*Instructor's Resource Guide for*

# CALCULUS: Early Transcendental Functions

THIRD EDITION

Larson / Hostetler / Edwards

Ann Rutledge Kraus
The Pennsylvania State University
The Behrend College

*with the assistance of*

Gerald A. Kraus
Gannon University

**Houghton Mifflin Company** Boston New York

Editor-in-Chief: Jack Shira
Managing Editor: Cathy Cantin
Development Manager: Maureen Ross
Development Editor: Laura Wheel
Assistant Editor: Rosalind Martin
Supervising Editor: Karen Carter
Senior Project Editor: Patty Bergin
Editorial Assistant: Meghan Lydon
Production Technology Supervisor: Gary Crespo
Executive Marketing Manager, Mathematics: Michael Busnach
Marketing Associate: Nicole Mollica
Senior Manufacturing Coordinator: Jane Spelman



Printed in the United States of America

ISBN: 0-618-22310-X

123456789-CRS-06 05 04 03 02

# PREFACE

This instructor's guide is a supplement to *Calculus: Early Transcendental Functions,* Third Edition, by Ron Larson, Robert P. Hostetler, and Bruce H. Edwards. All references to chapters, sections, theorems, definitions and examples apply to this text.

Part I of the guide contains teaching strategies. In Notes to the New Teacher and the Chapter Summaries, I have made comments and suggestions based on my own teaching experiences.

The second portion of the guide contains tests and test answer keys. The first three tests for each chapter are multiple choice and the last two are open ended. Each test is meant for a 50-minute class period. A graphing utility is required for test forms C and E for most chapters.

There are two final exams, one multiple choice and one open ended, for each semester. These tests are meant for a 90-minute class period.

Finally, in this portion of the guide there are four different versions of five Gateway Tests covering these topics: Algebra, Trigonometry, Exponential and Logarithmic Functions, Differentiation, and Integration.

Part III of this guide contains suggested solutions for the Special Features in the text: Chapter Openers, Explorations, Technology Boxes, and Section Projects.

I have made every effort to see that thc answers are correct. However, I would appreciate very much hearing about any errors or other suggestions for improvement.

I would like to thank the staff at Larson Texts, Inc. for preparing this manuscript for publication. I would especially like to thank my husband, Gerald A. Kraus, who wrote the solutions for the Chapter Opener problems and for many of the Section Projects, as well as providing input for other portions of this guide.

Ann Rutledge Kraus
The Pennsylvania State University
Erie, Pennsylvania 16563

# CONTENTS


| | | |
|---|---|---|
| **PART I** | **TEACHING STRATEGIES** | 1 |
| | Notes to the New Teacher | 2 |
| | Chapter Summaries | 4 |
| **PART II** | **TEST AND TEST ANSWER KEYS** | 27 |
| | Test Bank | 28 |
| | Answer Key to Chapter Tests | 202 |
| | Final Exams | 221 |
| | Answer Key to Final Exams | 237 |
| | Gateway Tests | 239 |
| | Answer Key to Gateway Tests | 259 |
| **PART III** | **SUGGESTED SOLUTIONS FOR SPECIAL FEATURES** | 264 |
| | Chapter Openers | 265 |
| | Explorations | 274 |
| | Technology Boxes | 300 |
| | Section Projects | 307 |

# PART I TEACHING STRATEGIES

**Notes to the New Teacher** .......... 2

**Chapter Summaries** .......... 4

**Chapter P** Preparation for Calculus .......... 4
**Chapter 1** Limits and Their Properties .......... 6
**Chapter 2** Differentiation .......... 8
**Chapter 3** Applications of Differentiation .......... 10
**Chapter 4** Integration .......... 11
**Chapter 5** Differential Equations .......... 13
**Chapter 6** Applications of Integration .......... 14
**Chapter 7** Integration Techniques, L'Hôpital's Rule, and Improper Integrals .......... 15
**Chapter 8** Infinite Series .......... 16
**Chapter 9** Conics, Parametric Equations, and Polar Coordinates .......... 17
**Chapter 10** Vectors and the Geometry of Space .......... 19
**Chapter 11** Vector-Valued Functions .......... 20
**Chapter 12** Functions of Several Variables .......... 21
**Chapter 13** Multiple Integration .......... 23
**Chapter 14** Vector Analysis .......... 24

In these notes you will find general suggestions to guide the new instructor in preparing for and conducting his or her class. For specific suggestions regarding teaching strategies, assignments, and topics that are traditionally difficult for students, refer to the chapter comments in this guide.

# NOTES TO THE NEW TEACHER

The key to having a good presentation is to prepare for class. First of all, read the text carefully. Then make notes (to be used in class) so you won't neglect to cover anything. Work out the examples that you intend to use in detail. You might want to use the even-numbered problems as examples and assign the odd-numbered problems.

Work out the assignments that you give. This will help you anticipate some of the students' difficulties. It will also give you some idea of how long an assignment takes so you can make adjustments on future assignments.

You may find it worthwhile to have your students write up and turn in several carefully chosen problems each week. This will give you a chance to look over the notation the student is using as well as his or her approach to the problems. If you insist on correct notation and an organized format, the students will be forced to write better mathematics.

There are chapter tests available in this guide. You may choose to use these or make up your own. Whichever you do, grade the tests immediately and fairly, and return them to the students, the sooner the better. Students like to know what their grades are throughout the semester, so establish a grading policy, put it in writing, and stick to it.

There are several other policies that you need to formulate before the semester begins: make-up exams for tests missed, class attendance, disturbance in class, and cheating. One way to discourage make-up exams is by stating that any test missed will be made up during the last class meeting of the semester. Encourage regular class attendance. Do not put up with class disturbances. Most students will respond to a quiet comment about the disturbance; however, occasionally you may have to ask a student to leave. Cheating occurs everywhere. You should look for it and proctor your exams carefully. Review your school's policy regarding cheating with your students, then enforce that policy if necessary. In this guide there are three multiple-choice tests for each chapter. Test A and Test B are comparable, and you may find it useful to use both in one class to discourage cheating.

It is often helpful to the new teacher to find someone in the department who cares a lot about teaching and who has had experience with the course you are teaching. This person can guide you as to the pace of the class and length of assignments.

If English is not your native language, you may find the following suggestions helpful.

- At the beginning of each class (or at the first class of the week), outline the material you plan to cover in the class (or in the week of classes).
- When you use a math term for the first time, write it on the board.
- Write on the board all key terms that you use.
- When you write definitions, theorems, and examples on the board, always label them as such.

Encourage class participation by taking questions during the classes. You'll be able to judge student comprehension and adjust accordingly. When students come to you for help, you may want to discuss good study habits, such as those outlined on the following page.

Above all, remember that you are in charge of the class, not your students. Good luck!

*Suggested Study Habits for Students*

- As you are reading the text or preparing homework assignments, make a note of anything you don't understand and ask about it in class.
- Take notes during class. After class, review your notes and rewrite them if necessary to fill in any missing details. If you don't understand a concept, question your instructor about it during the next class or during his or her office hours.
- As you start each chapter of the text, read *Chapter Openers* to begin to develop an understanding of what you'll learn in the chapter.
- Read the assigned material in each section before it is covered in class. After attending class, reread the assigned section to make sure that you understand the material.
- Remember that learning calculus is a step-by-step process. Always keep up with and complete your assignments because you must understand each topic in order to learn the next one.
- When you are working problems for homework assignments, show every step in your solution. Then, if you make an error, it will be easier for you or your instructor to find it.
- Work the problems in the homework assignments as though you were practicing for a test. First try to do the problems with your book and your notebook closed.
- Study with another student or small group of students, especially when preparing for a test.
- On the day of the test, allow yourself plenty of time to get to the testing location.
- Review your tests carefully, making sure you understand where and how you made errors.
- If you are having a problem, be sure to take advantage of your school's tutoring services for extra help.



**For additional teaching strategies, consult the following sources.**

Case, Bettye Anne. 1994. *You're the Professor, What Next?* Washington, DC: Mathematical Association of America.

Krantz, Steven G. 1993. *How to Teach Mathematics: a personal perspective.* Providence, RI: American Mathematical Society.

McKeachie, Wilbert James. 1994. *Teaching Tips, Strategies, Research, and Theory for College and University Teachers.* Lexington, MA: D. C. Heath and Company.

Rishel, Thomas W. 2000. *Teaching First, A Guide for New Mathematicians.* Washington, DC: Mathematical Association of America.

Schoenfeld, Alan. 1990. *A Source Book for College Mathematics Teaching.* Providence, RI: Mathematical Association of America.

Consult current math journals such as *The Mathematics Teacher, College Mathematics Journal, Computers in Mathematics and Science Teaching,* or *MAA-AMS Notes* for specific teaching strategies in a variety of areas.

# CHAPTER SUMMARIES

All calculus texts have more material in them than can be covered in three semesters. Therefore, you must make choices in the beginning of the semester. Do not assign all of the problems (or even all of the odd problems). There are far too many. Read the problems and choose your assignment based on the material for which you are going to hold your students responsible. At the end of each chapter is a set of review problems. While you might not have the time to assign these problems for class, you might suggest to your students that they do these problems when preparing for a chapter test.

# CHAPTER P
# Preparation for Calculus

## Section Topics

P.1 **Graphs and Models**—Sketch the graph of an equation. Find the intercepts of a graph. Test a graph for symmetry with respect to an axis and the origin. Find the points of intersection of two graphs. Interpret mathematical models for real-life data.

P.2 **Linear Models and Rates of Change**—Find the slope of a line passing through two points. Write the equation of a line with a given point and slope. Interpret slope as a ratio or as a rate in a real-life application. Sketch the graph of a linear equation in slope-intercept form. Write equations of lines that are parallel or perpendicular to a given line.

P.3 **Functions and Their Graphs**—Use function notation to represent and evaluate a function. Find the domain and range of a function. Sketch the graph of a function. Identify different types of transformations of functions. Classify functions and recognize combinations of functions.

P.4 **Fitting Models to Data**—Fit a linear model to a real-life data set. Fit a quadratic model to a real-life data set. Fit a trigonometric model to a real-life data set.

P.5 **Inverse Functions**—Verify that one function is the inverse function of another function. Determine whether a function has an inverse function. Develop properties of the six inverse trigonometric functions.

P.6 **Exponential and Logarithmic Functions**—Develop and use properties of exponential functions. Understand the definition of the number *e*. Understand the definition of the natural logarithmic function. Develop and use properties of the natural logarithmic function.

## Chapter Comments

Chapter P is a review chapter and, therefore, should be covered quickly. I usually spend about 3 or 4 days on this chapter, placing most of the emphasis on Sections 3 and 6. Of course, you cannot cover every single item that is in this chapter in that time, so this is a good opportunity to encourage your students to read the book. To convince your students of this, assign homework problems or give a quiz on some of the material that is in this chapter but that you do not go over in class. Although you will not hold your students responsible for everything in all 15 chapters, the tools in this chapter need to be readily at hand, i.e., memorized.

Sections P.1 and P.2 can be covered in a day. Students at this level of mathematics have graphed equations before so let them read about that information on their own. I discuss intercepts, emphasizing that they are points, not numbers, and, so, should be written as ordered pairs. I also discuss symmetry with respect to the

$x$-axis, the $y$-axis, and the origin. Be sure to do a problem like Example 5 in Section P.1. Students need to be able to find the points of intersection of graphs in order to calculate the area between two curves in Chapter 6.

In Section P.2, I discuss the slope of a line, the point-slope form of a line, equations of vertical and horizontal lines, and the slopes of parallel and perpendicular lines. You need to emphasize the point-slope form of a straight line because this is needed to write the equation of a tangent line in Chapter 2.

Students need to know everything in Section P.3, so carefully go over the definition of a function, domain and range, function notation, transformations, the terms algebraic and transcendental, and the composition of functions. Note that the authors assume a knowledge of trigonometric functions. If necessary to review these functions, refer to Section 3 of Appendix D, found on the website that accompanies this text at *college.hmco.com*. Because students need practice handling $\Delta x$, be sure to do an example calculating $f(x + \Delta x)$. Your students should know the graphs in Figure P.27.

Section P.4 introduces the idea of fitting models to data. Because a basic premise of science is that much of the physical world can be described mathematically, you and your students would benefit from looking at the three models presented in this section.

Students should be familiar with the inverses of functions from their Algebra courses, so perhaps the first part of Section 5 could be covered quickly. However, they probably are not familiar with the ideas of monotonic and one-to-one, so spend some time explaining these ideas and use the terminology frequently. Inverse trigonometric functions will not appeal to your students. They tend to get lazy when the trigonometry gets this involved. Be sure to do problems like those in Example 7. The concept of using a right triangle is an important tool for the integration in Chapters 4 and 7.

Although students should be familiar and comfortable with both the exponential and logarithmic functions, frequently they are not. Therefore, in Section 6, present these functions as though they are new ideas. Have your students commit to memory the graphs and properties of these functions. Your students will learn more about the number $e$ in Chapter 1 when limits are discussed.

The authors assume students have a working knowledge of inequalities, the formula for the distance between two points, absolute value, etc. If needed, you can find a review of these concepts in Appendix D.

# CHAPTER 1
# Limits and Their Properties

## Section Topics

1.1 **A Preview of Calculus**—Understand what calculus is and how it compares to precalculus. Understand that the tangent line problem is basic to calculus. Understand that the area problem is also basic to calculus.

1.2 **Finding Limits Graphically and Numerically**—Estimate a limit using a numerical or graphical approach. Learn different ways that a limit can fail to exist. Study and use a formal definition of a limit.

1.3 **Evaluating Limits Analytically**—Evaluate a limit using properties of limits. Develop and use a strategy for finding limits. Evaluate a limit using cancellation and rationalization techniques. Evaluate a limit using the Squeeze Theorem.

1.4 **Continuity and One-Sided Limits**—Determine continuity at a point and continuity on an open interval. Determine one-sided limits and continuity on a closed interval. Use properties of continuity. Understand and use the Intermediate Value Theorem.

1.5 **Infinite Limits**—Determine infinite limits from the left and from the right. Find and sketch the vertical asymptotes of the graph of a function.

## Chapter Comments

Section 1.1 gives a preview of calculus. On pages 61 and 62 of the textbook are examples of some of the concepts from precalculus extended to ideas that require the use of calculus. I suggest that you review these ideas with your students to give them a feel for where the course is heading.

The idea of a limit is central to calculus. So you should take the time to discuss the tangent line problem and/or the area problem in this section. Exercise 11 of Section 1.1 is yet another example of how limits will be used in calculus. A review of the formula for the distance between two points can be found in Section 2 of Appendix D.

The discussion of limits is difficult for most students the first time that they see it. For this reason you should carefully go over the examples and the informal definition of a limit presented in Section 1.2. Stress to your students that a limit exists only if the answer is a real number. Otherwise, we say the limit fails to exist, as shown in Examples 3, 4, and 5 of this section. You might want to work Exercise 99 of Section 1.2 with your students so that they are familiar with the limit definition of the number $e$. This exercise is similar to Example 3 of Section P.6. You may choose to omit the formal definition of a limit found in this section.

I carefully go over the properties of limits found in Section 1.3 because I want my students to be comfortable with the idea of a limit and also with the notation used for limits. By the time you get to Theorem 1.6, it should be obvious to your students that all of these properties amount to direct substitution.

When direct substitution for the limit of a quotient yields the indeterminate form $\frac{0}{0}$, I tell my students that we must rewrite the fraction using *legitimate* algebra. Then I do at least one problem using dividing out and another using rationalization. Exercises 60 and 62 in Section 1.3 are examples of other algebraic techniques needed for the limit problems. You need to go over the Squeeze Theorem, Theorem 1.8, with your students so that you can use it to prove $\lim_{x \to 0} \frac{\sin x}{x} = 1$. The proof of Theorem 1.8 and many other theorems can be found in Appendix B. Your students need to memorize the results in Theorem 1.9 as they will need these facts to do problems throughout the textbook. Most of your students will need help with Exercises 69–82 which involve the limits of transcendental functions.

Continuity, Section 1.4, is another idea that often puzzles students. However, if I describe a continuous function as one in which I can draw the entire graph without lifting my pencil, the idea seems to stay with them. Distinguishing between removable and non-removable discontinuities will help students to determine vertical asymptotes. An important theorem concerning the behavior of continuous functions, the Intermediate Value Theorem, is discussed in this section. Example 8 shows that this theorem is frequently used to locate the zeros of a function that is continuous on a closed interval.

To discuss infinite limits, Section 1.5, remind your students of the graph of the function $f(x) = 1/x$ studied in Section P.3. Be sure to make your students write a vertical asymptote as an equation, not just a number. For example, for the function $y = 1/x$, the vertical asymptote is $x = 0$.

# CHAPTER 2
# Differentiation

## Section Topics

**2.1** **The Derivative and the Tangent Line Problem**—Find the slope of the tangent line to a curve at a point. Use the limit definition to find the derivative of a function. Understand the relationship between differentiability and continuity.

**2.2** **Basic Differentiation Rules and Rates of Change**—Find the derivative of a function using the Constant Rule. Find the derivative of a function using the Power Rule. Find the derivative of a function using the Constant Multiple Rule. Find the derivative of a function using the Sum and Difference Rules. Find the derivative of the sine function and of the cosine function. Use derivatives to find rates of change.

**2.3** **The Product and Quotient Rules and Higher-Order Derivatives**—Find the derivative of a function using the Product Rule. Find the derivative of a function using the Quotient Rule. Find the derivative of a trigonometric function. Find a higher-order derivative of a function.

**2.4** **The Chain Rule**—Find the derivative of a composite function using the Chain Rule. Find the derivative of a function using the General Power Rule. Simplify the derivative of a function using algebra. Find the derivative of a trigonometric function using the Chain Rule.

**2.5** **Implicit Differentiation**—Distinguish between functions written in implicit form and explicit form. Use implicit differentiation to find the derivative of a function.

**2.6** **Derivatives of Inverse Functions**—Find the derivative of an inverse function. Differentiate an inverse trigonometric function. Review the basic differentiation formulas for elementary functions.

**2.7** **Related Rates**—Find a related rate. Use related rates to solve real-life problems.

**2.8** **Newton's Method**—Approximate a zero of a function using Newton's Method.

## Chapter Comments

The material presented in Chapter 2 forms the basis for the remainder of calculus. Much of it needs to be memorized, beginning with the definition of a derivative of a function found on page 115. Your students need to have a thorough understanding of the tangent line problem and they need to be able to find the equation of a tangent line. Frequently, students will use the function $f'(x)$ as the slope of the tangent line. They need to understand that $f'(x)$ is the formula for the slope and the actual value of the slope can be found by substituting into $f'(x)$ the appropriate value for $x$. On page 118 of Section 2.1, you will find a discussion of situations where the derivative fails to exist. These examples (or similar ones) should be discussed in class. On page 119 is a summary of the relationship between continuity and differentiability.

As you teach this chapter, vary your notations for the derivative. One time write $y'$, another time write $dy/dx$ or $f'(x)$. Terminology is also important: instead of saying "find the derivative," sometimes say, "differentiate." This would be an appropriate time, also, to talk a little about Leibnitz and Newton and the development of calculus.

Sections 2.2, 2.3, and 2.4 present a number of rules for differentiation. Have your students memorize the Product Rule and the Quotient Rule (Theorems 2.8 and 2.9) in words rather than symbols. Students tend to be lazy when it comes to trigonometry and therefore, you need to impress upon them that the formulas for the derivatives of the six trigonometric functions need to be memorized also. You will probably not have enough time in class to prove every one of these differentiation rules, so choose several to do in class and perhaps assign a few of the other proofs as homework.

The Chain Rule, in Section 2.4, will require two days of your class time. Students need lots of practice with this and the algebra involved in these problems. Many students can find the derivative of $f(x) = x^2\sqrt{1 - x^2}$ without much trouble, but simplifying the answer is often very difficult for them. Insist that they learn to factor and write the answer without negative exponents. Strive to get the answer in the form given in the back of the book. This will help them later on when the derivative is set equal to zero. Students will need to be able to differentiate $a^u$ in Chapter 8, so be sure to cover Theorem 2.15.

Implicit differentiation is often difficult for students. Have your students think of $y$ as a function of $x$ and therefore $y^3$ is $[f(x)]^3$. This way they can relate implicit differentiation to the Chain Rule studied in the previous section.

Section 2.6 discusses the derivative of an inverse function and, in particular, the derivative of the inverse trigonometric functions. Although the inverse trigonometric functions were reviewed in Section P.5, you will probably need to go over these again with your students. In particular, you will need to remind them of the restrictions on the ranges of these functions. Section 2.6 also contains a table of basic differentiation rules for elementary functions found on page 173. This table is also found on the inside cover on the front of the text.

Try to get your students to see that, Related Rates, discussed in Section 2.6, are another use of the Chain Rule.

Newton's Method, a technique for finding the zeros of a function, is a tool that students of future math and science courses will be expected to know. Note in Example 7 that the method does not always work. A calculator is needed.

# CHAPTER 3
# Applications of Differentiation

## Section Topics

**3.1** **Extrema on an Interval**—Understand the definition of extrema of a function on an interval. Understand the definition of relative extrema of a function on an open interval. Find extrema on a closed interval.

**3.2** **Rolle's Theorem and the Mean Value Theorem**—Understand and use Rolle's Theorem. Understand and use the Mean Value Theorem.

**3.3** **Increasing and Decreasing Functions and the First Derivative Test**—Determine intervals on which a function is increasing or decreasing. Apply the First Derivative Test to find relative extrema of a function.

**3.4** **Concavity and the Second Derivative Test**—Determine intervals on which a function is concave upward or concave downward. Find any points of inflection of the graph of a function. Apply the Second Derivative Test to find relative extrema of a function.

**3.5** **Limits at Infinity**—Determine (finite) limits at infinity. Determine the horizontal asymptotes, if any, of the graph of a function. Determine infinite limits at infinity.

**3.6** **A Summary of Curve Sketching**—Analyze and sketch the graph of a function.

**3.7** **Optimization Problems**—Solve applied minimum and maximum problems.

**3.8** **Differentials**—Understand the concept of a tangent line approximation. Compare the value of differential $dy$ with the actual change in $y$, $\Delta y$. Estimate a propagated error using a differential. Find the differential of a function using differentiation formulas.

## Chapter Comments

Chapter 3 considers some of the applications of a derivative. The first five sections in this chapter lead up to curve sketching in Section 3.6. Before beginning this material, you may want to quickly review with your students the ideas of intercepts, vertical asymptotes and symmetry with respect to the $x$-axis, the $y$-axis and the origin. These algebraic tools covered in Chapter P are needed, along with those from calculus, in order to analyze the graph of a function.

Don't get bogged down in Section 3.1. The students need to know the terminology, so stress the definitions of critical numbers and extrema. Section 3.2 covers Rolle's Theorem and the Mean Value Theorem. Rolle's Theorem is used to prove the Mean Value Theorem.

Sections 3.3 and 3.4 cover the bulk of the material necessary to analyze the graphs of functions. Go over these sections carefully.

In Section 3.5, limits at infinity, as well as horizontal asymptotes, are discussed. Be sure to require that your students write horizontal asymptotes as straight lines: ie., $y = 3$.

Section 3.6 contains a summary of all of the ideas involved in analyzing the graph of a function. Note that slant asymptotes are covered in the comment after Example 2. After this material is covered, each day until you finish the chapter, assign one or two carefully chosen graphing problems. Have your students make a list of all of the tools useful in sketching a graph: intercepts, asymptotes, symmetry, critical numbers, increasing and decreasing intervals, extrema, concavity and inflection points. Only after this list is completed should the function be sketched.

If your students know the material in Sections 3.1 through 3.6 well, then the optimization problems in Section 3.7 should go fairly well for them. Stress the difference between a primary and a secondary equation. Also, stress the need to have the function in one independent variable before attempting to maximize or minimize it.

You should be able to cover Differentials, Section 3.8, quickly. Students need to understand that a differential is a reasonable approximation for the actual difference in function values. Following Examples 2 and 7 there is a brief discussion of the tangent line approximation of the graph of a function at a point.

# CHAPTER 4
# Integration

## Section Topics

4.1 **Antiderivatives and Indefinite Integration**—Write the general solution of a differential equation. Use indefinite integral notation for antiderivatives. Use basic integration rules to find antiderivatives. Find a particular solution of a differential equation.

4.2 **Area**—Use sigma notation to write and evaluate a sum. Understand the concept of area. Approximate the area of a plane region. Find the area of a plane region using limits.

4.3 **Riemann Sums and Definite Integrals**—Understand the definition of a Riemann Sum. Evaluate a definite integral using limits. Evaluate a definite integral using properties of definite integrals.

4.4 **The Fundamental Theorem of Calculus**—Evaluate a definite integral using the Fundamental Theorem of Calculus. Understand and use the Mean Value Theorem for Integrals. Find the average value of a function over a closed interval. Understand and use the Second Fundamental Theorem of Calculus.

4.5 **Integration by Substitution**—Use pattern recognition to evaluate an indefinite integral. Use a change of variables to evaluate an indefinite integral. Use the General Power Rule for Integration to evaluate an indefinite integral. Use a change of variables to evaluate a definite integral. Evaluate a definite integral involving an even or odd function.

4.6 **Numerical Integration**—Approximate a definite integral using the Trapezoidal Rule. Approximate a definite integral using Simpson's Rule. Analyze the approximate error in the Trapezoidal Rule and in Simpson's Rule.

4.7 **The Natural Logarithmic Function: Integration**—Use the Log Rule for Integration to integrate a rational function. Integrate trigonometric functions.

4.8 **Inverse Trigonometric Functions: Integration**—Integrate functions whose antiderivatives involve inverse trigonometric functions. Use completing the square to integrate a function. Review the basic integration formulas involving elementary functions.

4.9 **Hyperbolic Functions**—Develop properties of hyperbolic functions. Differentiate and integrate hyperbolic functions. Develop properties of inverse hyperbolic functions. Differentiate and integrate functions involving inverse hyperbolic functions.

## Chapter Comments

The first section of this chapter stresses correct terminology and notation. Recognizing the different ways to express the same algebraic expression is very important here. Algebra will get a workout. As you introduce antiderivatives, insist that your students write the proper differential when they write an integral problem. Deduct points on homework and quizzes for incorrect notation.

Some of your students may never have seen sigma notation before, so introduce it as though it is a new topic. The idea of the limit of a sum as $n$ approaches infinity should go quickly if the students learned the rules for limits at infinity from Section 3.5. If you are pressed for time you may want to go over area using a limit definition (as done in Section 4.2) briefly. For example, you could do one problem to show the idea and the technique and then move on to Section 4.3.

The amount of time spent on Riemann Sums will depend on your schedule and your class. The definite integral is defined in Section 4.3, as is area. Be sure that your students set up the limits for area properly. Do not accept integrals for area with the upper and lower limits interchanged.

Students should know the Fundamental Theorem of Calculus by name and have it memorized. The Mean Value Theorem for Integrals and the Average Value of a Function are covered in Section 4.4. Students usually find the Second Fundamental Theorem of Calculus suprising and interesting.

It is very important for your students to be comfortable with all of the integration techniques covered in Section 4.5. Allow two days to cover these ideas and go over each carefully. Be sure to do some change of variable problems, a technique that many students are reluctant to use.

To introduce the approximation techniques of Simpson's Rule and Trapezoidal Rule, present an integral problem for which there is no antiderivative that is an elementary function, such as $\int_{\pi/4}^{\pi/2} \sin x^2\, dx$. This is a good opportunity to convince your students that not all functions have antiderivatives that are elementary functions. If you are pressed for time, the development of Simpson's Rule could be eliminated, but not the rule itself.

Integrals that yield a natural logarithmic function, discussed in Section 4.7, are hard for the student to see at first. You must tell your students that whenever the problem is $\int f(x)/g(x)\, dx$, there are two things to consider before anything else:

(1) if the degree of $f(x) \geq$ the degree of $g(x)$, then perform algebraic long division to rewrite the integrand;

(2) investigate the possibility of $u'/u$.

Notice that with Theorem 4.20 we can now integrate all six trigonometric functions. These integration rules, on page 347, are important; have your students memorize them. The problems in the exercises for Section 4.7 bring out a number of important concepts for integration. Be sure to allow enough time to go over these thoroughly.

In Section 4.8 there are some examples that demonstrate important tools that will be needed for future work with integration. The rewriting of the integrand as the sum of two quotients is shown in Example 3. Completing the square on a quadratic expression is demonstrated in Examples 4 and 5. These examples should be done in class. Again, remind your students of the need to divide if the degree of the polynomial in the numerator is greater than or equal to the degree of the polynomial in the denominator. The problems in Section 4.8 contain a significant amount of information, reviewing all that we know about integration so far. You should plan to spend quite a bit of time covering them. Exercise 29 would be a good example to do with your students in class after they have attempted Exercises 1–27 on their own.

Section 4.9 discusses hyperbolic functions and could be covered lightly if you are pressed for time. Do not omit completely, however, because the students will see these functions again. A familiar example of a hyperbolic cosine function is the St. Louis Arch, discussed in the Section Project on page 366.

# CHAPTER 5
# Differential Equations

## Sections Topics

**5.1** **Differential Equations: Growth and Decay**—Use separation of variables to solve a simple differential equation. Use exponential functions to model growth and decay in applied problems.

**5.2** **Differential Equations: Separation of Variables**—Use initial conditions to find particular solutions of differential equations. Use a slope field to sketch solutions of a differential equation. Solve differential equations by separation of variables. Solve homogeneous differential equations. Use Euler's Method to approximate a solution of a differential equation. Use a differential equation to model and solve an applied problem.

**5.3** **First-Order Linear Differential Equations**—Solve a first-order linear differential equation. Solve a Bernoulli differential equation. Use linear differential equations to solve applied problems.

## Chapter Comments

Chapter 5 introduces differential equations. Be sure to do at least one example on exponential growth or decay so that the students get some exposure to an application of the exponential function. Students frequently become frustrated checking their answers to homework problems in differential equations. Be sure to point out that there are many correct ways that an answer to a differential equation could be given. As the students become more familiar with this work, they will have a better idea of what is expected for an answer.

Separation of variables and homogeneous differential equations, as well as general and particular solutions, are covered in Section 5.2. Also discussed in this section is Euler's Method, a way of solving initial value problems using an approximation technique.

Section 5.3 considers first-order linear differential equations and Bernoulli equations.

# CHAPTER 6
# Applications of Integration

## Section Topics

6.1 **Area of a Region Between Two Curves**—Find the area of a region between two curves using integration. Find the area of a region between intersecting curves using integration. Describe integration as an accumulation process.

6.2 **Volume: The Disk Method**—Find the volume of a solid of revolution using the disk method. Find the volume of a solid of revolution using the washer method. Find the volume of a solid with known cross sections.

6.3 **Volume: The Shell Method**—Find the volume of a solid of revolution using the shell method. Compare the uses of the disk method and the shell method.

6.4 **Arc Length and Surfaces of Revolution**—Find the arc length of a smooth curve. Find the area of a surface of revolution.

6.5 **Work**—Find the work done by a constant force. Find the work done by a variable force.

6.6 **Moments, Centers of Mass, and Centroids**—Understand the definition of mass. Find the center of mass in a one-dimensional system. Find the center of mass in a two-dimensional system. Find the center of mass of a planar lamina. Use the Theorem of Pappus to find the volume of a solid of revolution.

6.7 **Fluid Pressure and Fluid Force**—Find fluid pressure and fluid force.

## Chapter Comments

Chapter 6 covers many applications to the definite integral. There are far too many problems in this chapter to cover in a normal calculus course. You would be better off assigning just a few carefully chosen problems from each section covered.

Because you may not have the time to cover each topic in this chapter, you should cover the area between two curves carefully so that the students get used to the idea of using a representative rectangle in the given region. Remind your students that they are expected to know how to sketch a graph (Chapter 3) and how to find points of intersection (Chapter P).

Both disk and shell methods, Sections 6.2 and 6.3, need to be discussed to do a thorough treatment of volumes of solids of revolution. Be sure to go over Examples 3, 4 and 5 of Section 6.3 to convince your students that it is necessary to know both methods.

Arc length, Section 6.4, should be covered because this is built upon in later chapters (Chapters 10 and 12).

The remaining topics in this chapter should be covered as time allows.

# CHAPTER 7
# Integration Techniques, L'Hôpital's Rule, and Improper Integrals

## Section Topics

**7.1** **Basic Integration Rules**—Review procedures for fitting an integrand to one of the basic integration rules.

**7.2** **Integration by Parts**—Find an antiderivative using integration by parts. Use a tabular method for integration by parts.

**7.3** **Trigonometric Integrals**—Solve trigonometric integrals involving powers of sine and cosine. Solve trigonometric integrals involving powers of secant and tangent. Solve trigonometric integrals involving sine-cosine products with different angles.

**7.4** **Trigonometric Substitution**—Use trigonometric substitution to solve an integral. Use integrals to model and solve real-life applications.

**7.5** **Partial Fractions**—Understand the concept of a partial fraction decomposition. Use partial fraction decomposition with linear factors to integrate rational functions. Use partial fraction decomposition with quadratic factors to integrate rational functions.

**7.6** **Integration by Tables and Other Integration Techniques**—Evaluate an indefinite integral using a table of integrals. Evaluate an indefinite integral using reduction formulas. Evaluate an indefinite integral involving rational functions of sine and cosine.

**7.7** **Indeterminate Forms and L'Hôpital's Rule**—Recognize limits that produce indeterminate forms. Apply L'Hôpital's Rule to evaluate a limit.

**7.8** **Improper Integrals**—Evaluate an improper integral that has an infinite limit of integration. Evaluate an improper integral that has an infinite discontinuity.

## Chapter Comments

This chapter and the following one will take up a lot of time. Be prepared to use two class days for most of the sections in this chapter. The first section is a review of all of the integration covered up to this point. Your students should have memorized the list of basic integration rules found on page 484. Besides these basic rules, this section also reviews techniques necessary for integration, such as completing the square. Be sure to go over the examples in this section.

When you cover integration by parts in Section 7.2, be sure to do one of the integrals $\int \sec^3 x\, dx$ or $\int \csc^3 x\, dx$. It is important to point out that "parts" is the way to integrate $\int \sec^m x\, dx$, where $m$ is positive and odd.

Sections 7.3 and 7.4 will give many of your students trouble because they do not have their trigonometry memorized. Insist that they learn it and deduct points when the trigonometry is not correct. In Section 7.4, do some problems converting the limits of integration when a new variable is introduced into the problem, such as Example 4 on page 509.

Section 7.5 will take 2 days to cover because most students will need help reviewing how to find partial fractions and how to solve simultaneous equations. Choose your in-class examples carefully in order to review these techniques. I usually spend the first day on linear factors and the second day working with the quadratic factors.

In Section 7.6 is the technique for integrating rational functions of sine and cosine (page 527). This technique is interesting and worth doing, but do not dwell on it.

When you cover indeterminate forms and L'Hôpital's Rule in Section 7.7, be sure to do Example 5 on page 534 because that limit turns out to be the number *e*. Some books use this limit as a definition of *e* and your students need to recognize it, as it comes up in problems later on.

Improper integrals, Section 7.8, incorporates the ideas of the previous section.

# CHAPTER 8
# Infinite Series

## Section Topics

**8.1** **Sequences**—List the terms of a sequence. Determine whether a sequence converges or diverges. Write a formula for the *n*th term of a sequence. Use properties of monotonic sequences and bounded sequences.

**8.2** **Series and Convergence**—Understand the definition of a convergent infinite series. Use properties of infinite geometric series. Use the *n*th Term Test for Divergence of an infinite series.

**8.3** **The Integral Test and *p*-Series**—Use the Integral Test to determine whether an infinite series converges or diverges. Use properties of *p*-series and harmonic series.

**8.4** **Comparisons of Series**—Use the Direct Comparison Test to determine whether a series converges or diverges. Use the Limit Comparison Test to determine whether a series converges or diverges.

**8.5** **Alternating Series**—Use the Alternating Series Test to determine whether an infinite series converges. Use the Alternating Series Remainder to approximate the sum of an alternating series. Classify a convergent series as absolutely or conditionally convergent. Rearrange an infinite series to obtain a different sum.

**8.6** **The Ratio and Root Tests**—Use the Ratio Test to determine whether a series converges or diverges. Use the Root Test to determine whether a series converges or diverges. Review the tests for convergence and divergence of an infinite series.

**8.7** **Taylor Polynomials and Approximations**—Find and compare polynomial approximations of an elementary function to the elementary function. Find Taylor and Maclaurin polynomial approximations of elementary functions. Use the remainder of a Taylor polynomial.

**8.8** **Power Series**—Understand the definition of a power series. Find the radius and interval of convergence of a power series. Determine the endpoint convergence of a power series. Differentiate and integrate a power series.

**8.9** **Representation of Functions by Power Series**—Find a geometric power series that represents a function. Construct a power series using series operations.

**8.10** **Taylor and Maclaurin Series**—Find a Taylor or Maclaurin series for a function. Find a binomial series. Use a basic list of Taylor series to find other Taylor series.

## Chapter Comments

You may want to think of this chapter as two parts. Part I (Sections 8.1 through 8.6) covers sequences and series of constant terms and Part II (Sections 8.7 through 8.10) covers series with variable terms. Part I should be covered quickly so that most of your time in this chapter is spent in Part II.

In Sections 8.1 through 8.6 there are many different kinds of series and many different tests for convergence or divergence. Be sure to go over each of these carefully. It is a good idea to review the basic facts of each test each day before covering the new material for that day. This provides a review for the students and also allows them to see the similarities and differences among tests. The table on page 602 in Section 8.6 is a good way to compare the various tests. Be sure to go over with your students the guidelines for choosing the appropriate test found on page 601.

The *n*th-Term Test for Divergence, Theorem 8.9 on page 571, is frequently misunderstood. Your students need to know that it proves divergence only and that it says absolutely nothing about convergence.

Sections 8.7 through 8.10 often seem difficult for students so allow extra time for these sections. You will need to go over the material slowly and do lots of examples. Students should be able to find the coefficients of a Taylor or Maclaurin polynomial, write a Taylor Series, derive a Taylor Series from a basic list, find the radius of convergence and the interval of convergence. Checking the endpoints should be a matter of recalling Sections 8.2 through 8.6.

# CHAPTER 9
# Conics, Parametric Equations, and Polar Coordinates

## Section Topics

9.1 **Conics and Calculus**—Understand the definition of a conic section. Analyze and write equations of parabolas using properties of parabolas. Analyze and write equations of ellipses using properties of ellipses. Analyze and write equations of hyperbolas using properties of hyperbolas.

9.2 **Plane Curves and Parametric Equations**—Sketch the graph of a curve given by a set of parametric equations. Eliminate the parameter in a set of parametric equations. Find a set of parametric equations to represent a curve. Understand two classic calculus problems: Tautochrone and brachistochrone problems.

9.3 **Parametric Equations and Calculus**—Find the slope of a tangent line to a curve given by a set of parametric equations. Find the arc length of a curve given by a set of parametric equations. Find the area of a surface of revolution (parametric form).

9.4 **Polar Coordinates and Polar Graphs**—Understand the polar coordinate system. Rewrite rectangular equations in polar form and vice versa. Sketch the graph of an equation given in polar form. Find the slope of a tangent line to a polar graph. Identify several types of special polar graphs.

9.5 **Area and Arc Length in Polar Coordinates**—Find the area of a region bounded by a polar graph. Find the points of intersection of two polar graphs. Find the arc length of a polar graph. Find the area of a surface of revolution (polar form).

9.6 **Polar Equations of Conics and Kepler's Laws**—Analyze and write polar equations of conics. Understand and use Kepler's Laws of planetary motion.

## Chapter Comments

For each of the conics (parabola, circle, ellipse, and hyperbola), your students should be able to write the equation in standard form, identify the center, radius, vertices, foci, directrix, axes, or asymptotes, and sketch the graph from these facts. Many of these concepts will be used when quadric surfaces are discussed in Section 10.6. A review of the circle can be found in Section D.2 of Appendix D.

Eccentricity for ellipses is discussed on page 655 and for hyperbolas on page 658. For a discussion of rotation of axes and the use of the discriminant to determine which conic the general second-degree equation represents, see Appendix E.

Be sure to do some of the real life applications in Section 9.1, such as the length of a parabolic cable used for a bridge, Exercise 39 on page 661.

When you discuss parametric equations in Section 9.2, be sure to point out to your students that a parametric representation for a curve is not unique. Therefore, their correct answers may not agree with those provided in the back of the book. The process of writing parametric equations for a given rectangular equation should convince your students of this. The opposite process, eliminating the parameter, sometimes involves adjusting the domain, as in Example 2 of Section 9.2. When graphing in parametric form, have your students note the orientation. For example, $x = 2\cos\theta$ and $y = 2\sin\theta$ graphs a circle counterclockwise, but $x = 2\sin\theta$ and $y = 2\cos\theta$ graphs the same circle clockwise.

Parametric form of the derivative is discussed in Section 9.3. Be sure to point out to your students that the slope of the tangent line is $\frac{dy}{dx}$ and not $\frac{dy}{dt}$. Students very often have difficulty with the higher order derivatives with respect to $x$. For example, to find $\frac{d^2y}{dx^2}$, they forget to divide $\frac{d}{dt}\left(\frac{dy}{dx}\right)$ by $\frac{dx}{dt}$. Arc length for parametric equations of a curve is a concept that will be expanded in 3-space (Section 11.5).

Polar coordinates, discussed in Section 9.4, use only radian measure for the angle $\theta$. Point out to your students that when plotting points in polar coordinates, it is usually easier to find the angle first and then mark off $r$ units on the terminal side of the angle. Note, too, that contrary to rectangular coordinates, a point can be represented many ways in polar coordinates. Students should know the equations for coordinate conversion found in Theorem 9.10 on page 685.

Curve sketching in polar coordinates can be time consuming and tedious. Don't get bogged down in this. Your students should easily recognize limaçons and rose curves from the equation and be able to sketch them quickly. Use a graphing calculator and note the orientation.

Area, discussed in Section 9.5, is an important idea for a good understanding of polar coordinates. The difficulty here is in finding the limits of integration. Note that there may be points of intersection of two curves given in polar coordinates that do not show up from solving simultaneously. Therefore, it is necessary to graph the curves.

Section 9.6 on conics in polar form may be omitted if time is a problem in your course.

# CHAPTER 10
# Vectors and the Geometry of Space

## Section Topics

**10.1** **Vectors in the Plane**—Write the component form of a vector. Perform vector operations and interpret the results geometrically. Write a vector as a linear combination of standard unit vectors. Use vectors to solve problems involving force or velocity.

**10.2** **Space Coordinates and Vectors in Space**—Understand the three-dimensional rectangular coordinate system. Analyze vectors in space. Use three-dimensional vectors to solve real-life problems.

**10.3** **The Dot Product of Two Vectors**—Use properties of the dot product of two vectors. Find the angle between two vectors using the dot product. Find the direction consines of a vector in space. Find the projection of a vector onto another vector. Use vectors to find the work done by a constant force.

**10.4** **The Cross Product of Two Vectors in Space**—Find the cross product of two vectors in space. Use the triple scalar product of three vectors in space.

**10.5** **Lines and Planes in Space**—Write a set of parametric equations for a line in space. Write a linear equation to represent a plane in space. Sketch the plane given by a linear equation. Find the distance between points, planes, and lines in space.

**10.6** **Surfaces in Space**—Recognize and write equations for cylindrical surfaces. Recognize and write equations for quadratic surfaces. Recognize and write equations for surfaces of revolution.

**10.7** **Cylindrical and Spherical Coordinates**—Use cylindrical coordinates to represent surfaces in space. Use spherical coordinates to represent surfaces in space.

## Chapter Comments

All of the ideas in this chapter need to be discussed in order for your students to have a good understanding of vectors. Point out to your students that vectors are not little pointed arrows, but that a directed line segment is just our way of representing a vector geometrically. Also note that this geometric representation of a vector does not have location. It does have direction and magnitude. On page 737 in the text is a Note about the words perpendicular, orthogonal, and normal. This is worth discussing with your students because sometimes it seems that these words are used interchangeably.

The dot product of two vectors is sometimes referred to as scalar multiplication and the cross product as vector multiplication. The reason for this is because the dot product is a *scalar* and the cross product is a *vector*. Point out this distinction to your students. The way to find a cross product is to use a 3 by 3 determinant. You will probably have to show your students how to calculate this.

When discussing lines and planes in space, Section 10.5, point out to your students that direction numbers are not unique. If $a, b, c$ is a set of direction numbers for a line or a plane, then $ka, kb, kc$ where $k \in R, k \neq 0$, is also a set of direction numbers for that line or plane.

The distance formulas in Section 10.5 between a point and a plane and between a point and a line need not be memorized. However, go over these so that the students know where they are when they need to look them up.

# CHAPTER 11
# Vector-Valued Functions

## Section Topics

11.1 **Vector-Valued Functions**—Analyze and sketch a space curve given by a vector-valued function. Extend the concept of limits and continuity to vector-valued functions.

11.2 **Differentiation and Integration of Vector-Valued Functions**—Differentiate a vector-valued function. Integrate a vector-valued function.

11.3 **Velocity and Acceleration**—Describe the velocity and acceleration associated with a vector-valued function. Use a vector-valued function to analyze projectile motion.

11.4 **Tangent Vectors and Normal Vectors**—Find a unit tangent vector at a point on a space curve. Find the tangential and normal components of acceleration.

11.5 **Arc Length and Curvature**—Find the arc length of a space curve. Use the arc length parameter to describe a plane curve or space curve. Find the curvature of a curve at a point on the curve. Use a vector-valued function to find frictional force.

## Chapter Comments

In discussing vector-valued functions with your students, be sure to distinguish them from real-valued functions. A vector-valued function is a vector, whereas a real-valued function is a real number. Remind your students that the parameterization of a curve is not unique. As discussed on page 786 of the text, the choice of a parameter determines the orientation of the curve.

Sections 11.1 and 11.2 of this chapter go over the domain, limit, continuity, differentiation and integration of vector-valued functions. Go over all of this carefully because it is the basis for the applications discussed in Sections 11.3, 11.4 and 11.5. Be sure to point out that when integrating a vector-valued function, the constant of integration is a vector, not a real number.

Go over all of the ideas presented in Sections 11.3, 11.4 and 11.5. Some, such as arc length, will be familiar to your students. Choose your assignments carefully in these sections as the problems are lengthy. Be sure to assign problems with a mix of functions, algebraic as well as transcendental.

If you are pressed for time, it is not necessary to cover every formula for curvature. However, do not omit this idea entirely.

# CHAPTER 12
# Functions of Several Variables

## Section Topics

**12.1** **Introduction to Functions of Several Variables**—Understand the notation for a function of several variables. Sketch the graph of a function of two variables. Sketch level curves for a function of two variables. Sketch level surfaces for a function of three variables. Use computer graphics to sketch the graph of a function of two variables.

**12.2** **Limits and Continuity**—Understand the definition of a neighborhood in the plane. Understand the definition of a limit of a function of two variables. Extend the concept of continuity to a function of two variables. Extend the concept of continuity to a function of three variables.

**12.3** **Partial Derivatives**—Find and use a partial derivative of a function of two variables. Find and use a partial derivative of a function of three variables. Find higher order partial derivatives of a function of two or three variables.

**12.4** **Differentials**—Understand the concepts of increments and differentials. Extend the concept of differentiability to a function of two variables. Use a differential as an approximation.

**12.5** **Chain Rules for Functions of Several Variables**—Use the Chain Rules for functions of several variables. Find partial derivatives implicitly.

**12.6** **Directional Derivatives and Gradients**—Find and use directional derivatives of a function of two variables. Find the gradient of a function of two variables. Use the gradient of a function of two variables in applications. Find directional derivatives and gradients for functions of three variables.

**12.7** **Tangent Planes and Normal Lines**—Find equations of tangent planes and normal lines to surfaces. Find the angle of inclination of a plane in space. Compare the gradients $\nabla f(x, y)$ and $\nabla F(x, y, z)$.

**12.8** **Extrema of Functions of Two Variables**—Find absolute and relative extrema of a function of two variables. Use the Second Partials Test to find relative extrema of a function of two variables.

**12.9** **Applications of Extrema of Functions of Two Variables**—Solve optimization problems involving functions of several variables. Use the method of least squares.

**12.10** **Lagrange Multipliers**—Understand the Method of Lagrange Multipliers. Use Lagrange Multipliers to solve constrained optimization problems. Use the Method of Lagrange Multipliers with two constraints.

## Chapter Comments

The terminology used for functions of several variables parallels functions of one variable. So a discussion of domain, independent and dependent variables, composite functions, and polynomial and rational functions should move quickly. Graphing functions of several variables is difficult for most students so be sure to note the comment after Example 2 on using traces in planes parallel to the coordinate planes on page 840.

Discussion of level curves and contour maps is often meaningful to your students since many have seen a topographic map depicting high mountains or deep oceans.

Unless you are taking a very theoretic approach to the course, don't get bogged down in the $\varepsilon$-$\delta$ definition of the limit of a function of two variables in Section 12.2. Point out that limits of functions of several variables have the same properties regarding sums, differences, products and quotients as do limits of functions of one variable. Go over Example 4 on page 853 with your students and be sure to point out that in order for a limit of a function of several variables to exist, the limit must be the same along *all* possible approaches.

A discussion of continuity for functions of several variables should parallel your discussion of continuity from Chapter 1.

Be sure to have your students memorize the definitions of partial derivatives for a function of several variables. Otherwise, a partial derivative won't mean anything to them. If possible, show a film depicting a geometric interpretation of a partial derivative when applied to a function of two independent variables.

Hopefully, your students already understand that for a function of a single variable, a differential is an approximation for a change in the function value. This concept is generalized in Section 12.4. Examples, such as 3 and 4 on pages 871 and 872, showing how the total differential is used to approximate a change in the function value, are important.

Be careful of notation when discussing Chain Rule in Section 12.5. Many students are careless about the notation and easily mix up $dy/dx$ and $\partial y/\partial x$. I have found that implicit differentiation seems confusing to many students. You may want to go slowly and carefully over this idea.

Sections 12.6 through 12.10 consider some applications of the partial derivative. As you begin Section 12.6 with the discussion of the directional derivative and the gradient, be sure that your students understand that a gradient is a vector; hence, the boldface print, $\nabla f$. It is important for your students to understand the difference between $\nabla f(x, y)$, a vector in the $xy$-plane, and $\nabla F(x, y, z)$, a vector in space. For a function of two independent variables, the gradient is normal to the level curves in the same plane as the curves, whereas for a function of three independent variables, the gradient, $\nabla F$, is normal to the surface $F(x, y, z) = 0$. This concept is brought out in Section 12.7 with the discussion of tangent planes and normal lines to a surface.

Section 12.8 examines extrema in space. The ideas and terminology again parallel what was done in the plane. Be sure to note that critical points occur where *one* partial derivative is undefined or when *both* partials are zero. The idea of a saddle point may be new to your students. The Second Partials Test for extrema is great when it works. However, point out to your students that like the Second Derivative Test for functions of one variable, it can fail. Example 4 on page 909 demonstrates this.

The method of least squares, discussed in Section 12.9 could be skipped if time is a problem. In order to do LaGrange Multipliers, Section 12.10, your students need to solve simultaneous equations. Encourage them to be creative in this regard.

# CHAPTER 13
# Multiple Integration

## Section Topics

**13.1** **Iterated Integrals and Area in the Plane**—Evaluate an iterated integral. Use an iterated integral to find the area of a plane region.

**13.2** **Double Integrals and Volume**—Use a double integral to represent the volume of a solid region. Use properties of double integrals. Evaluate a double integral as an iterated integral.

**13.3** **Change of Variables: Polar Coordinates**—Write and evaluate double integrals in polar coordinates.

**13.4** **Center of Mass and Moments of Inertia**—Find the mass of a planar lamina using a double integral. Find the center of mass of a planar lamina using double integrals. Find moments of inertia using double integrals.

**13.5** **Surface Area**—Use a double integral to find the area of a surface.

**13.6** **Triple Integrals and Applications**—Use a triple integral to find the volume of a solid region. Find the center of mass and moments of inertia of a solid region.

**13.7** **Triple Integrals in Cylindrical and Spherical Coordinates**—Write and evaluate a triple integral in cylindrical coordinates. Write and evaluate a triple integral in spherical coordinates.

**13.8** **Change of Variables: Jacobians**—Understand the concept of a Jacobian. Use a Jacobian to change variables in a double integral.

## Chapter Comments

I have found that the most difficult part of multiple integration for students is in setting up the limits. Go slowly and carefully over this *every time.* Point out that the variable of integration cannot appear in either limit of integration and that the outside limits must be constant with respect to *all* variables. It is important to stress early that the order of integration does not affect the value of the integral (Example 5 on page 940). However, very often it does effect the difficulty of the problem (Example 4 on page 950). Sketching the region is a necessary part of making the decision on the order of integration.

Because polar coordinates were already covered in Chapter 9, Section 13.3 should be easy for your students and so move quickly through it.

You may have to choose which of the applications in Sections 13.4 and 13.5 that you want to discuss because you probably will not have enough time for all of them. Surface area problems are often easier when done in polar coordinates.

The problems on triple integrals in Section 13.7 are lengthy. However, take the time to carefully set these up and work them out in detail. Encourage your students to consider the different orders of integration, looking for the simplest method of solving the problem.

Cylindrical and spherical coordinates, Section 13.7, are worth discussing well since many multiple integral problems are far easier done in these coordinate systems rather than in rectangular coordinates.

The Jacobian is discussed in Section 13.8. Example 1 shows why the extra factor appears when area is calculated in polar coordinates. Be sure to assign Exercise 29 which asks for the Jacobian for the change of variables from rectangular coordinates to spherical coordinates.

# CHAPTER 14
# Vector Analysis

## Section Topics

**14.1** **Vector Fields**—Understand the concept of a vector field. Decide whether a vector field is conservative. Find the curl of a vector field. Find the divergence of a vector field.

**14.2** **Line Integrals**—Understand and use the concept of a piecewise smooth curve. Write and evaluate a line integral. Write and evaluate a line integral of a vector field. Write and evaluate a line integral in differential form.

**14.3** **Conservative Vector Fields and Independence of Path**—Understand and use the Fundamental Theorem of Line Integrals. Understand the concept of independence of path. Understand the concept of conservation of energy.

**14.4** **Green's Theorem**—Use Green's Theorem to evaluate a line integral. Use alternative forms of Green's Theorem.

**14.5** **Parametric Surfaces**—Understand the definition of a parametric surface. Find a set of parametric equations to represent a surface. Find a normal vector and a tangent plane to a parametric surface. Find the area of a parametric surface.

**14.6** **Surface Integrals**—Evaluate a surface integral as a double integral. Evaluate a surface integral for a parametric surface. Determine the orientation of a surface. Understand the concept of a flux integral.

**14.7** **Divergence Theorem**—Understand and use the Divergence Theorem. Use the Divergence Theorem to calculate flux.

**14.8** **Stokes's Theorem**—Understand and use Stoke's Theorem. Use curl to analyze the motion of a rotating liquid.

## Chapter Comments

Chapter 14 is divided into two parts. Sections 14.1 through 14.4 consider situations in which the integration is done over a plane region bounded by curves. Sections 14.5 through 14.8 consider integrals over regions in space bounded by surfaces. Most students find this material difficult, so go over these topics slowly and carefully. Section 14.1 contains a lot of new ideas and terminology. Take two days if necessary to get your students familiar with these ideas. Be sure to make it clear that the divergence of a vector field $F$ is a *scalar* while **curl F** is a *vector.*

When using parametric forms of line integrals, be sure that the parameters are chosen with *increasing* parameter values that keep the specified direction along the curve. Confusion often arises when the direction of the path is not considered (Example 7 on page 1026). Students need to understand that a line integral is independent of the parameterization of a curve $C$ provided $C$ is given the same orientation by all sets of parametric equations defining $C$. The discussion of work is an application of line integrals. Be sure to include it.

Conservative vector fields and potential functions, discussed in Section 14.3, are ideas that your students will see again in differential equations. It is important that your students know the conditions necessary for the Fundamental Theorem of Line Integrals and for path independence (Example 1 on page 1032).

Green's Theorem, in Section 14.4, is a good way to integrate line integrals over closed curves. Using this theorem should be fairly easy for your students if they are able to calculate the area between two curves. Be sure to note the alternate forms of Green's Theorem on page 1047.

Surface integrals, like other multiple integrals, are tedious and time consuming. However, they should be worked out in detail. Both the Divergence Theorem and Stokes's Theorem are higher dimension analogues of Green's Theorem.

# PART II TEST AND TEST ANSWER KEYS

**Test Bank** . . . . . . . . . . 28

| | | |
|---|---|---|
| **Chapter P** | Preparation for Calculus | 28 |
| **Chapter 1** | Limits and Their Properties | 43 |
| **Chapter 2** | Differentiation | 56 |
| **Chapter 3** | Applications of Differentiation | 69 |
| **Chapter 4** | Integration | 81 |
| **Chapter 5** | Differential Equations | 95 |
| **Chapter 6** | Applications of Integration | 103 |
| **Chapter 7** | Integration Techniques, L'Hôpital's Rule, and Improper Integrals | 111 |
| **Chapter 8** | Infinite Series | 122 |
| **Chapter 9** | Conics, Parametric Equations, and Polar Coordinates | 135 |
| **Chapter 10** | Vectors and the Geometry of Space | 151 |
| **Chapter 11** | Vector-Valued Functions | 165 |
| **Chapter 12** | Functions of Several Variables | 174 |
| **Chapter 13** | Multiple Integration | 183 |
| **Chapter 14** | Vector Analysis | 192 |

**Answer Key to Chapter Tests** . . . . . . . . . . 202

**Final Exams** . . . . . . . . . . 221

**Answer Key to Final Exams** . . . . . . . . . . 237

**Gateway Tests** . . . . . . . . . . 239

**Answer Key to Gateway Tests** . . . . . . . . . . 259

**Test Form A** **Name** ______________________ **Date** ______________

**Chapter P** **Class** ______________________ **Section** ______________

**1.** Find all intercepts of the graph of $y = \dfrac{x+2}{x-3}$.

(a) $(-2, 0)$ (b) $(-2, 0), (3, 0)$ (c) $\left(0, \frac{2}{3}\right), (3, 0)$

(d) $(-2, 0), \left(0, -\frac{2}{3}\right)$ (e) None of these

**2.** Determine if the graph of $y = \dfrac{x}{x^2 - 4}$ is symmetrical with respect to the $x$-axis, the $y$-axis, or the origin.

(a) About the $x$-axis (b) About the $y$-axis (c) About the origin

(d) All of these (e) None of these

**3.** Find all points of intersection of the graphs of $x^2 - 2x - y = 6$ and $x - y = -4$.

(a) $(0, -6), (0, 4)$ (b) $(10, 14), (13, 17)$ (c) $(5, 9), (-2, 2)$

(d) $(-5, -1), (2, 6)$ (e) None of these

**4.** Which of the following is a sketch of the graph of the equation $y = x^3 + 1$?

(a)

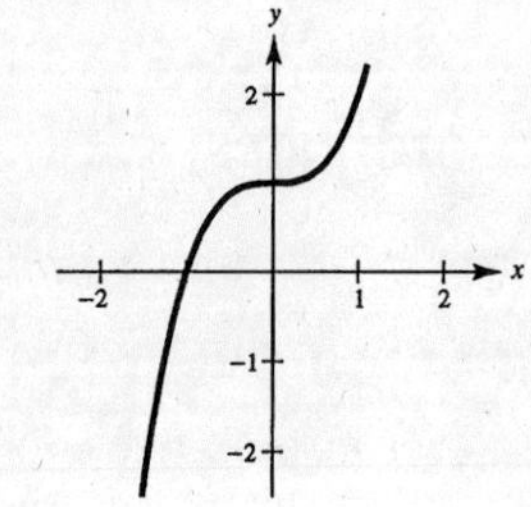


(b)

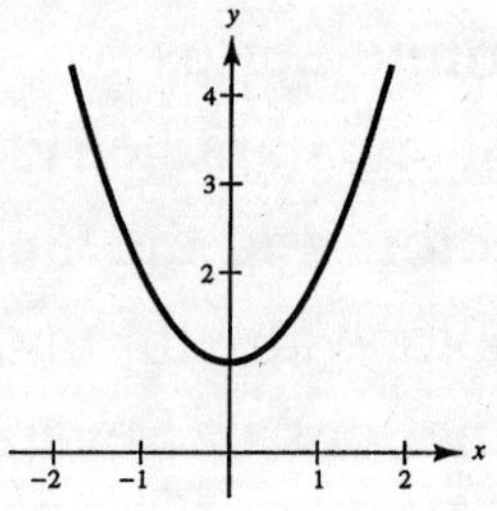


(c)

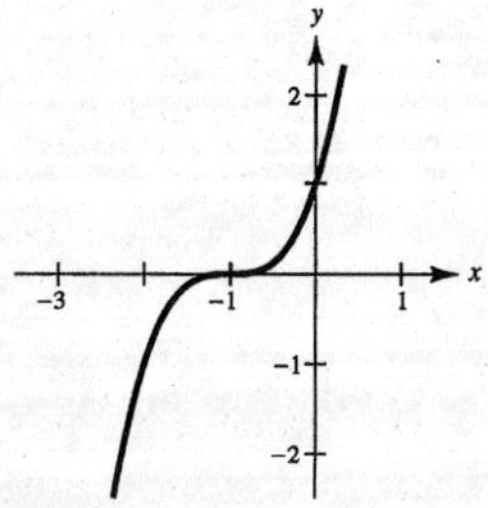


(d)

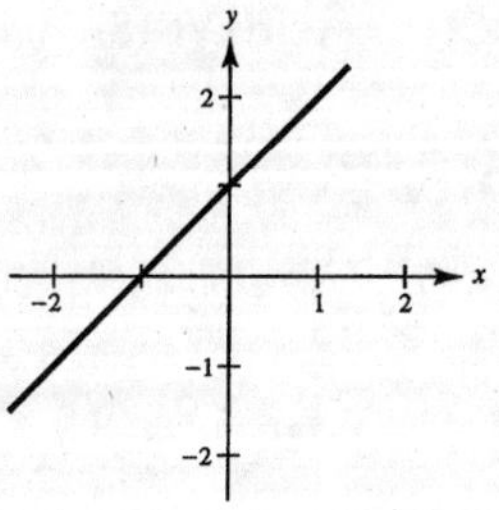


(e) None of these

**5.** Find an equation for the line passing through the point $(4, -1)$ and perpendicular to the line $2x - 3y = 3$.

(a) $y = \frac{2}{3}x - 1$ (b) $3x + 2y + 2 = 0$ (c) $2x + 3y = 10$

(d) $3x + 2y = 10$ (e) None of these

**6.** Find the domain of $f(x) = \dfrac{1}{\sqrt{3 - 2x}}$.

(a) $\left(-\infty, \dfrac{3}{2}\right)$
(b) $\left[\dfrac{3}{2}, \infty\right)$
(c) $\left(\dfrac{3}{2}, \infty\right)$
(d) $\left(-\infty, \dfrac{3}{2}\right) \cup \left(\dfrac{3}{2}, \infty\right)$
(e) None of these

**7.** Find $f(x + \Delta x)$ for $f(x) = x^3 + 1$.

(a) $x^3 + 1 + \Delta x$
(b) $x^3 + 3x^2(\Delta x) + 3x(\Delta x)^2 + (\Delta x)^3 + 1$
(c) $x^3 + (\Delta x)^3 + 1$
(d) $\Delta^3 x^6 + 1$
(e) None of these

**8.** If $f(x) = \dfrac{1}{\sqrt{x}}$ and $g(x) = 1 - x^2$, find $f(g(x))$.

(a) $\dfrac{1 - x^2}{\sqrt{x}}$
(b) $\dfrac{1}{\sqrt{1 - x^2}}$
(c) $1 - \dfrac{1}{x}$
(d) $\dfrac{1}{\sqrt{x}} + 1 - x^2$
(e) None of these

**9.** If the point $\left(-3, \frac{1}{2}\right)$ lies on the graph of the equation $2x + ky = -11$, find the value of $k$.

(a) $-\frac{5}{2}$
(b) $-34$
(c) $-\frac{17}{2}$
(d) $-10$
(e) None of these

**10.** Which of the following equations expresses $y$ as a function of $x$?

(a) $3y + 2x - 9 = 17$
(b) $2x^2y + x = 4y$
(c) Both *a* and *b*
(d) Neither *a* nor *b*
(e) $3y^2 - x^2 = 5$

**11.** Determine which function is neither even nor odd.

(a) $f(x) = \tan x$
(b) $f(x) = 3x^5 + 5x^3 + 1$
(c) $f(x) = \dfrac{3}{x^2}$
(d) $f(x) = \sqrt{x^2 + 1}$
(e) Both *a* and *b*

**12.** Describe the transformation needed to sketch the graph of $y = \dfrac{1}{x - 2}$ using the graph of $f(x) = \dfrac{1}{x}$.

(a) Shift $f(x)$ two units to the right.
(b) Shift $f(x)$ two units to the left.
(c) Shift $f(x)$ two units upward.
(d) Shift $f(x)$ two units downward.
(e) Reflect $f(x)$ about the $x$-axis.

**13.** Use the vertical line test to determine which of the following graphs represent $y$ as a function of $x$.

(a)

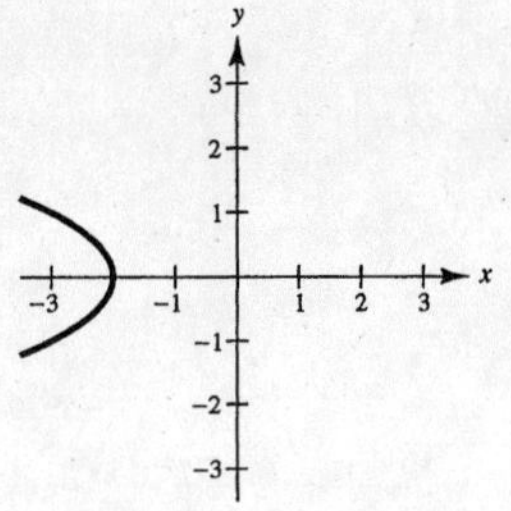

(b)

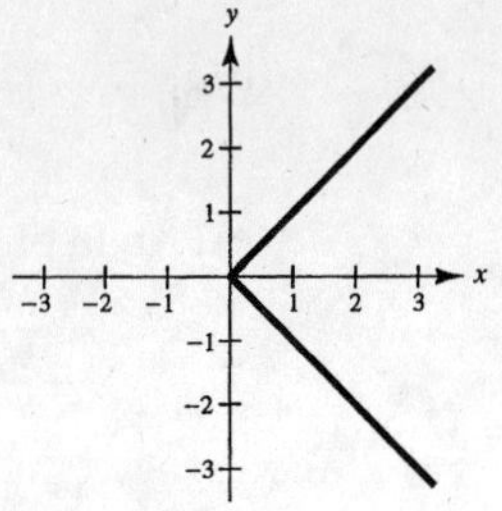

(c)

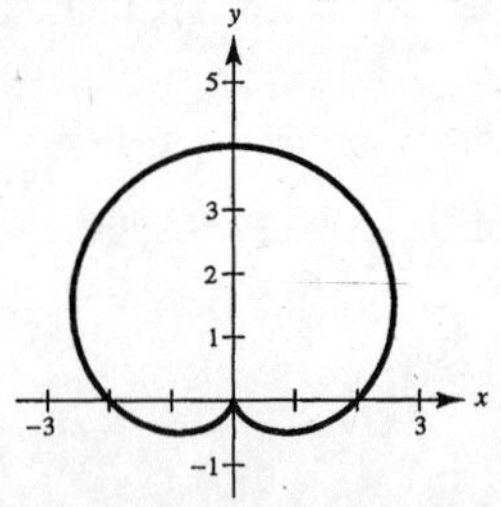

(d)

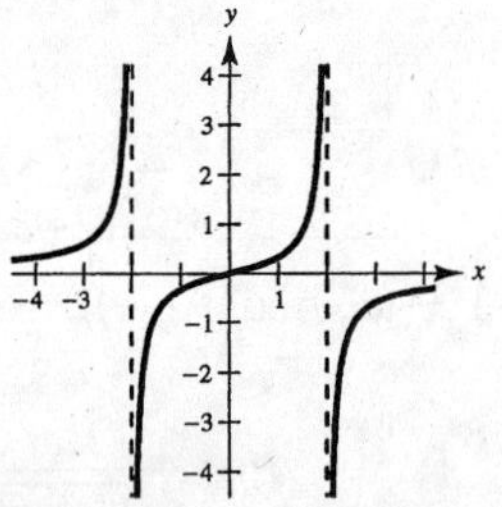

(e) None of these

**14.** Let $f(x) = \begin{cases} \dfrac{1}{x} & x < 0 \\ 2x + 1, & x \geq 0 \end{cases}$. Find $f(3)$.

(a) $\dfrac{1}{3}$ (b) 1 (c) 7

(d) Undefined (e) $\dfrac{22}{3}$

**15.** During the first and second quarters of the year, a business had sales of \$150,000 and \$185,000, respectively. If the growth of sales follows a linear pattern, what will sales be during the fourth quarter?

(a) \$220,000 (b) \$235,000 (c) \$335,000

(d) \$255,000 (e) None of these

**16.** In order for a company to realize a profit in the manufacture and sale of a certain item, the revenue, $R$, for selling $x$ items must be greater than the cost, $C$, of producing $x$ items. If $R = 79.99x$ and $C = 61x + 1050$, for what values of $x$ will this product return a profit?

(a) $x \geq 55$ (b) $x \geq 8$ (c) $x \geq 18$

(d) $x \geq 56$ (e) None of these

**17.** Determine whether $f(x) = \dfrac{x - b}{a}$ is one-to-one; if it is, find $f^{-1}$.

(a) $f$ is not one-to-one. (b) $ax + b$ (c) $\dfrac{a}{x - b}$

(d) $\dfrac{x - a}{b}$ (e) None of these

**18.** Let $f(x) = \sqrt{3x^3 - 1}$. Calculate $f^{-1}(x)$.

(a) $\dfrac{1}{\sqrt{3x^3 - 1}},\ x \geq 0$ (b) $\sqrt[3]{\dfrac{x^2 + 1}{3}},\ x \geq 0$ (c) $\sqrt{\dfrac{3}{x^3} - 1},\ x \geq 0$

(d) $\sqrt[3]{\dfrac{x^2}{3} + 1},\ x \geq 0$ (e) None of these

**19.** Solve for $x$: $\ln(5x + 1) + \ln x = \ln 4$.

(a) $-1, \frac{4}{5}$ (b) $\frac{4}{5}$ (c) $\frac{3}{5}, 4$ (d) $e^4, e^{3/5}$ (e) None of these

**20.** Given $y = \arcsin \dfrac{1}{x}$, find $\cot y$.

(a) $\sqrt{1 - x^2}$ (b) $x$ (c) $\sqrt{x^2 - 1}$ (d) $\dfrac{1}{\sqrt{1 - x^2}}$ (e) None of these

**Test Form B** **Name** ____________ **Date** ____________

**Chapter P** **Class** ____________ **Section** ____________

1. Find all intercepts of the graph of $y = \dfrac{x - 1}{x + 3}$.

(a) $(1, 0), \left(0, -\frac{1}{3}\right)$ (b) $(1, 0)$ (c) $(-3, 0), (1, 0)$

(d) $(-3, 0), \left(0, -\frac{1}{3}\right)$ (e) None of these

2. Determine if the graph of $y = \dfrac{x^2}{x^2 - 4}$ is symmetrical with respect to the $x$-axis, the $y$-axis, or the origin.

(a) About the $x$-axis (b) About the $y$-axis (c) About the origin

(d) All of these (e) None of these

3. Find all points of intersection of the graphs of $x^2 + 3x - y = 3$ and $x + y = 2$.

(a) $(5, -3), (1, 1)$ (b) $(0, -3), (0, 2)$ (c) $(-5, -3), (1, 1)$

(d) $(-5, 7), (1, 1)$ (e) None of these

4. Which of the following is a sketch of the graph of the equation $y = (x - 1)^3$?

(a)

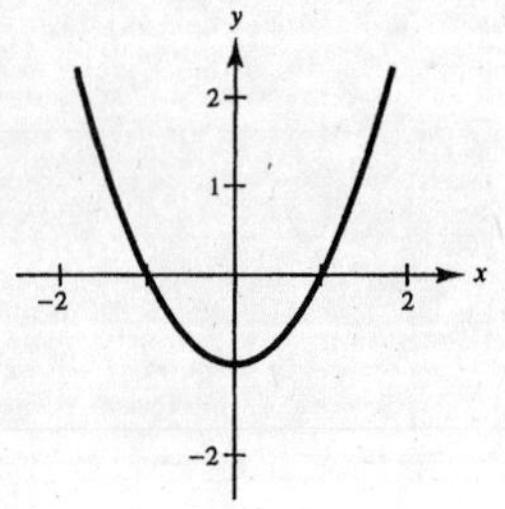

(b)

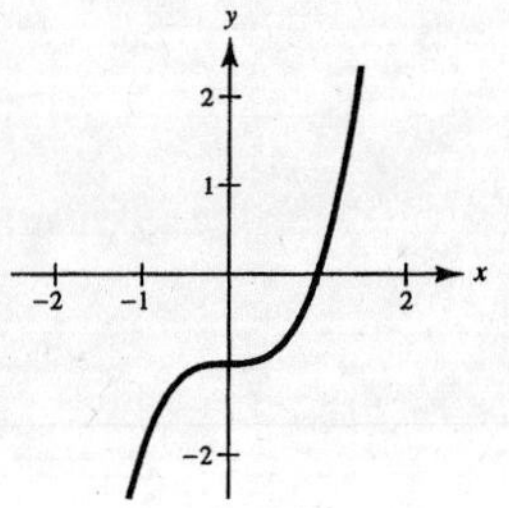

(c)

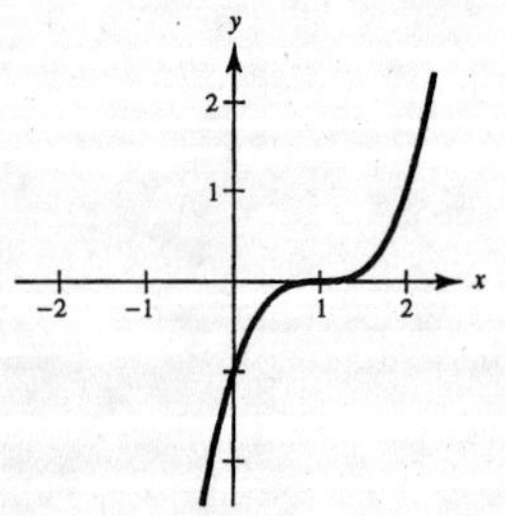

(d)

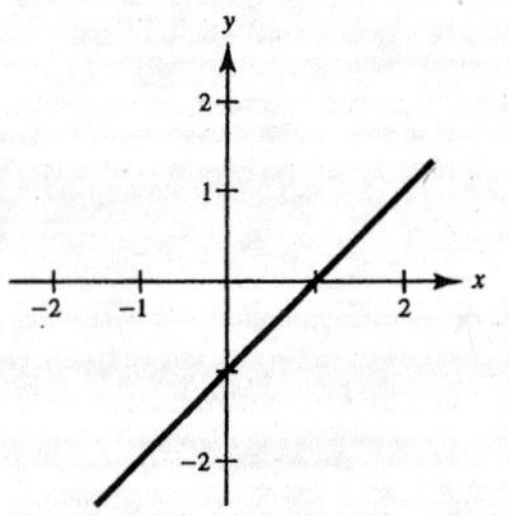

(e) None of these

5. Find an equation for the line passing through the point $(4, -1)$ and parallel to the line $2x - 3y = 3$.

(a) $2x - 3y = 11$ (b) $2x - 3y = -5$ (c) $3x - 2y = -5$

(d) $y = \frac{2}{3}x - 1$ (e) None of these

**6.** Find the domain of $f(x) = \dfrac{1}{\sqrt{3 + 2x}}$.

(a) $\left(-\infty, -\frac{3}{2}\right)$ (b) $\left[-\frac{3}{2}, \infty\right)$ (c) $\left(-\frac{3}{2}, \infty\right)$

(d) $\left(-\infty, -\frac{3}{2}\right) \cup \left(-\frac{3}{2}, \infty\right)$ (e) None of these

**7.** Find $f(x + \Delta x)$ for $f(x) = x^2 - 2x - 3$.

(a) $x^2 - x - 3 + \Delta x$ (b) $x^2 + 2x(\Delta x) + (\Delta x)^2 - 2x - 2\,\Delta x - 3$

(c) $x^2 - 2x - 3 + \Delta x$ (d) 5

(e) None of these

**8.** If $f(x) = 1 - x^2$ and $g(x) = \dfrac{1}{\sqrt{x}}$, find $f(g(x))$.

(a) $\dfrac{1 - x^2}{\sqrt{x}}$ (b) $\dfrac{1}{\sqrt{1 - x^2}}$ (c) $1 - \dfrac{1}{x}$

(d) $\dfrac{1}{\sqrt{x}} + 1 - x^2$ (e) None of these

**9.** If the point $(-1, 1)$ lies on the graph of the equation $kx^2 - xy + y^2 = 5$, find the value of $k$.

(a) 7 (b) 3 (c) 5

(d) $-3$ (e) None of these

**10.** In which of the following equations is $y$ a function of $x$?

(a) $2x + 3y - 1 = 0$ (b) $x^2 + 3y^2 = 7$ (c) $2x^2y = 7$

(d) Both *a* and *b* (e) Both *a* and *c*

**11.** Determine the even function.

(a) $f(x) = \sin x$ (b) $f(x) = \dfrac{x^3}{x^2 + 1}$ (c) $f(x) = 3x^4 + 5x^2 - 1$

(d) $f(x) = \sqrt{x^3 + 1}$ (e) None of these

**12.** Describe the transformation needed to sketch the graph of $y = \dfrac{1}{x} + 2$ using the graph of $f(x) = \dfrac{1}{x}$.

(a) Shift $f(x)$ two units to the right.

(b) Shift $f(x)$ two units to the left.

(c) Shift $f(x)$ two units upward.

(d) Shift $f(x)$ two units downward.

(e) Reflect $f(x)$ about the $x$-axis.

**13.** Use the vertical line test to determine which of the following graphs does *not* represent $y$ as a function of $x$.

(a)

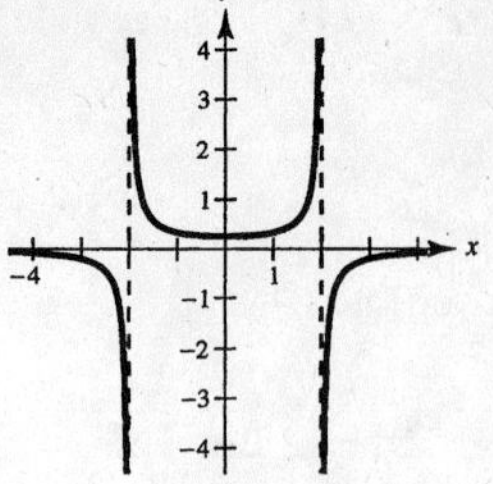


(b)

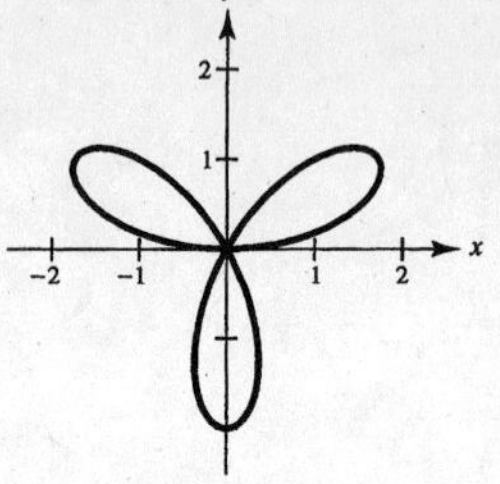


(c)

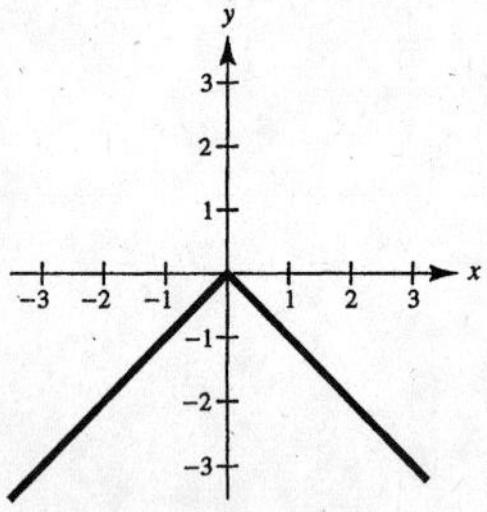


(d)

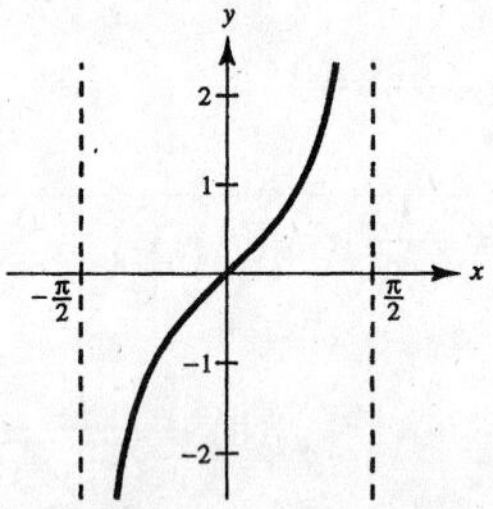


(e) Both *a* and *d*

**14.** Let $f(x) = \begin{cases} x^2 - 5, & x < 2 \\ 3x + 1, & x \geq 2 \end{cases}$. Find $f(1)$.

(a) $-4$ (b) $-2$ (c) 4

(d) 2 (e) 0

**15.** A business had annual retail sales of \$124,000 in 1993 and \$211,000 in 1996. Assuming that the annual increase in sales follows a linear pattern, predict the retail for 2001.

(a) \$356,000 (b) \$435,000 (c) \$646,000

(d) \$298,000 (e) \$327,000

**16.** In order for a company to realize a profit in the manufacture and sale of a certain item, the revenue, $R$, for selling $x$ items must be greater than the cost, $C$, of producing $x$ items. If $R = 69.99x$ and $C = 59x + 850$, for what values of $x$ will this product return a profit?

(a) $x \geq 78$ (b) $x \geq 15$ (c) $x \geq 85$

(d) $x \geq 13$ (e) None of these

**17.** Determine whether the function $f(x) = \dfrac{7}{x + 2}$ is one-to-one. If it is, find its inverse.

(a) $f$ is not one-to-one. (b) $f^{-1}(x) = \dfrac{x + 2}{7}$ (c) $f^{-1}(x) = \dfrac{7 - 2x}{x}$

(d) $f^{-1}(x) = \dfrac{7}{x + 2}$ (e) None of these

**18.** Let $f(x) = \sqrt[3]{1 - x}$. Calculate $f^{-1}(x)$.

(a) $1 - x^3$ (b) $\dfrac{1}{\sqrt[3]{1 - x}}$ (c) $(1 - x)^3$ (d) $\sqrt[3]{x - 1}$ (e) None of these

**19.** Solve for $x$: $2^{3x-1} = 5$.

(a) $\dfrac{1}{3}\ln 10$ (b) $\dfrac{7}{6}$ (c) $\dfrac{1}{3}[\ln 2.5 + 1]$

(d) $\dfrac{1}{3}\left[\dfrac{\ln 5}{\ln 2} + 1\right]$ (e) None of these

**20.** Given $y = \arctan \dfrac{1}{x}$, find $\cos y$.

(a) $\dfrac{x}{\sqrt{1 + x^2}}$ (b) $\dfrac{x}{1 + x^2}$ (c) $\dfrac{1}{\sqrt{1 + x^2}}$ (d) $x$ (e) None of these

**Test Form C** **Name** ____________ **Date** ________

**Chapter P** **Class** ____________ **Section** ________

**A graphing calculator/utility is recommended for this test.**

1. Find the $x$-intercepts for the graph of $y = 3x^5 + 5x^4 - 2x^3$.

   (a) $(0, 0), (-2, 0), \left(\frac{1}{3}, 0\right)$ (b) $(0, 0)$ (c) $(0, 0), (2, 0), (-1, 0)$

   (d) $\left(-2, \frac{1}{3}\right)$ (e) There are no $x$-intercepts.

2. Use a graphing utility to graph the equation $y = -x^4 + 3x^3 + 20$.

   (a)

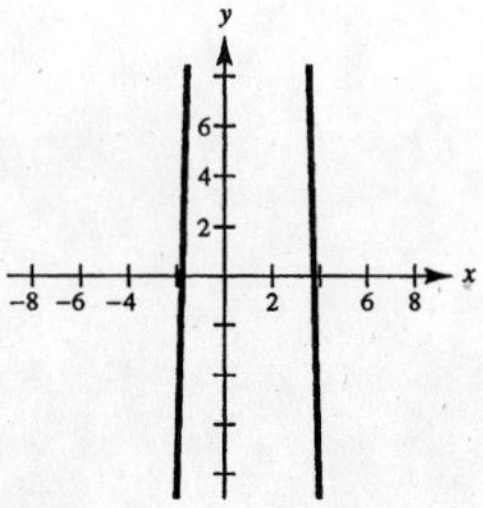


   (b)

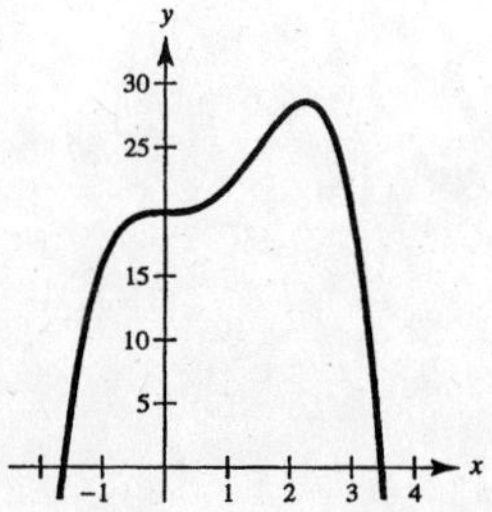


   (c)

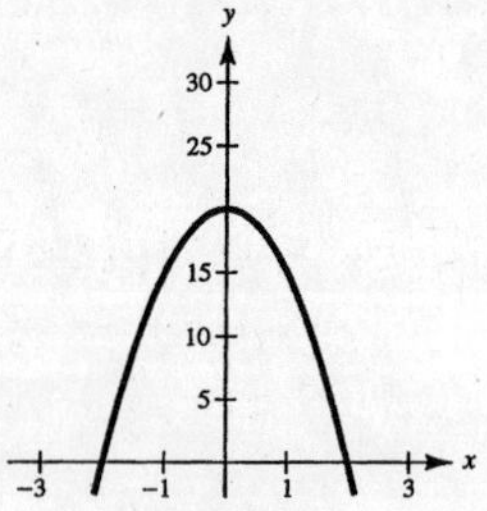


   (d)

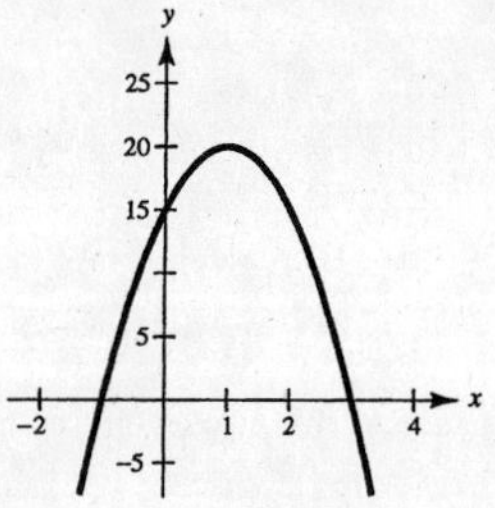


   (e) None of these

3. Identify the type(s) of symmetry: $x^4y^2 + 2x^2y - 1 = 0$.

   (a) About the $x$-axis (b) About the $y$-axis (c) About the origin

   (d) Both $a$ and $b$ (e) None of these

4. Find all points of intersection of the graphs of $x^2 + 3x - y = 3$ and $x + y = 2$.

   (a) $(2, 0)$ and $(-1, 3)$ (b) $(5, -3)$ and $(-1, 3)$ (c) $(1, 1)$ and $(-5, 7)$

   (d) $(5, 37)$ and $(-1, -5)$ (e) None of these

5. A business had annual retail sales of \$110,000 in 1993 and \$224,000 in 1996. Assuming that the annual increase in sales followed a linear pattern, what was the retail sales in 1995?

   (a) \$182,000 (b) \$195,000 (c) \$188,000

   (d) \$186,000 (e) None of these

**6.** Find an equation for the horizontal line that passes through the point $(-3, 2)$.

(a) $x = 2$ (b) $y = 2$ (c) $x = -3$

(d) $y = -3$ (e) None of these

**7.** Write an equation satisfied by the graph obtained by shifting $y = 2x - 5$ three units to the left.

(a) $y = 2x - 8$ (b) $y = 2x + 1$ (c) $y = 2x - 11$

(d) $y = 2x - 2$ (e) None of these

**8.** Let $f(x) = \dfrac{1}{\sqrt{x}}$ and $g(x) = 2x + 3$. Find the domain of $(f \circ g)(x)$.

(a) $x \geq -\dfrac{3}{2}$ (b) $x > 0$ (c) $x < \dfrac{2}{3}$

(d) $x > -\dfrac{3}{2}$ (e) None of these

**9.** Find the zero(s) of the function $f(x) = \dfrac{1}{x - 3} + \dfrac{1}{x - 4}$.

(a) 3 and 4 (b) 0 (c) $\dfrac{7}{2}$

(d) $\dfrac{2}{7}$ (e) None of these

**10.** Given $f(x) = |3x + 1| - 5$, find $f(x + 1) - f(x)$.

(a) 3 (b) $-5$ (c) $|3x + 4| - |3x + 1| - 10$

(d) $|3x + 4| - |3x + 1|$ (e) None of these

**11.** Find $\dfrac{f(x + \Delta x) - f(x)}{\Delta x}$ for $f(x) = 8x^2 + 1$.

(a) $8(\Delta x)^2 + 1$ (b) $8\,\Delta x + \dfrac{1}{\Delta x}$ (c) $16x + 8\,\Delta x$

(d) $16x(\Delta x) + 8(\Delta x)^2$ (e) None of these

**12.** Determine the odd function.

(a) $f(x) = x^5 + x^3 + x + 1$ (b) $f(x) = \dfrac{x^3}{x^2 + 1}$ (c) $f(x) = 3x^2 + 5x - 1$

(d) $f(x) = \cos x$ (e) None of these

**13.** Find the point that does not lie on the line determined by the points $(-5, 2)$ and $(1, -3)$.

(a) $(0, -4)$ (b) $(7, -8)$ (c) $(-11, 7)$

(d) $\left(-2, -\frac{1}{2}\right)$ (e) $(13, -13)$

**14.** Find the equation of the line that passes through the point (4, 2) and is perpendicular to the line that passes through the points (9, 7) and (11, 4).

(a) $y = \frac{2}{3}(x - 9) + 7$ (b) $y = -\frac{3}{2}(x - 9) + 7$ (c) $y = -\frac{3}{2}(x - 4) + 2$

(d) $y = \frac{2}{3}(x - 4) + 2$ (e) None of these

**15.** Let $f(x) = \begin{cases} 1 - 2x, & x < 1 \\ -x^2, & x \geq 1 \end{cases}$. Find $f(5)$.

(a) $-9$ (b) $-25$ (c) $-17$

(d) $-34$ (e) $-1$

**16.** An open box is to be made from a rectangular piece of material 9 inches by 12 inches by cutting equal squares from each corner and turning up the sides. Let $x$ be the length of each side of the square cut out of each corner. Write the volume $V$ of the box as a function of $x$.

(a) $V = x^3$ (b) $V = 108x$

(c) $V = x(9 - x)(12 - x)$ (d) $V = x(9 - 2x)(12 - 2x)$

(e) None of these

**17.** If $\ln a = 2.565$ and $\ln b = 3.045$, find $\ln\left(\frac{a}{b}\right)$.

(a) $-0.48$ (b) 0.842 (c) 5.61 (d) 0.48 (e) None of these

**18.** Use a graphing utility to graph the function $f(x) = \ln(x^2 - 3x + 2)$ and use the graph to state the domain of $f$.

(a) $(-\infty, \infty)$ (b) $(-\infty, 1), (2, \infty)$ (c) $(-\infty, 0), (0, \infty)$

(d) $\left(-\infty, \frac{3}{2}\right), \left(\frac{3}{2}, \infty\right)$ (e) None of these

**19.** Determine the inverse function of the function $f(x) = -x$.

(a) $x$ (b) $\frac{1}{x}$ (c) $-x$ (d) $-\frac{1}{x}$ (e) None of these

**20.** Use your calculator to find the value of arcsec(1.459). Round your answer to four decimal places.

(a) 46.7332 (b) 1.0939 (c) Does not exist

(d) 0.8156 (e) None of these

**Test Form D** **Name** ________________ **Date** __________

**Chapter P** **Class** ________________ **Section** __________

**1.** Show that $y = \dfrac{x}{x^2 + 1}$ is symmetric with respect to the origin.

**2.** Find the intercepts: $y = \dfrac{2x - 1}{3 - x}$.

**3.** Find all points of intersection: $y = -x^2 + 4x$ and $y = x^2$.

**4.** Find an equation in general form for the straight line that passes through the point $(-1, 4)$ and is perpendicular to the line $2x + 3y = 6$.

**5.** If $g(x) = x^2 + 3x - 1$, find $\dfrac{g(x + \Delta x) - g(x)}{\Delta x}$.

**6.** If $f(x) = \dfrac{1}{\sqrt{x}}$ and $g(x) = x^2 - 5$, find $g(f(x))$.

**7.** Find the domain: $f(x) = \dfrac{1}{x^2 - 2x - 2}$.

**8.** Sketch a graph of $y = x^3 - 1$.

**9.** Sketch the graph of the equation $4x - 2y + 8 = 0$.

**10.** In which of the following equations is $y$ a function of $x$?

**a.** $3x + 2y - 7 = 0$ **b.** $5x^2y = 9 - 2x$ **c.** $3x^2 - 4y^2 = 9$

**d.** $x = 3y^2 - 1$ **e.** None of these

**11.** Determine whether the function is even, odd, or neither. Justify your answer.

$f(x) = -x^5 + 3x^3 - 2x + 1$

**12.** Use the graph of $f(x) = |x|$ to sketch the graph of $y = |x - 1| + 3$.

**13.** Find the slope and $y$-intercept of the line given by the equation $5x + 4y - 12 = 0$.

**14.** Let $f(x) = \begin{cases} x^2 - 4, & x < 2 \\ 3 - 2x, & x \ge 2 \end{cases}$. Evaluate:

**a.** $f(0)$ **b.** $f(2)$ **c.** $f(3)$

**15.** A student working for a telemarket company gets paid \$3 per hour plus \$1.50 for each sale. Let $x$ represent the number of sales the student has in an 8-hour day.

**a.** Write a linear equation giving the day's salary $S$ in terms of $x$.

**b.** Use the linear equation to calculate the student's salary on Wednesday if the student makes 14 sales that day.

**c.** Use the linear equation to calculate the number of sales per day the student would have to make in order to earn at least \$100 a day.

**16.** Determine whether $f(x) = \dfrac{1}{ax + b}$ is one-to-one; if it is, find $f^{-1}$.

**17.** Write an algebraic expression for tan[arcsin $x$]. $\left(-\dfrac{\pi}{2} \leq \arcsin x \leq \dfrac{\pi}{2}\right)$

**18.** Let $f(x) = x^3 + 5$. Find $f^{-1}$.

**19.** Solve for $x$: $\ln(5x - 1) - \ln x = 3$.

**20.** Solve for $x$: $(x - 2)^{3/2} = 8$.

**Test Form E** Name ______________ Date __________

**Chapter P** Class ______________ Section __________

**A graphing calculator/utility is recommended for this test.**

**1.** Given the equation: $x^2 + y^2 + 4x - 6y + 12 = 0$.

**a.** Solve for $y$.

**b.** Use a graphing utility to graph the resulting equations on the same set of axes and sketch the graph.

**2.** Create an equation whose graph has intercepts at $(-5, 0)$, $(0, 0)$, and $(5, 0)$.

**3.** **a.** Use a graphing utility to graph the equation $y = x^4 - 3x^2 + 2$.

**b.** Identify the intercepts of the graph.

**c.** Test for symmetry.

**4.** Identify the type(s) of symmetry: $x^2 + xy + y^2 = 0$.

**5.** Find all points of intersection of the graphs of $x^2 - 2x - y = 6$ and $x - y = -4$.

**6.** Sketch the graph of the equation: $4x - 2y + 8 = 0$.

**7.** Let $f(x) = \begin{cases} |x|, & x < 2 \\ x - 3, & x \geq 2 \end{cases}$. Evaluate:

**a.** $f(-3)$

**b.** $f(-2)$

**c.** $f(0)$

**d.** $f(2)$

**8.** Let $f(x) = x + \sqrt{x + 2}$.

**a.** Use a graphing utility to graph $y = f(x)$.

**b.** Estimate the domain and range from the graph.

**c.** Estimate the coordinates of any intercepts of the graph.

**d.** Find the intercepts analytically.

**9.** Use a graphing utility to estimate the zero(s) of $f(x) = x^3 - 5x + 2$.

**10.** In which of the following is $y$ a function of $x$?

**a.** $y = 3x^2 - 9$

**b.** $x^2 + y^2 = 7$

**c.** $x^2 - y^2 = 2$

**d.** $3x + 2y = 5$

**e.** $|x| = y$

**11.** Given $f(x) = |3x - 6|$, find $f(0) - f(3)$.

**12.** Find $\dfrac{g(x + \Delta x) - g(x)}{\Delta x}$ for $g(x) = x^2 + 3x - 1$.

**13.** Determine whether the function $f(x) = -x^4 + 2x^2 - 1$ is even, odd, or neither. Justify your answer.

**14.** Use the graph of $f$ shown below to sketch the graph of $y = f(x) - 2$.

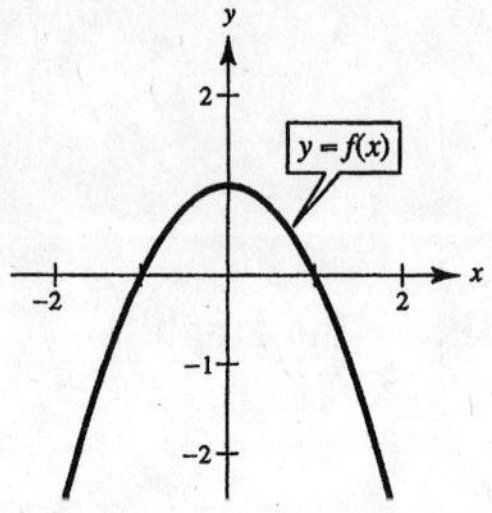


**15.** For the functions $f(x) = x - 2$ and $g(x) = \dfrac{x + 5}{3}$, find $g(f(x))$.

**16.** A business had annual retail sales of $224,000 in 1993 and $186,500 in 1996. Assume the annual decrease in sales follows a linear pattern.

**a.** Write a linear equation giving sales $S$ in terms of the year $t$ where $t = 0$ corresponds to 1993.

**b.** Use a graphing utility to graph the equation.

**c.** Use the graph to estimate the annual retail sales for 1998 to two digits of precision (the nearest multiple of $10,000).

**d.** Use the graph to estimate the first year when there will be no sales.

**17.** If $\ln a = 2.565$ and $\ln b = 3.045$, find:

**a.** $\ln(ab)$

**b.** $\ln \sqrt{a}$

**18.** Use a graphing utility to graph the function $f(x) = x \ln x$. Use the graph to determine whether the function is one-to-one over its entire domain.

**19.** Find the exact value: $\cos\left[\arctan\left(-\frac{3}{10}\right)\right]$.

**20.** Solve for $x$: $e^{4x+1} = 3$.

**Test Form A** **Name** ____________ **Date** ____________

**Chapter 1** **Class** ____________ **Section** ____________

**1.** Find the limit: $\lim_{x \to 2} (3x^2 + 5)$.

(a) 41 (b) 17 (c) 11

(d) 0 (e) None of these

**2.** Given $\lim_{x \to 2} (2x - 1) = 3$. Find $\delta$ such that $|(2x - 1) - 3| < 0.01$ whenever $0 < |x - 2| < \delta$.

(a) 3 (b) 0.05 (c) 0.03

(d) 0.005 (e) None of these

**3.** Use the graph to find $\lim_{x \to 1} f(x)$ if $f(x) = \begin{cases} 3 - x, & x \neq 1 \\ 1, & x = 1 \end{cases}$.

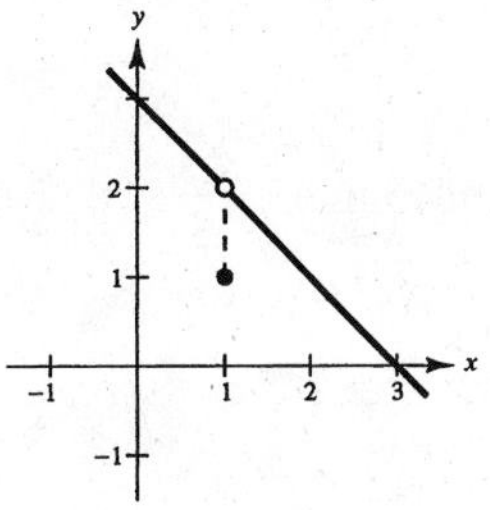


(a) 2 (b) 1

(c) $\frac{3}{2}$ (d) Does not exist

(e) None of these

**4.** Find the limit: $\lim_{x \to -1} \dfrac{x^2 + 3x + 2}{x^2 + 1}$.

(a) 0 (b) $\infty$ (c) $-1$

(d) Does not exist (e) None of these

**5.** Find the limit: $\lim_{x \to 1} (\ln x^4)$.

(a) 0 (b) 4 (c) 1

(d) $4e$ (e) None of these

**6.** If $\lim_{x \to c} f(x) = -\frac{1}{2}$ and $\lim_{x \to c} g(x) = \frac{2}{3}$, find the limit: $\lim_{x \to c} \dfrac{f(x)}{g(x)}$.

(a) $-\frac{1}{3}$ (b) $\frac{1}{3}$ (c) $-\frac{3}{4}$

(d) $-3$ (e) None of these

7. Find the limit: $\lim_{x \to 2} \frac{x - 2}{x^2 - 4}$.

(a) 0 (b) $\frac{1}{4}$ (c) $\infty$

(d) 1 (e) None of these

8. Find the limit: $\lim_{x \to 0} \frac{\sqrt{x + 4} - 2}{x}$.

(a) 0 (b) $\frac{1}{4}$ (c) $\infty$

(d) 1 (e) None of these

9. Find the limit: $\lim_{x \to 3} \frac{x - 3}{|x - 3|}$.

(a) 0 (b) 1 (c) 3

(d) Does not exist (e) None of these

10. Find the limit: $\lim_{x \to 2} \sec \frac{\pi x}{3}$.

(a) $-2$ (b) $\frac{2}{\sqrt{3}}$ (c) $-\frac{\sqrt{3}}{2}$

(d) $\frac{1}{2}$ (e) None of these

11. Find the limit: $\lim_{x \to 0} \frac{x}{\tan x}$.

(a) 0 (b) $\frac{\pi}{4}$ (c) 1

(d) Does not exist (e) None of these

12. Find the limit: $\lim_{x \to 3^+} \sqrt{2x - 5}$.

(a) 1 (b) 0 (c) $2i$

(d) Does not exist (e) None of these

13. Find the limit: $\lim_{x \to 2^-} \frac{1}{x - 2}$.

(a) $\infty$ (b) $-\infty$ (c) 0

(d) $-\frac{1}{4}$ (e) None of these

**14.** Find the limit: $\lim_{x \to 2} \frac{1}{(x-2)^2}$.

(a) $\infty$ (b) $-\infty$ (c) 0

(d) $\frac{1}{4}$ (e) None of these

**15.** Find the limit: $\lim_{x \to 0} \frac{e^{2x}-1}{e^x-1}$.

(a) $e^2$ (b) 0 (c) 1

(d) 2 (e) None of these

**16.** At which value(s) of $x$ is $f(x) = \frac{x^2-2x-3}{x-2}$ discontinuous?

(a) 2 (b) $-1, 2, 3$ (c) 1

(d) $-1, \frac{3}{2}, 2, 3$ (e) None of these

**17.** Let $f(x) = \frac{1}{x+1}$ and $g(x) = x^2 - 5$. Find all values of $x$ for which $f(g(x))$ is discontinuous.

(a) $-1$ (b) $-1, \pm\sqrt{5}$ (c) $\pm\sqrt{5}$

(d) $-2, 2$ (e) None of these

**18.** Determine the value of $c$ so that $f(x)$ is continuous on the entire real line when $f(x) = \begin{cases} x-2, & x \le 5 \\ cx-3, & x > 5 \end{cases}$.

(a) 0 (b) $\frac{6}{5}$ (c) 1

(d) $\frac{5}{6}$ (e) None of these

**19.** Find all vertical asymptotes of $f(x) = \frac{x-3}{x+2}$.

(a) $x = -2, x = 3$ (b) $x = -2$ (c) $x = 3$

(d) $x = 1$ (e) None of these

**20.** Find all vertical asymptotes of $g(x) = \frac{2x+3}{2x^2+x-3}$.

(a) $x = -\frac{3}{2}, x = 1$ (b) $x = -\frac{3}{2}$ (c) $x = 1$

(d) $y = 1$ (e) None of these

**Test Form B** Name ______________ Date ______________

**Chapter 1** Class ______________ Section ______________

1. Find the limit: $\lim_{x\to -3} (-2x^2 + 1)$.

(a) 37 (b) 19 (c) $-17$

(d) $\pm\sqrt{2}$ (e) None of these

2. Given $\lim_{x\to 1} (2x + 1) = 3$. Find $\delta$ such that $|(2x + 1) - 3| < 0.01$ whenever $0 < |x - 1| < \delta$.

(a) 3 (b) 0.05 (c) 0.005

(d) 1 (e) None of these

3. Use the graph at the right to find $\lim_{x\to -1} f(x)$ for $f(x) = \dfrac{1}{x+1}$.

(a) 0 (b) 1

(c) $\infty$ (d) Does not exist

(e) None of these

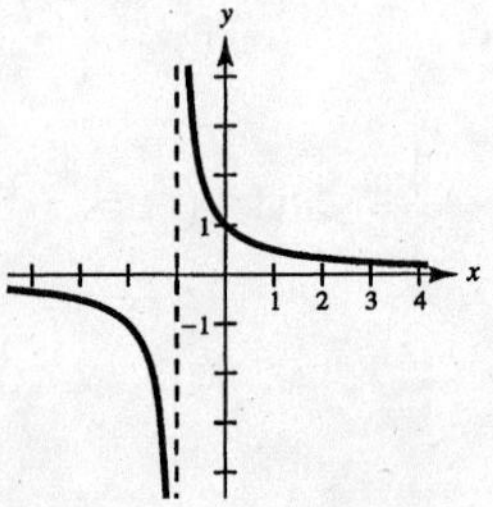


4. Find the limit: $\lim_{x\to -1} \dfrac{x^2 + 2x + 3}{x^2 + 1}$.

(a) 0 (b) 1 (c) $\infty$

(d) Does not exist (e) None of these

5. Find the limit: $\lim_{x\to e} \dfrac{\ln(\ln x)}{\ln x}$.

(a) ln (b) 0 (c) 1

(d) $e$ (e) None of these

6. If $\lim_{x\to c} f(x) = -\frac{1}{2}$ and $\lim_{x\to c} g(x) = \frac{2}{3}$, find $\lim_{x\to c} [f(x)g(x)]$.

(a) $\frac{1}{6}$ (b) $-\frac{1}{3}$ (c) 1

(d) Does not exist (e) None of these

7. Find the limit: $\lim_{x\to -1} \dfrac{x^2 - 5x - 6}{x + 1}$.

(a) 0 (b) $-7$ (c) $-\infty$

(d) $\infty$ (e) None of these

8. Find the limit: $\lim_{x\to 2} \dfrac{x-2}{|x-2|}$.

(a) 0 (b) 1 (c) 2

(d) Does not exist (e) None of these

9. Find the limit: $\lim_{x\to 0} \dfrac{\sqrt{x+9}-3}{x}$.

(a) 0 (b) 1 (c) $\infty$

(d) $\dfrac{1}{3}$ (e) None of these

10. Find the limit: $\lim_{x\to 5} \csc \dfrac{\pi x}{4}$.

(a) 1 (b) $-1$ (c) $-\sqrt{2}$

(d) $-\dfrac{1}{\sqrt{2}}$ (e) None of these

11. Find the limit: $\lim_{x\to 0} \dfrac{1-\cos^2 x}{x}$.

(a) 1 (b) 0 (c) $\infty$

(d) Does not exist (e) None of these

12. Find the limit: $\lim_{x\to 2^-} \sqrt{2x-3}$.

(a) $1, -1$ (b) 1 (c) $-1$

(d) $\frac{1}{2}$ (e) None of these

13. Find the limit: $\lim_{x\to 0^+} \dfrac{1}{x}$.

(a) $+\infty$ (b) 0 (c) $-\infty$

(d) Does not exist (e) None of these

14. Find the limit: $\lim_{x\to 0} e^{[(\sin x)/x]}$.

(a) 1 (b) $\dfrac{1}{e}$ (c) $e$

(d) Does not exist. (e) None of these

15. Find the limit: $\lim_{x\to 1} \left(2 - \dfrac{5}{(x-1)^2}\right)$.

(a) $-\infty$ (b) $+\infty$ (c) $-3$

(d) 2 (e) None of these

16. At which values of $x$ is $f(x) = \dfrac{x-4}{x^2-x-2}$ discontinuous?

(a) 4 (b) $-1, 2, 4$ (c) $-1, 2$

(d) $-1, 2, 4, -2$ (e) None of these

17. Let $f(x) = \dfrac{1}{|x|}$ and $g(x) = x - 1$. Find all values of $x$ for which $f(g(x))$ is discontinuous.

(a) 0 (b) 1 (c) 0, 1

(d) $-1, 1$ (e) None of these

18. Determine the value of $c$ so that $f(x)$ is continuous on the entire real line when $f(x) = \begin{cases} x+3, & x \le -1 \\ 2x - c, & x > -1 \end{cases}$.

(a) $-4$ (b) 4 (c) 0

(d) $-1$ (e) None of these

19. Find all vertical asymptotes of $f(x) = \dfrac{2x-1}{x+3}$.

(a) $x = 2$ (b) $x = \dfrac{1}{2}, x = -3$ (c) $x = -3$

(d) $x = \dfrac{1}{2}$ (e) None of these

20. Find all vertical asymptotes of $f(x) = \dfrac{x-2}{x^2-4}$.

(a) $x = -2, x = 2$ (b) $x = -2$ (c) $x = 0$

(d) $x = 2$ (e) None of these

**Test Form C** Name ____________ Date ________

**Chapter 1** Class ____________ Section ________

**A graphing calculator/utility is recommended for this test.**

1. Use a graphing utility to graph the function: $f(x) = -x^2 + 4x$ and then estimate $\lim_{x\to 2} f(x)$ (if it exists).

   (a) 0 (b) 12 (c) 4

   (d) $-12$ (e) None of these

2. Use the graph to find $\lim_{x\to -1} f(x)$ (if it exists).

   (a) 1

   (b) $-2$

   (c) The limit does not exist.

   (d) $-1$

   (e) $-3$

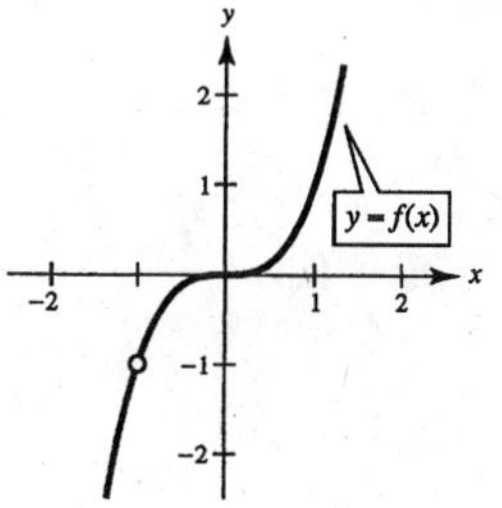


3. Find the limit: $\lim_{x\to 1} \frac{x^2 + x - 2}{x - 3}$.

   (a) 3 (b) 0 (c) $-\frac{1}{2}$

   (d) The limit does not exist. (e) None of these

4. Find the limit: $\lim_{x\to \pi/2} \frac{\sin x}{x}$.

   (a) 0 (b) $\frac{2}{\pi}$ (c) $-\frac{\pi}{2}$

   (d) $\frac{2\sqrt{2}}{\pi}$ (e) None of these

5. Use a graphing utility to graph the function $f(x) = \frac{\ln x}{x - 1}$ and then estimate $\lim_{x\to 1} f(x)$.

   (a) 0 (b) 1 (c) The limit does not exist.

   (d) $e$ (e) None of these

6. Find the limit: $\lim_{x\to 0} \frac{e^{3x} - 1}{e^x - 1}$.

   (a) 3 (b) The limit does not exist. (c) 0

   (d) 1 (e) None of these

**7.** Find the limit: $\lim_{x\to 0} \dfrac{\dfrac{1}{x+3} - \dfrac{1}{3}}{x}$.

(a) $-\dfrac{1}{9}$ (b) 0 (c) $\dfrac{1}{9}$

(d) The limit does not exist. (e) None of these

**8.** Find the limit: $\lim_{x\to 0} \dfrac{1 - \cos^2 x}{x}$.

(a) 1 (b) 0 (c) $\infty$

(d) Does not exist (e) None of these

**9.** Match the graph with the correct function.

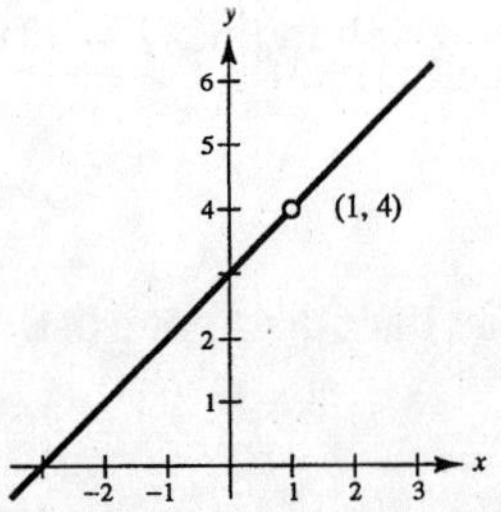


(a) $f(x) = \dfrac{x+3}{x-1}$ (b) $f(x) = x + 3$

(c) $f(x) = \dfrac{x-1}{x^2+2x-3}$ (d) $f(x) = \dfrac{x^2+2x-3}{x-1}$

(e) None of these

**10.** Find the $x$-values (if any) for which $f$ is not continuous. $f(x) = \begin{cases} \dfrac{1}{x-3}, & x \le 5 \\ \dfrac{1}{2}, & x > 5 \end{cases}$

(a) 5 (b) $\frac{1}{2}$

(c) 3 (d) 3, 5

(e) None of these

**11.** Use the graph to find the $x$-values (if any) at which $f$ is not continuous.

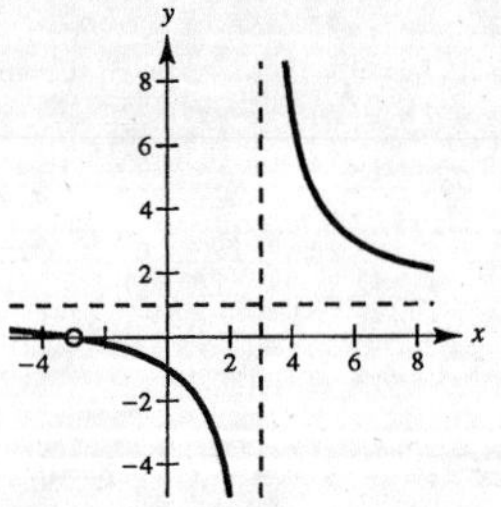


(a) 3 (b) $-3, 3$

(c) $-2, -3$ (d) 1, 3

(e) None of these

**12.** Use a graphing utility to graph $f(x) = x^3 - 2x - 5$. Then use this graph to find the interval for which the Intermediate Value Theorem guarantees the existence of at least one number $c$ in that interval for which $f(c) = 0$.

(a) $[-1, 1]$ (b) $[1, 2]$ (c) $[2, 3]$

(d) $[3, 4]$ (e) None of these

**13.** Use the graph to find the interval(s) for which the function $f$ is continuous.

(a) $(-\infty, 2)$ and $(2, \infty)$ (b) $(-\infty, \infty)$

(c) $(-\infty, 1)$ and $(1, \infty)$ (d) $(1, 2)$

(e) None of these

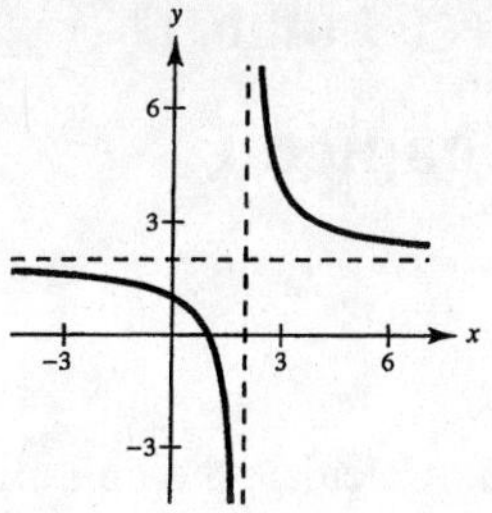


**14.** Find the limit: $\lim_{x\to(1/2)+} \sqrt{2x-1}$.

(a) 0 (b) 2 (c) 1

(d) The limit does not exist. (e) None of these

**15.** Find the limit: $\lim_{x\to 6^-} \frac{|3x-18|}{6-x}$.

(a) $-1$ (b) 1 (c) 3

(d) $-3$ (e) None of these

**16.** Which of the following statements is *not* true of $f(x) = \sqrt{x^2 - 25}$?

(a) $f$ is continuous at $x = 10$. (b) $f$ is continuous on the interval $(-\infty, -5]$.

(c) $f$ is continuous on the interval $[5, \infty)$. (d) $f$ is continuous on the interval $[-5, 5]$.

(e) $f$ is not continuous at $x = 0$.

**17.** Find the limit: $\lim_{x\to 1^-} \frac{-2}{x-1}$.

(a) $\infty$ (b) $-\infty$ (c) 0

(d) The limit does not exist. (e) None of these

**18.** Find the limit: $\lim_{x\to 3^-} \frac{x^2-3x+2}{x^2-5x+6}$.

(a) $\frac{1}{3}$ (b) $+\infty$ (c) $-\infty$

(d) 1 (e) None of these

**19.** Find the vertical asymptote(s): $f(x) = \frac{x-2}{x^2-3x-10}$

(a) $x = -2, x = 5$ (b) $y = 1$ (c) $y = 0$

(d) $x = 5$ (e) $x = -5$

**20.** $f(x)$ decreases without bound as $x$ approaches what value from the right? $f(x) = \frac{4}{(x-3)(5-x)}$

(a) 5 (b) $-3$ (c) $-5$

(d) 3 (e) None of these

**Test Form D** Name ______________ Date ________

**Chapter 1** Class ______________ Section ________

1. Calculate the limit: $\lim_{x\to 2} (2x^2 - 6x + 1)$.

2. If $\lim_{x\to 3} (3x - 2) = 7$, find $\delta$ such that $|(3x - 2) - 7| < 0.003$ whenever $0 < |x - 3| < \delta$.

3. Find the limit: $\lim_{x\to 1} f(x)$ if $f(x) = \begin{cases} x^2 + 4, & x \neq 1 \\ 2, & x = 1 \end{cases}$.

4. Find the limit: $\lim_{x\to -1} \dfrac{x^2 + 3x + 2}{x^2 + 1}$.

5. Find the limit: $\lim_{x\to \pi/6} e^{\sin x}$.

6. If $\lim_{x\to c} f(x) = -\frac{1}{2}$ and $\lim_{x\to c} g(x) = \frac{2}{3}$, find $\lim_{x\to c} [f(x) - g(x)]$.

7. Find the limit: $\lim_{x\to -2} \dfrac{x + 2}{x^3 + 8}$.

8. Find the limit: $\lim_{\Delta x\to 0} \dfrac{\sqrt{x + \Delta x} - \sqrt{x}}{\Delta x}$.

9. Find the limit: $\lim_{x\to 1} \dfrac{x - 1}{|x - 1|}$.

10. Find the limit: $\lim_{x\to 5} \cot \dfrac{\pi x}{6}$.

11. Find the limit: $\lim_{x\to 0} \dfrac{x}{\sin 3x}$.

12. Find the limit: $\lim_{x\to 2^+} \sqrt{2x - 1}$.

13. Find the limit: $\lim_{x\to -1^-} \dfrac{1}{x + 1}$.

14. Find the limit: $\lim_{x\to 3} \dfrac{3}{x^2 - 6x + 9}$.

15. Find the limit: $\lim_{x\to 0} \left(3x + 2 + \dfrac{1}{x^2}\right)$.

**16.** Find the value(s) of $x$ for which $f(x) = \dfrac{x-2}{x^2-4}$ is discontinuous and label these discontinuities as removable or nonremovable.

**17.** Let $f(x) = \dfrac{5}{x-1}$ and $g(x) = x^4$.

**a.** Find $f(g(x))$.

**b.** Find all values of $x$ for which $f(g(x))$ is discontinuous.

**18.** Determine the value of $c$ so that $f(x)$ is continuous on the entire real line if $f(x) = \begin{cases} x^2, & x \le 3 \\ \dfrac{c}{x}, & x > 3 \end{cases}$.

**19.** Find all vertical asymptotes of $f(x)$ if $f(x) = \dfrac{2x-2}{(x-1)(x^2+x-1)}$.

**20.** Evaluate the function $f(x) = \dfrac{\ln(1+x)}{x}$ at $x = 0.1, 0.01, 0.001$, and $-0.1, -0.01, -0.001$.

Then use the result to estimate the limit: $\displaystyle\lim_{x \to 0} \frac{\ln(1+x)}{x}$.

**Test Form E** **Name** ______________ **Date** ______________

**Chapter 1** **Class** ______________ **Section** ______________

**A graphing calculator/utility is recommended for this test.**

**1.** Use the graph to find $\lim_{x\to 0} f(x)$ (if it exists).

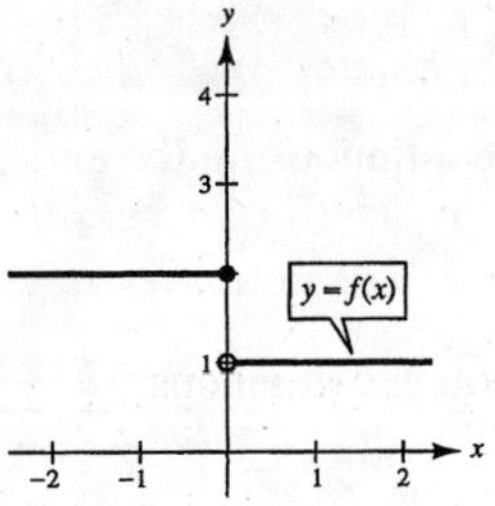


**2.** Determine whether the statement is true or false. If it is false, give an example to show that it is false.
If $\lim_{x\to 3} f(x) = 9$, then $f(3) = 9$.

**3.** Calculate the limit: $\lim_{x\to 2} (2x^2 - 6x + 1)$.

**4.** Find the limit: $\lim_{x\to 0} \dfrac{1 - e^x}{1 + e^x}$.

**5.** Find the limit: $\lim_{x\to \pi} \dfrac{x}{\cos x}$.

**6.** Use a graphing utility to graph the function $f(x) = -x^3 + x + 5$ and then estimate $\lim_{x\to -1} f(x)$.

**7.** Let $f(x) = \dfrac{x^2 - 4}{x - 2}$.

**a.** Use a graphing utility to graph the function.

**b.** Use the graph to estimate $\lim_{x\to 2} f(x)$.

**c.** Find the limit by analytical methods.

**8.** Let $f(x) = \dfrac{x^3 + 2x^2}{x + 2}$.

**a.** Find $\lim_{x\to -2} f(x)$ (if it exists).

**b.** Identify another function that agrees with $f(x)$ at all but one point.

**c.** Sketch the graph of $f(x)$.

**9.** Find the limit: $\lim_{\Delta x\to 0} \dfrac{\sqrt{(x + \Delta x) + 2} - \sqrt{x + 2}}{\Delta x}$.

**10.** Find the limit: $\lim_{\Delta x\to 0} \dfrac{(x + \Delta x)^2 - 2(x + \Delta x) - (x^2 - 2x)}{\Delta x}$.

**11.** Find the $x$-values (if any) for which $f$ is not continuous.

$$f(x) = \begin{cases} 3x + 2, & x < -1 \\ 2x^2 - 3x + 6, & x \geq -1 \end{cases}$$

**12.** Use the graph to find the $x$-values (if any) for which $f$ is not continuous.

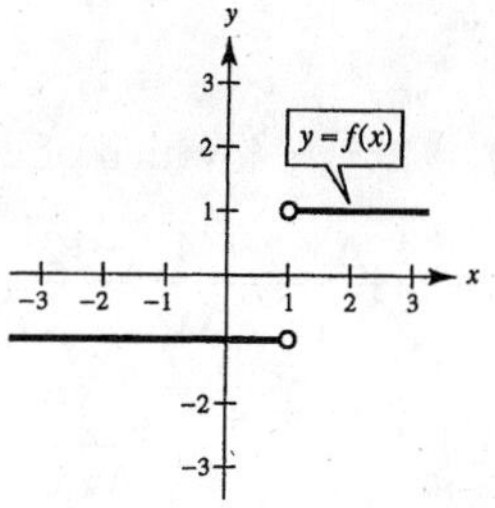


**13.** Find the interval(s) for which the function is continuous.

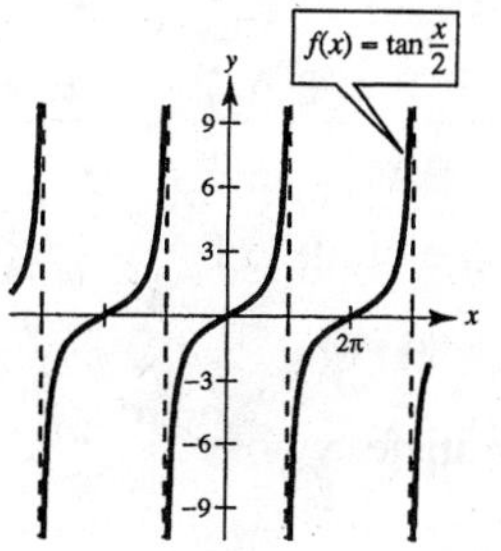


**14.** Use the Intermediate Value Theorem to show that the function $f(x) = x^4 - 2x^2 + 3x$ has a zero in the interval $[-2, -1]$.

**15.** Use a graphing utility to graph $f(x) = \dfrac{x + 2}{x^2 - 4}$. Then use the graph to determine $x$-values at which the function is not continuous.

**16.** Let $f(x) = \begin{cases} x^2 + 1, & x \leq 0 \\ 2x - 3, & x > 0 \end{cases}$. Find each limit (if it exists).

**a.** $\lim_{x \to 0^-} f(x)$ **b.** $\lim_{x \to 0^+} f(x)$ **c.** $\lim_{x \to 0} f(x)$

**17.** Use a graphing utility to find the limit: $\lim_{x \to 0} \dfrac{\sin 3x}{x}$. Then verify your answer analytically.

**18.** Find all vertical asymptotes of $f(x) = \dfrac{x^2 - x - 2}{x^2 + x - 6}$.

**19.** Use a graphing utility to graph the function $f(x) = \ln\left(\dfrac{1}{x^2 + x - 6}\right)$. Then use the graph to evaluate the limit:

$$\lim_{x \to 2^+} \left[\ln\left(\frac{1}{x^2 + x - 6}\right)\right].$$

**20.** Find the limit: $\lim_{x \to 1} \dfrac{-2}{(1 - x)^2}$.

**Test Form A** **Name** ______________ **Date** ______________

**Chapter 2** **Class** ______________ **Section** ______________

1. If $f(x) = 2x^2 + 4$, which of the following will calculate the derivative of $f$?

(a) $\dfrac{[2(x + \Delta x)^2 + 4] - (2x^2 + 4)}{\Delta x}$

(b) $\displaystyle\lim_{\Delta x \to 0} \frac{(2x^2 + 4 + \Delta x) - (2x^2 + 4)}{\Delta x}$

(c) $\displaystyle\lim_{\Delta x \to 0} \frac{[2(x + \Delta x)^2 + 4] - (2x^2 + 4)}{\Delta x}$

(d) $\dfrac{(2x^2 + 4 + \Delta x) - (2x^2 + 4)}{\Delta x}$

(e) None of these

2. Differentiate: $y = \dfrac{1 + \cos x}{1 - \cos x}$.

(a) $-1$ (b) $-2 \csc x$ (c) $2 \csc x$

(d) $\dfrac{-2 \sin x}{(1 - \cos x)^2}$ (e) None of these

3. Find $dy/dx$ for $y = x^3\sqrt{x + 1}$.

(a) $\dfrac{3x^2}{2\sqrt{x + 1}}$ (b) $\dfrac{x^2(7x + 6)}{2\sqrt{x + 1}}$ (c) $3x^2\sqrt{x + 1}$

(d) $\dfrac{7x^3 + x^2}{2\sqrt{x + 1}}$ (e) None of these

4. Find $f'(x)$ for $f(x) = \sqrt{4 + e^{2x}}$.

(a) $\dfrac{e^{2x}}{\sqrt{4 + e^{2x}}}$ (b) $\dfrac{1}{2\sqrt{2e^{2x}}}$ (c) $\dfrac{xe^{2x-1}}{\sqrt{4 + e^{2x}}}$

(d) $e^x$ (e) None of these

5. The position equation for the movement of a particle is given by $s = (t^2 - 1)^3$ when $s$ is measured in feet and $t$ is measured in seconds. Find the acceleration at two seconds.

(a) 342 units/sec$^2$ (b) 18 units/sec$^2$ (c) 288 units/sec$^2$

(d) 90 units/sec$^2$ (e) None of these

**6.** Find $\frac{dy}{dx}$ for $xe^y + 1 = xy$.

(a) $0$
(b) $\dfrac{y - e^y}{xe^y - x}$
(c) $\dfrac{y}{e^y - x}$
(d) $\dfrac{e^y}{xe^y - 1}$
(e) None of these

**7.** Find $y'$ if $y = \sin(x + y)$.

(a) $0$
(b) $\dfrac{\cos(x + y)}{1 - \cos(x + y)}$
(c) $\cos(x + y)$
(d) $1$
(e) None of these

**8.** Differentiate: $y = \sec^2 x + \tan^2 x$.

(a) $0$
(b) $\tan x + \sec^4 x$
(c) $\sec^2 x(\sec^2 x + \tan^2 x)$
(d) $4 \sec^2 x \tan x$
(e) None of these

**9.** Find the derivative: $s(t) = \csc \frac{t}{2}$.

(a) $-\csc \frac{t}{2} \cot \frac{t}{2}$
(b) $-\frac{1}{2} \cot^2 \frac{t}{2}$
(c) $\frac{1}{2} \csc \frac{t}{2} \cot \frac{t}{2}$
(d) $\frac{1}{2} \cot^2 \frac{t}{2}$
(e) None of these

**10.** Find an equation for the tangent line to the graph of $f(x) = 2x^2 - 2x + 3$ at the point where $x = 1$.

(a) $y = 2x - 2$
(b) $y = 4x^2 - 6x + 5$
(c) $y = 2x + 1$
(d) $y = 4x^2 - 6x + 2$
(e) None of these

**11.** Find all points on the graph of $f(x) = -x^3 + 3x^2 - 2$ at which there is a horizontal tangent line.

(a) $(0, -2), (2, 2)$
(b) $(0, -2)$
(c) $(1, 0), (0, -2)$
(d) $(2, 2)$
(e) None of these

**12.** Find the instantaneous rate of change of $w$ with respect to $z$ if $w = \frac{7}{3z^2}$.

(a) $\dfrac{7}{6z}$
(b) $\dfrac{14}{3}z$
(c) $-\dfrac{14}{3z}$
(d) $-\dfrac{14}{3z^3}$
(e) None of these

13. Suppose the position equation for a moving object is given by $s(t) = 3t^2 + 2t + 5$ where $s$ is measured in meters and $t$ is measured in seconds. Find the velocity of the object when $t = 2$.

(a) 13 m/sec
(b) 14 m/sec
(c) 10 m/sec
(d) 6 m/sec
(e) None of these

14. A point moves along the curve $y = 2x^2 + 1$ in such a way that the $y$ value is decreasing at the rate of 2 units per second. At what rate is $x$ changing when $x = \frac{3}{2}$?

(a) increasing $\frac{1}{3}$ unit/sec
(b) decreasing $\frac{1}{3}$ unit/sec
(c) decreasing $\frac{7}{2}$ unit/sec
(d) increasing $\frac{7}{2}$ unit/sec
(e) None of these

15. Find the derivative: $f(x) = \ln \dfrac{\sqrt{x^2+1}}{x(2x^3-1)^2}$.

(a) $\dfrac{x}{x^2+1} - \dfrac{1}{x} + \dfrac{12x^2}{2x^3-1}$
(b) $\dfrac{x}{x^2+1} - \dfrac{1}{x} + \dfrac{6x^2}{2x^3-1}$
(c) $\dfrac{1}{(x^2+1)^{1/2}(4x^2)(2x^3-1)}$
(d) $\dfrac{x}{x^2+1} - \dfrac{1}{x} - \dfrac{12x^2}{2x^3-1}$
(e) None of these

16. Differentiate: $y = x^{1-x}$.

(a) $(1-x)x^{-x}$
(b) $x^{1-x}\left[\dfrac{1-x}{x} - \ln x\right]$
(c) $(x-1)x^{-x}$
(d) $x^{1-x}\left(-\dfrac{1}{x}\right)$
(e) None of these

17. Find $y'$ if $y = \dfrac{x^3}{3^x}$.

(a) $\dfrac{x}{3^{x-2}}$
(b) $\dfrac{3x^2}{3^x(\ln 3)}$
(c) $\dfrac{x^2(9-x^2)}{3^{x+1}}$
(d) $\dfrac{x^2[3 - x(\ln 3)]}{3^x}$
(e) None of these

18. Find $\dfrac{dy}{dx}$ for $y = \arctan \dfrac{x}{2}$.

(a) $\dfrac{4}{4+x^2}$
(b) $\dfrac{4}{1+x^2}$
(c) $\dfrac{1}{\sqrt{4-x^2}}$
(d) $\dfrac{1}{2}\sec^2\left(\dfrac{x}{2}\right)$
(e) $\dfrac{2}{4+x^2}$

**Test Form B** **Name** ____________ **Date** ____________

**Chapter 2** **Class** ____________ **Section** ____________

1. If $f(x) = -x^2 + x$, which of the following will calculate the derivative of $f(x)$?

(a) $\lim\limits_{\Delta x \to 0} \dfrac{(-x^2 + x + \Delta x) - (-x^2 + x)}{\Delta x}$

(b) $\lim\limits_{\Delta x \to 0} \dfrac{[-(x + \Delta x)^2 + (x + \Delta x)] - (-x^2 + x)}{\Delta x}$

(c) $\dfrac{[-(x + \Delta x)^2 + (x + \Delta x)] - (-x^2 + x)}{\Delta x}$

(d) $\dfrac{(-x^2 + x + \Delta x) - (-x^2 + x)}{\Delta x}$

(e) None of these

2. Differentiate: $y = \dfrac{3x}{x^2 + 1}$.

(a) $\dfrac{3}{1 + x^2}$ (b) $\dfrac{3}{2x}$ (c) $\dfrac{3x^2 - 3}{(1 + x^2)^3}$

(d) $\dfrac{3(1 - x^2)}{(1 + x^2)^2}$ (e) None of these

3. Find $dy/dx$ for $y = \sqrt{x}(3x - 1)$.

(a) $\dfrac{9x - 1}{2\sqrt{x}}$ (b) $\dfrac{9}{2}\sqrt{x} - 1$ (c) $3\sqrt{x}$

(d) $\dfrac{3}{2\sqrt{x}}$ (e) None of these

4. Find $f'(x)$ for $f(x) = \dfrac{2}{2x + e^{2x}}$.

(a) 0 (b) $\dfrac{1}{1 + e^{2x}}$ (c) $\dfrac{-4(1 + e^{2x})}{(2x + e^{2x})^2}$

(d) $\dfrac{1 + xe^{2x-1}}{(2x + e^{2x})^2}$ (e) None of these

5. Find $\dfrac{dy}{dx}$ for $\ln(xy) = x + y$.

(a) $-\dfrac{y}{x}$ (b) $e^{x+y}$ (c) $\dfrac{xy}{1 - xy}$

(d) $\dfrac{xy - y}{x - xy}$ (e) None of these

**6.** Find $y'$ for $x = \tan(x + y)$.

(a) $-\sin^2(x + y)$ (b) $\sec^2(x + y)$ (c) $-\tan^2(x + y)$

(d) $\dfrac{1 - \sec^2 x}{\sec^2 y}$ (e) None of these

**7.** Differentiate: $y = \csc^2 \theta + \cot^2 \theta$.

(a) $\cot \theta + \csc^4 \theta$ (b) $0$ (c) $-4 \csc^2 \theta \cot \theta$

(d) $-\csc^2 \theta(\csc^2 \theta + \cot^2 \theta)$ (e) None of these

**8.** Find the derivative: $s(t) = \sec \sqrt{t}$.

(a) $\tan^2 \sqrt{t}$ (b) $\dfrac{\sec \sqrt{t} \tan \sqrt{t}}{2\sqrt{t}}$ (c) $\sec \dfrac{1}{2\sqrt{t}} \tan \dfrac{1}{2\sqrt{t}}$

(d) $\sec \sqrt{t} \tan \sqrt{t}$ (e) None of these

**9.** A particle moves along the curve given by $y = \sqrt{t^3 + 1}$. Find the acceleration when $t = 2$ seconds.

(a) 3 units/sec$^2$ (b) $\frac{2}{3}$ units/sec$^2$ (c) $-\frac{1}{108}$ units/sec$^2$

(d) $-\frac{1}{9}$ units/sec$^2$ (e) None of these

**10.** Find an equation for the tangent line to the graph of $f(x) = -2x^2 + 2x + 3$ at the point where $x = 1$.

(a) $y = -4x + 2$ (b) $2x + y - 1 = 0$ (c) $y = -4x^2 + 2x + 1$

(d) $2x + y = 5$ (e) None of these

**11.** Find point(s) on the graph of the function $f(x) = x^3 - 2$ where the slope is 3.

(a) $(1, 3), (-1, 3)$ (b) $(1, -1), (-1, -3)$ (c) $(\sqrt[3]{2}, 0)$

(d) $(1, 3)$ (e) None of these

**12.** Find the instantaneous rate of change of $w$ with respect to $z$ for $w = \dfrac{1}{z} + \dfrac{z}{2}$.

(a) $\dfrac{3}{2}$ (b) $-2$ (c) $\dfrac{z^2 - 2}{2z^2}$

(d) $\dfrac{-1}{z^2}$ (e) None of these

**13.** Suppose the position equation for a moving object is given by $s(t) = 3t^2 - 2t + 5$ where $s$ is measured in meters and $t$ is measured in seconds. Find the velocity of the object when $t = 2$.

(a) 13 m/sec (b) 6 m/sec (c) 10 m/sec

(d) 14 m/sec (e) None of these

**14.** A point moves along the curve $y = 2x^2 - 1$ in such a way that the $y$ value is decreasing at the rate of 2 units per second. At what rate is $x$ changing when $x = -\frac{3}{2}$?

(a) decreasing $\frac{7}{2}$ unit/sec
(b) increasing $\frac{7}{2}$ unit/sec
(c) increasing $\frac{1}{3}$ unit/sec
(d) decreasing $\frac{1}{3}$ unit/sec
(e) None of these

**15.** Find the derivative: $f(x) = \ln \dfrac{x^2\sqrt{4x+1}}{(x^3+5)^3}$.

(a) $\dfrac{x}{9x^2(x^3+5)^2\sqrt{4x+1}}$
(b) $\dfrac{2}{x} + \dfrac{2}{4x+1} - \dfrac{9x^2}{x^3+5}$
(c) $\dfrac{2}{x} + \dfrac{1}{2(4x+1)} - \dfrac{3}{x^3+5}$
(d) $\dfrac{2}{x} - \dfrac{2}{4x+1} - \dfrac{9x^2}{x^3+5}$
(e) None of these

**16.** Differentiate: $y = x^{e^x}$

(a) $e^x x^{e^{x-1}}$
(b) $e^x$
(c) $x^{e^x}\left[\dfrac{e^x}{x} + (\ln x)(e^x)\right]$
(d) $xe^x + e^x$
(e) None of these

**17.** Find $y'$ if $y = 3^x x^3$.

(a) $3^x x^2[3 + (\ln 3)x]$
(b) $2x^2 3^{x-1}$
(c) $3^{x-1}x^2[9 + x^2]$
(d) $9x^2$
(e) None of these

**18.** Find $\dfrac{dy}{dx}$ for $y = \arcsin \dfrac{x}{3}$.

(a) $\dfrac{1}{\sqrt{9-x^2}}$
(b) $\dfrac{3}{\sqrt{9-x^2}}$
(c) $\dfrac{x}{\sqrt{9-x^2}}$
(d) $\dfrac{3}{\sqrt{x^2-9}}$
(e) $\dfrac{3}{\sqrt{9+x^2}}$

**Test Form C** Name ____________ Date ____________

**Chapter 2** Class ____________ Section ____________

**A graphing calculator/utility is recommended for this test.**

**1.** Determine whether the slope at the indicated point is positive, negative, or zero.

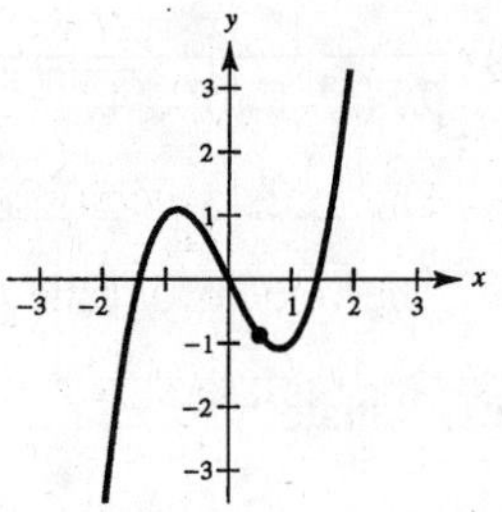

(a) Zero

(b) No slope

(c) Positive

(d) Negative

(e) None of these

**2.** Find the slope of the graph of $f(x) = x^2 - 2x$ at the point $(a, f(a))$.

(a) 0 (b) $2a - 2$ (c) $f(a)$

(d) $a^2 - 2a$ (e) None of these

**3.** Use the graph to determine all $x$-values at which the function is not differentiable.

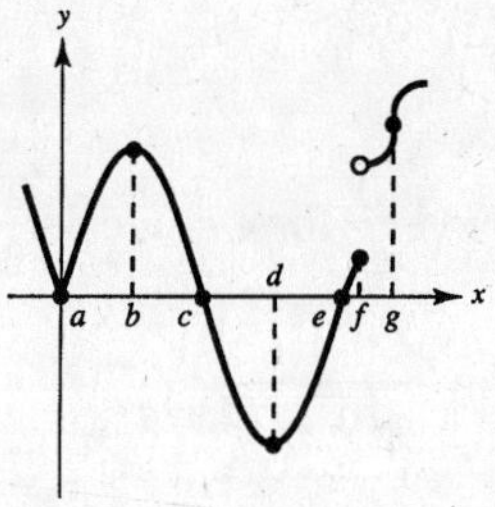

(a) $a, b, c, d, e, f,$ and $g$

(b) $b$ and $d$

(c) $a$ and $f$

(d) $a, f,$ and $g$

(e) None of these

**4.** Use a graphing utility to graph $f(x) = \dfrac{3x^2 - 8}{x^2 - 4}$ and its derivative, $f'$, on the same coordinate axes. Then use the graph to describe the behavior of $f$ at that value of $x$ where $f'(x) = 0$.

(a) $f(x) = 0$ (b) $f$ increases without bound. (c) $f$ has a horizontal tangent line.

(d) $f$ has no tangent line. (e) None of these

**5.** Find the value of the derivative of the function $f(t) = \dfrac{t^3 + 2}{t}$ at the point $(-2, 3)$.

(a) $-\dfrac{9}{2}$ (b) $-\dfrac{7}{2}$ (c) 12

(d) $-\dfrac{11}{16}$ (e) None of these

**6.** The graph at the right represents the graph of the derivative of which of the following functions?

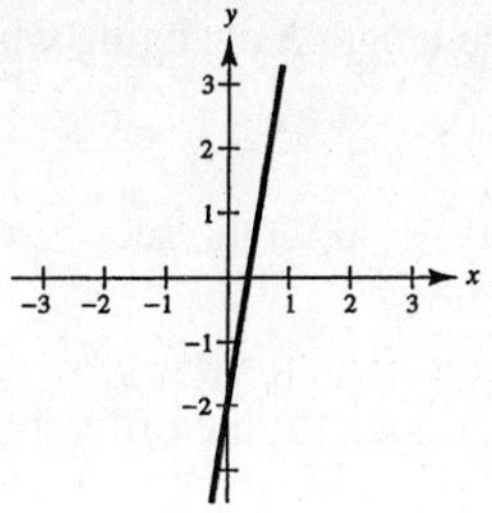


(a) $f(x) = 2x^2 + 1$

(b) $f(x) = 2x - 3$

(c) $f(x) = 3x^2 - 2x - 1$

(d) $f(x) = x^3 + x^2$

(e) None of these

**7.** The position function for a particular object is $s = -\frac{35}{2}t^2 + 58t + 91$. Which statement is true?

(a) The initial velocity is $-35$

(b) The velocity is a constant

(c) The velocity at time $t = 1$ is 23

(d) The initial position is $-\frac{35}{2}$

(e) None of these

**8.** Find $\frac{dy}{d\theta}$ for $y = \csc \theta - \cot \theta$.

(a) 0

(b) $-\cot^2 \theta + \csc \theta \cot \theta$

(c) $\sec \theta \tan \theta - \sec^2 \theta$

(d) $-\csc \theta \cot \theta + \csc^2 \theta$

(e) None of these

**9.** Find an equation of the tangent line to the graph of $f(\theta) = \tan \theta$ at the point $\left(\frac{\pi}{4}, 1\right)$.

(a) $4x - 4y = \pi - 4$

(b) $4\sqrt{2}x - 4y = \pi - 4$

(c) $4x - 2y = \pi - 2$

(d) $y = x$

(e) None of these

**10.** Let $f(3) = 0$, $f'(3) = 6$, $g(3) = 1$ and $g'(3) = \frac{1}{3}$. Find $h'(3)$ if $h(x) = \frac{f(x)}{g(x)}$.

(a) 18

(b) 6

(c) $-6$

(d) $-2$

(e) None of these

**11.** Find the derivative: $f(x) = \ln(e^{-x^2})$.

(a) $e^{x^2}$

(b) $-2xe^{-x^2}$

(c) $-2x$

(d) $-2xe^{2x^2}$

(e) None of these

**12.** Find the derivative: $f(\theta) = \sqrt{\sin 2\theta}$.

(a) $\frac{\cos 2\theta}{\sqrt{\sin 2\theta}}$

(b) $\sqrt{\sec 2\theta}$

(c) $\frac{\cos 2\theta}{2\sqrt{\sin 2\theta}}$

(d) $\cos \theta$

(e) None of these

**13.** Determine the slope of the graph of the relation $2x^2 - 3xy + y^3 = -1$ at the point $(2, -3)$.

(a) $-\frac{17}{21}$

(b) $\frac{5}{7}$

(c) $-\frac{1}{3}$

(d) $\frac{4}{3}$

(e) None of these

**14.** A machine is rolling a metal cylinder under pressure. The radius of the cylinder is decreasing at a constant rate of 0.05 inches per second and the volume $V$ is $128\pi$ cubic inches. At what rate is the length $h$ changing when the radius $r$ is 1.8 inches? [Hint: $V = \pi r^2 h$]

(a) $-2.195$ in./sec (b) 39.51 in./sec (c) 2.195 in./sec

(d) $-43.90$ in./sec (e) None of these

**15.** Use logarithmic differentiation to find $\frac{dy}{dx}$: $y = \frac{(x+2)\sqrt{1-x^2}}{4x^3}$.

(a) $\frac{-x}{12x^2\sqrt{1-x^2}}$ (b) $\frac{1}{x+2} - \frac{x}{1-x^2} - \frac{3}{x}$

(c) $\frac{(x+2)\sqrt{1-x^2}}{4x^3}\left[\frac{1}{x+2} - \frac{x}{1-x^2} - \frac{3}{x}\right]$ (d) $\frac{1}{x+2} - \frac{2x}{1-x^2} - \frac{3}{x}$

(e) None of these

**16.** Find $y'$: $y = \arctan\sqrt{t}$.

(a) $\frac{\sec^2\sqrt{t}}{2\sqrt{t}}$ (b) $\frac{1}{2\sqrt{t}(1+t)}$ (c) $\frac{1}{2\sqrt{t}(1-t)}$

(d) $\frac{1}{1+t}$ (e) None of these

**17.** Find $f'(x)$ for $f(x) = 2/(2x + e^{2x})$.

(a) 0 (b) $\frac{1}{1+e^{2x}}$ (c) $\frac{-4(1+e^{2x})}{(2x+e^{2x})^2}$

(d) $\frac{1+xe^{2x-1}}{(2x+e^{2x})^2}$ (e) None of these

**18.** Find $\frac{dy}{dx}$ for $y = x^x$.

(a) $x(x^{x-1})$ (b) $x^x$ (c) $x^{x-1}$

(d) $x^x(1 + \ln x)$ (e) $1 + \ln x$

**Test Form D** **Name** ______________ **Date** ____________

**Chapter 2** **Class** ______________ **Section** ____________

1. Use the definition of a derivative to calculate the derivative of $f(x) = \frac{1}{x}$.

2. Differentiate: $y = \frac{2x}{1 - 3x^2}$.

3. Find $\frac{dy}{dx}$ for $y = \sqrt{2x + 1}\,(x^3)$.

4. Find $f'(x)$ for $f(x) = \cot^3\sqrt{x}$.

5. The position equation for the movement of a particle is given by $s = (t^3 + 1)^2$ where $s$ is measured in feet and $t$ is measured in seconds. Find the acceleration of this particle at 1 second.

6. Find $y'$ if $y = \frac{x}{x + y}$.

7. Find $\frac{dy}{dx}$ for $ye^x - x = y^2$.

8. Find the derivative: $f(\theta) = \sec \theta^2$.

9. Differentiate: $y = \sin^2 x - \cos^2 x$.

10. Find an equation for the tangent line to the graph of $f(x) = \arctan x$ at the point where $x = 1$.

11. Find the values of $x$ for all points on the graph of $f(x) = x^3 - 2x^2 + 5x - 16$ at which the slope of the tangent line is 4.

12. Find the instantaneous rate of change of $R$ with respect to $x$ if $R = 2x^2 + \frac{1}{x}$.

13. An object is thrown (straight down) from the top of a 220-foot building with an initial velocity of 26 feet per second.

    **a.** Write the position equation for the movement described.

    **b.** What is the velocity at 1 second?

14. As a balloon in the shape of a sphere is being blown up, the volume is increasing at the rate of 4 cubic inches per second. At what rate is the radius increasing when the radius is 1 inch?

**15.** Find $\frac{dy}{dx}$ if $y = \ln \frac{\sqrt{x}}{5 - x}$.

**16.** Differentiate: $y = x^x$.

**17.** Find $y'$: $y = e^{1/x}$.

**18.** Find $y'$ if $y = x^e e^x$.

**Test Form E** **Name** ______________ **Date** ____________

**Chapter 2** **Class** ______________ **Section** ____________

**A graphing calculator/utility is recommended for this test.**

**1.** At each point indicated on the graph, determine whether the value of the derivative is positive, negative, zero, or the function has no derivative.

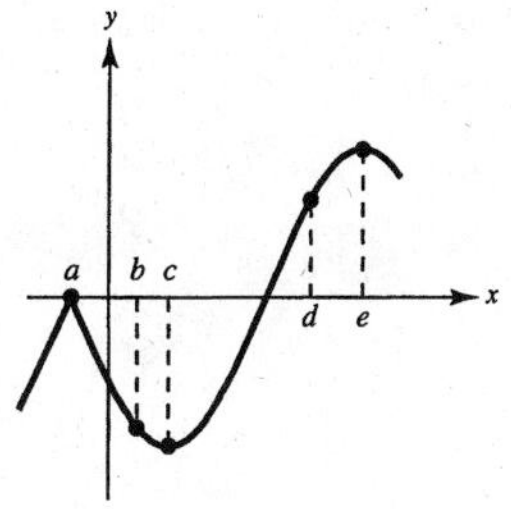


**2.** Let $f(x) = \dfrac{4}{x}$.

**a.** Use the definition of the derivative to calculate the derivative of $f$.

**b.** Find the slope of the tangent line to the graph of $f$ at the point $(-2, -2)$.

**c.** Write an equation of the tangent line in part **b.**

**d.** Use a graphing utility to graph $f$ and the tangent line on the same axes. Then sketch the graphs.

**3.** Find the point(s) on the graph of $y = \dfrac{1}{x}$ where the graph is parallel to the line $4x + 9y = 3$.

**4.** A coin is dropped from a height of 750 feet. The height, $s$, (measured in feet), at time, $t$ (measured in seconds), is given by $s = -16t^2 + 750$.

**a.** Find the average velocity on the interval $[1, 3]$.

**b.** Find the instantaneous velocity when $t = 3$.

**c.** How long does it take for the coin to hit the ground?

**d.** Find the velocity of the coin when it hits the ground.

**5.** The volume of a right circular cone of radius $r$ and height $r$ is given by $V = \dfrac{\pi}{3}r^3$. How fast is the volume changing with respect to changes in $r$ when the radius is equal to 2 feet?

**6.** Differentiate: $f(x) = \dfrac{1}{(4 + e^{2x})^4}$.

**7.** Evaluate the derivative for the function $f(t) = \dfrac{t}{\cos t}$ at the point $\left(\dfrac{\pi}{3}, \dfrac{2\pi}{3}\right)$.

8. Let $f(x) = x^5 - 5x$.

a. Calculate $f'(x)$.

b. Use a graphing utility to graph $f$ and $f'$ on the same axes. Sketch the graphs.

c. Use the graph to determine those point(s) where $f$ has a horizontal tangent line.

d. Give the value of $f'$ at each of the points found in part c.

9. The graphs of a function $f$ and its derivative $f'$ are given on the same coordinate axes. Label the graphs as $f$ or $f'$ and state the reasons for your choice.

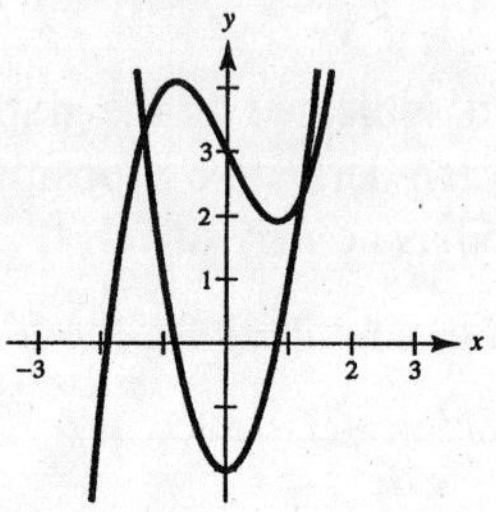


10. Find $f''(x)$: $f(x) = \sqrt{2x^2 + 5}$.

11. Find the point(s) (if any) of horizontal tangent lines: $x^2 + xy + y^2 = 6$.

12. A balloon rises at the rate of 8 feet per second from a point on the ground 60 feet from an observer. Find the rate of change of the angle of elevation when the balloon is 25 feet above the ground.

13. Find the derivative: $y = \ln(x^2 + 6x + 9)$.

14. Use logarithmic differentiation to find $\dfrac{dy}{dx}$: $y = \dfrac{x^3\sqrt{2x+3}}{(x-2)^2}$.

15. Find the derivative: $h(t) = \arccos t^2$.

**Test Form A** **Name** ____________________ **Date** ____________

**Chapter 3** **Class** ____________________ **Section** ____________

1. Find all open intervals on which the function $f(x) = \frac{x^2}{x^2 + 4}$ is decreasing.

(a) $(0, \infty)$ (b) $(-2, 2)$ (c) $(-\infty, 0)$

(d) $(-\infty, \infty)$ (e) None of these

2. Find all critical numbers for the function $f(x) = x \ln x$.

(a) There are none. (b) $-e$ (c) $-1$

(d) $\frac{1}{e}$ (e) None of these

3. Find the values of $x$ that give relative extrema for the function $f(x) = 3x^5 - 5x^3$.

(a) Relative maximum: $x = 0$; Relative minimum: $x = \sqrt{5/3}$

(b) Relative maximum: $x = -1$; Relative minimum: $x = 1$

(c) Relative maxima: $x = \pm 1$; Relative minimum: $x = 0$

(d) Relative maximum: $x = 0$; Relative minima: $x = \pm 1$

(e) None of these

4. Find all intervals on which the graph of the function is concave upward: $f(x) = \frac{x^2 + 1}{x^2}$.

(a) $(-\infty, \infty)$ (b) $(-\infty, -1)$ and $(1, \infty)$ (c) $(-\infty, 0)$ and $(0, \infty)$

(d) $(1, \infty)$ (e) None of these

5. Let $f''(x) = 4x^3 - 2x$ and let $f(x)$ have critical numbers $-1$, 0, and 1. Use the Second Derivative Test to determine if any of the critical numbers gives a relative maximum.

(a) $-1$ (b) 0 (c) 1

(d) $-1$ and 1 (e) None of these

6. Find $\lim_{x \to \infty} \frac{2x^3 + 6x^2 + 5}{3 + x^3}$.

(a) $\frac{2}{3}$ (b) $\infty$ (c) 1

(d) 2 (e) None of these

**7.** Which of the following functions has a horizontal asymptote at $y = 2$?

(a) $\dfrac{x - 2}{3x - 5}$ (b) $\dfrac{2x}{\sqrt{x - 2}}$ (c) $\dfrac{2x^2 - 6x + 1}{1 + x^2}$

(d) $\dfrac{2x - 1}{x^2 + 1}$ (e) None of these

**8.** Which of the following is the correct sketch of the graph of the function $f(x) = \dfrac{1}{(x - 2)^2}$?

(a)

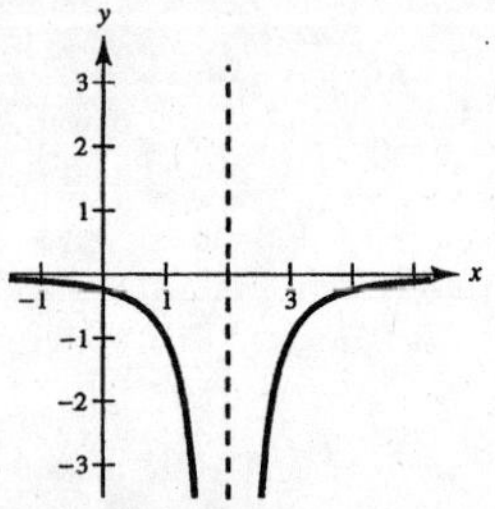


(b)

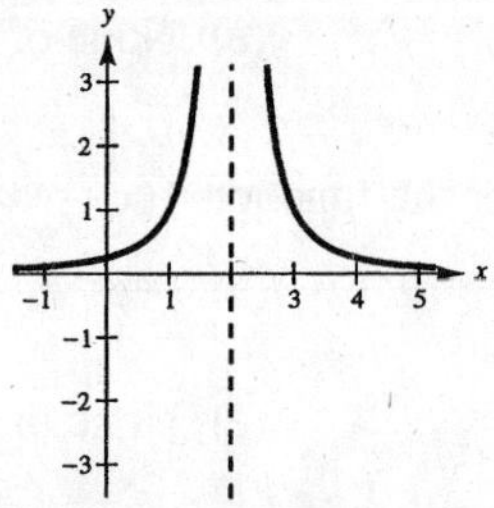


(c)

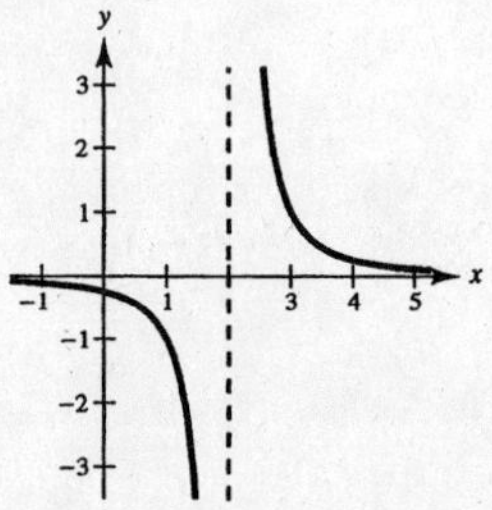


(d)

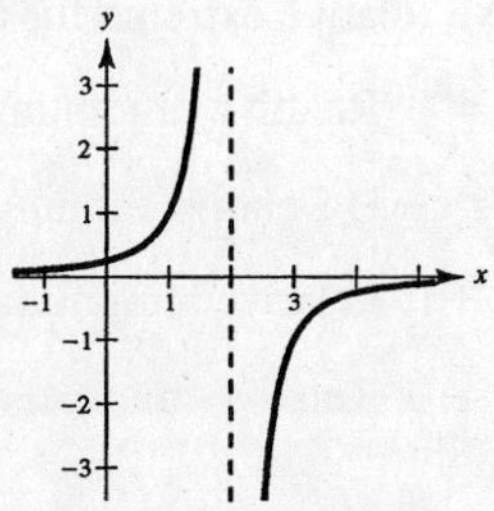


(e) None of these

**9.** Find all points of inflection: $f(x) = \frac{1}{12}x^4 - 2x^2 + 15$.

(a) $(2, 0)$ (b) $(2, 0), (-2, 0)$ (c) $(0, 15)$

(d) $\left(2, \frac{25}{3}\right), \left(-2, \frac{25}{3}\right)$ (e) None of these

**10.** The management of a large store wishes to add a fenced-in rectangular storage yard of 20,000 square feet, using the building as one side of the yard. Find the minimum amount of fencing that must be used to enclose the remaining 3 sides of the yard.

(a) 400 ft (b) 200 ft (c) 20,000 ft

(d) 500 ft (e) None of these

**11.** Which of the following is the correct sketch of the graph of the function $y = x^3 - 12x + 20$?

(a)

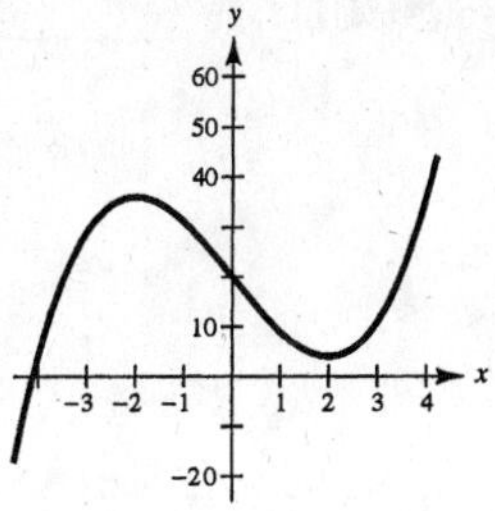


(b)

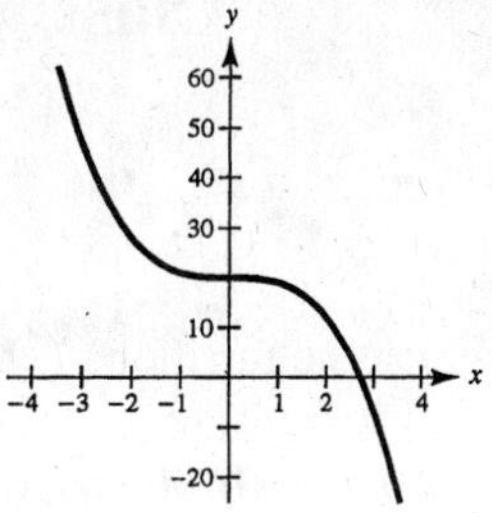


(c)

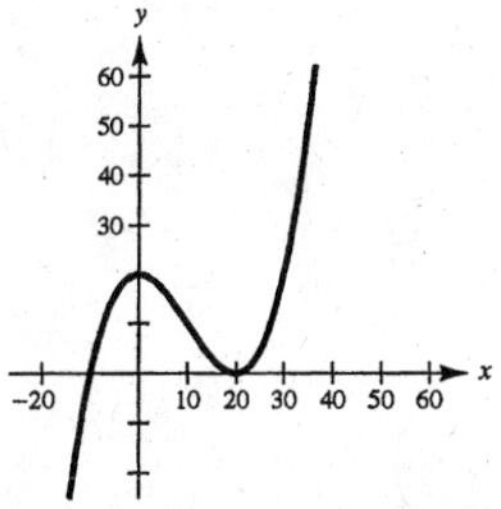


(d)

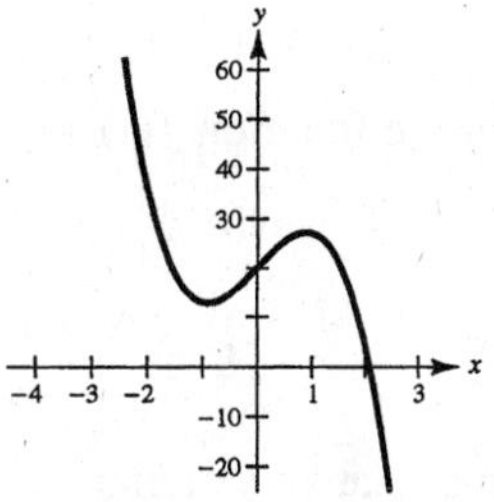


(e) None of these

**12.** State why Rolle's Theorem does not apply to the function $f(x) = \dfrac{2}{(x+1)^2}$ on the interval $[-2, 0]$.

(a) $f$ is not continuous on $[-2, 0]$.

(b) $f(-2) \neq f(0)$

(c) $f$ is not differentiable at $x = -1$.

(d) Both *a* and *c*.

(e) None of these

**13.** Find all extrema in the interval $[0, 2\pi]$ if $y = x + \sin x$.

(a) $\left(-1, -1 + \dfrac{3\pi}{2}\right)$, $(0, 0)$

(b) $(2\pi, 2\pi)$, $(0, 0)$

(c) $(2\pi, 2\pi)$ $(\pi, \pi)$

(d) $(\pi, \pi)$ $(0, 0)$

(e) None of these

**14.** The side of a cube is measured to be 3.0 inches. If the measurement is correct to within 0.01 inch, use differentials to estimate the propagated error in the volume of the cube.

(a) $\pm 0.000001$ in.$^3$

(b) $\pm 0.06$ in.$^3$

(c) $\pm 0.027$ in.$^3$

(d) $\pm 0.27$ in.$^3$

(e) None of these

**Test Form B** **Name** ______________ **Date** ______________

**Chapter 3** **Class** ______________ **Section** ______________

1. Find all open intervals on which the function $f(x) = \dfrac{x}{x^2 + x - 2}$ is decreasing.

(a) $(-\infty, \infty)$ (b) $(-\infty, 0)$ (c) $(-\infty, -2)$ and $(1, \infty)$

(d) $(-\infty, -2), (-2, 1)$ and $(1, \infty)$ (e) None of these

2. Find all critical numbers for the function $f(x) = (9 - x^2)^{3/5}$.

(a) 0 (b) 3 (c) $-3, 3$

(d) $-3, 0, 3$ (e) None of these

3. Find the values of $x$ that give relative extrema for the function $f(x) = (x + 1)^2(x - 2)$.

(a) Relative maximum: $x = -1$; relative minimum: $x = 1$

(b) Relative maxima: $x = 1, x = 3$; Relative minimum: $x = -1$

(c) Relative minimum: $x = 2$

(d) Relative maximum: $x = -1$; Relative minimum: $x = 2$

(e) None of these

4. Find all intervals on which the graph of the function is concave upward: $f(x) = \dfrac{x - 1}{x + 3}$.

(a) $(-\infty, \infty)$ (b) $(-\infty, -3)$ (c) $(1, \infty)$

(d) $(-3, \infty)$ (e) None of these

5. Let $f''(x) = 3x^2 - 4$ and let $f(x)$ have critical numbers $-2$, 0, and 2. Use the Second Derivative Test to determine which critical numbers, if any, gives a relative maximum.

(a) $-2$ (b) 2 (c) 0

(d) $-2$ and 2 (e) None of these

6. Find $\displaystyle\lim_{x\to\infty} \frac{\sqrt{4x^2 - 1}}{x^2}$.

(a) 4 (b) 0 (c) 2

(d) $\infty$ (e) None of these

**7.** Which of the following functions has a horizontal asymptote at $y = -\frac{1}{2}$?

(a) $\dfrac{x^3}{1 - 2x^3}$ (b) $\dfrac{x}{\sqrt{2x + 1}}$ (c) $\dfrac{2x^2 - 6x + 1}{1 + x^2}$

(d) $\dfrac{x - 1}{2x^2 + 1}$ (e) None of these

**8.** Find the value of $x$ at which $f'(x)$ is largest on the interval $[1, 5]$ for $f(x) = 2 - 4x + 7 \ln x$.

(a) $\frac{7}{4}$ (b) 5 (c) 1 (d) $\frac{7}{2}$ (e) None of these

**9.** Find all points of inflection: $f(x) = x^3 - 12x$.

(a) $(0, 0), (\pm\sqrt{12}, 0)$ (b) $(0, 0)$ (c) $(2, 0), (-2, 0)$

(d) $(2, -16), (-2, 16)$ (e) None of these

**10.** A farmer has 160 feet of fencing to enclose 2 adjacent rectangular pig pens. What dimensions should be used so that the enclosed area will be a maximum?

(a) $4\sqrt{15}$ ft by $\frac{8}{5}\sqrt{15}$ ft (b) 40 ft by $\frac{80}{3}$ ft (c) 20 ft by $\frac{80}{3}$ ft

(d) 40 ft by 40 ft (e) None of these

**11.** State why the Mean Value Theorem does not apply to the function $f(x) = \dfrac{2}{(x + 1)^2}$ on the interval $[-3, 0]$.

(a) $f(-3) \neq f(0)$ (b) $f$ is not continuous at $x = -1$.

(c) $f$ is not defined at $x = -3$ and $x = 0$. (d) Both $a$ and $b$

(e) None of these

**12.** Which of the following is the correct sketch of the graph of the function $y = 8x^3 - 2x^4$?

(a)

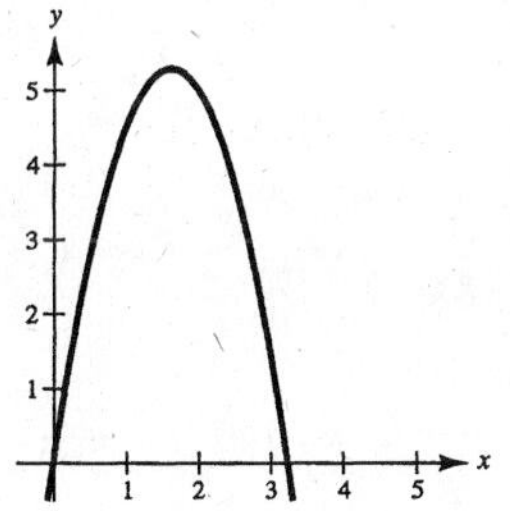


(b)

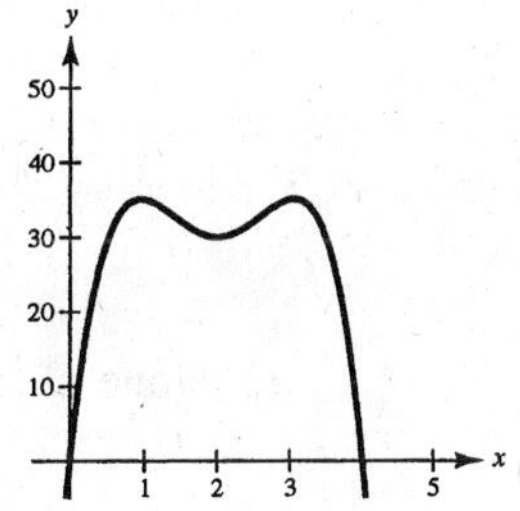


(c)

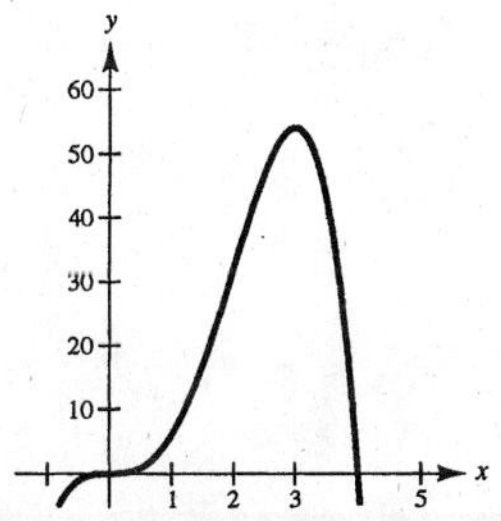


(d)

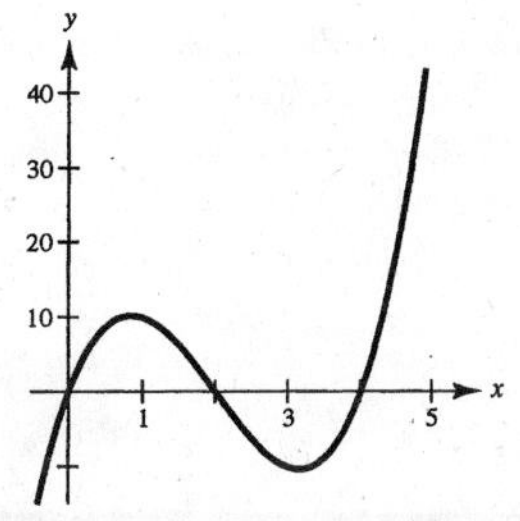


(e) None of these

**13.** Find all extrema in the interval $[0, 2\pi]$ for $y = x - \cos x$.

(a) $\left(\frac{3\pi}{2}, \frac{3\pi}{2}\right), (2\pi, 2\pi - 1)$ (b) $(\pi, \pi + 1), (0, 0)$ (c) $(2\pi, 2\pi - 1), (0, 0)$

(d) $(2\pi, 2\pi - 1), (\pi, \pi + 1)$ (e) None of these

**14.** The radius of a sphere is measured to be 3.0 inches. If the measurement is correct to within 0.01 inch, use differentials to estimate the propagated error in the volume of the sphere.

(a) $\pm 0.000001$ in.$^3$ (b) $\pm 0.36\pi$ in.$^3$ (c) $\pm 0.036\pi$ in.$^3$

(d) $\pm 0.06$ in.$^3$ (e) None of these

**Test Form C** Name ____________ Date ____________

**Chapter 3** Class ____________ Section ____________

**A graphing calculator/utility is recommended for this test.**

1. Determine from the graph whether $f$ possesses extrema on the inverval $(a, b)$.

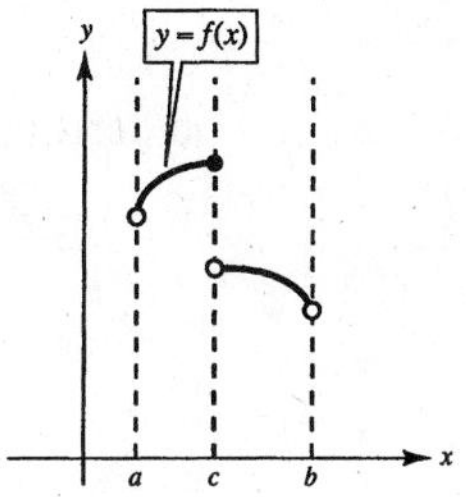


   (a) Maximum at $x = c$, minimum at $x = b$
   (b) Maximum at $x = c$, no minimum
   (c) No maximum, minimum at $x = b$
   (d) No extrema
   (e) None of these

2. Use a graphing utility to graph $f(x) = x^{2/3}$. State why Rolle's Theorem does not apply to $f$ on the interval $[-1, 1]$.
   (a) $f$ is not continuous on $[-1, 1]$.
   (b) $f$ is not defined on the entire interval.
   (c) $f$ is not differentiable at $x = 0$.
   (d) The interval $[-1, 1]$ is not closed.
   (e) None of these

3. Find the open intervals on which $f(x) = 4 \ln x - \frac{1}{2}x$ is increasing or decreasing.
   (a) Increasing $(-\infty, 8)$; decreasing $(8, \infty)$
   (b) Increasing $(-\infty, \infty)$
   (c) Increasing $(0, 2)$; decreasing $(2, \infty)$
   (d) Increasing $(0, 8)$; decreasing $(8, \infty)$
   (e) None of these

4. Let $f(x) = (x + 2)^3 - 4$. The point $(-2, -4)$ is ____________.
   (a) An absolute maximum
   (b) An absolute minimum
   (c) A critical point but not an extremum
   (d) Not a critical point
   (e) None of these

5. Given that $f(x) = -x^2 + 12x - 28$ has a relative maximum at $x = 6$, choose the correct statement.
   (a) $f'$ is negative on the interval $(-\infty, 6)$
   (b) $f'$ is positive on the interval $(-\infty, \infty)$
   (c) $f'$ is negative on the interval $(6, \infty)$
   (d) $f'$ is positive on the interval $(6, \infty)$
   (e) None of these

**6.** Use a graphing utility to graph $f(x) = -\frac{1}{(x+1)^2}$. Use the graph to determine the open intervals where the graph of the function is concave upward or concave downward.

(a) Concave downward: $(-\infty, \infty)$

(b) Concave downward: $(-\infty, -1)$; concave upward: $(-1, \infty)$

(c) Concave downward: $(-\infty, -1)$ and $(-1, \infty)$

(d) Concave upward: $(-\infty, -1)$ and $(-1, \infty)$

(e) None of these

**7.** The figure given in the graph is the second derivative of a polynomial function, $f$. Choose a graph of $f$.

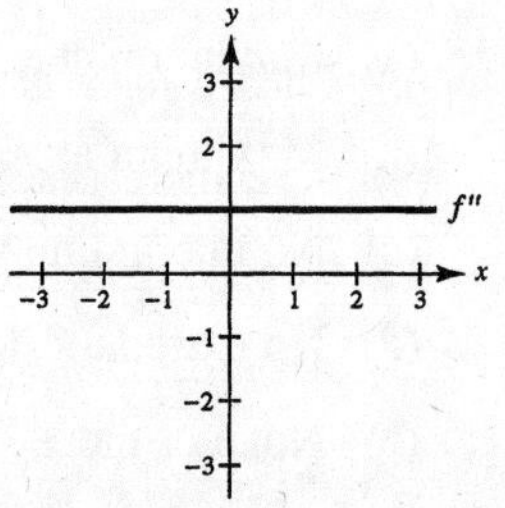


(a)

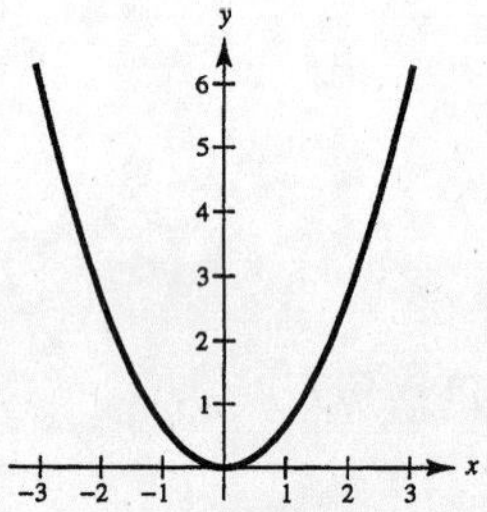


(b)

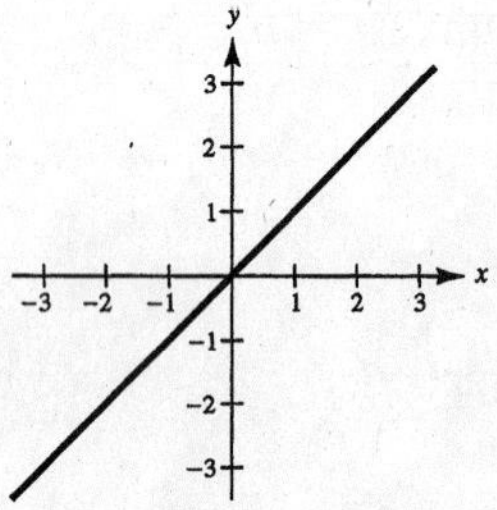


(c)

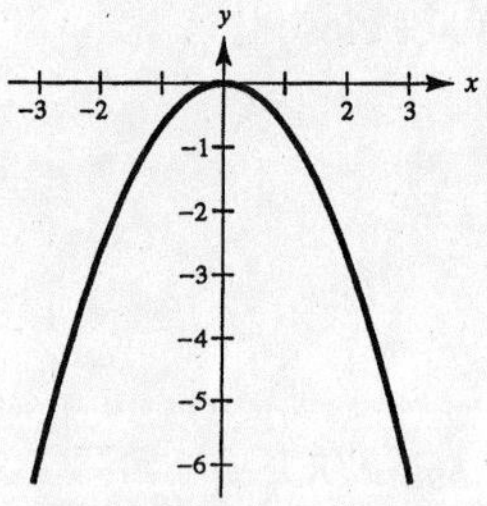


(d)

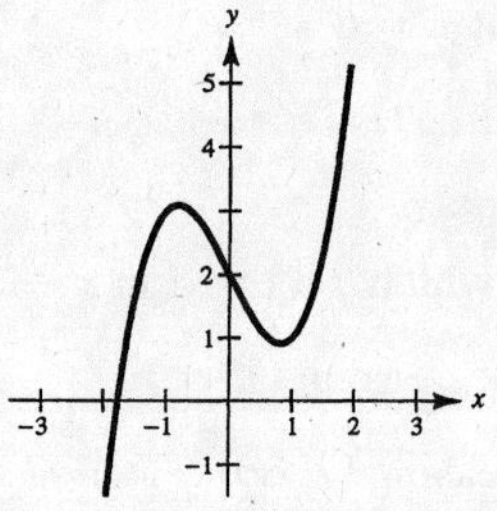


(e) None of these

**8.** Let $f(x)$ be a polynomial function such that $f(-2) = 5$, $f'(-2) = 0$, and $f''(-2) = 3$. The point $(-2, 5)$ is a __________________ the graph of $f$.

(a) Relative maximum (b) Relative minimum (c) Intercept

(d) Point of inflection (e) None of these

**9.** Use a graphing utility to graph $f(x) = \frac{2x+5}{1-x}$. Use the graph to find the limit: $\lim_{x\to\infty} \frac{2x+5}{1-x}$.

(a) 0 (b) $\infty$ (c) 2

(d) 5 (e) None of these

**10.** Find the limit: $\lim_{x\to\infty} \frac{a - bx^4}{cx^4 + x^2}$.

(a) 0 (b) $\infty$ (c) $-\frac{b}{c}$

(d) $\frac{a}{c}$ (e) None of these

**11.** Consider $f(x) = \frac{x^2}{x^2 + a}$, $a > 0$. Use a graphing utility to determine the effect on the graph of $f$ if $a$ is varied.

(a) Each $y$ value is multiplied by $a$.

(b) As $a$ increases, the vertical tangent lines move further from the origin.

(c) The graph of the curve is shifted $\sqrt{a}$ units to the left.

(d) As $a$ increases, the curve approaches its asymptotes more slowly.

(e) None of these

**12.** Determine the function whose graph has vertical asymptotes at $x = \pm 2$ and a horizontal asymptote at $y = 0$.

(a) $f(x) = \frac{x}{(x-2)^2}$ (b) $f(x) = \frac{x}{x^2 + 4}$ (c) $f(x) = \frac{3x}{x^2 - 4}$

(d) $f(x) = x(x - 2)(x + 2)$ (e) None of these

**13.** Find the point on the graph of $y = \sqrt{x + 1}$ closest to the point $(3, 0)$.

(a) $(0, 1)$ (b) $\left(\frac{5}{2}, \sqrt{\frac{7}{2}}\right)$ (c) $(3, 2)$

(d) $\left(2, \sqrt{3}\right)$ (e) None of these

**14.** Find all points of inflection for the function $f(x) = \ln(1 + x^2)$.

(a) $\left(\sqrt{\frac{1}{2}}, \ln\frac{3}{2}\right), \left(\sqrt{\frac{1}{2}}, \ln\frac{3}{2}\right)$ (b) $\left(-\sqrt{2}, \ln 3\right), \left(\sqrt{2}, \ln 3\right)$ (c) $(0, 0)$

(d) $\left(-\sqrt{\frac{1}{6}}, \ln\frac{7}{6}, \sqrt{\frac{1}{6}}, \ln\frac{7}{6}\right)$ (e) None of these

**15.** Find $dy$ for $y = \sec 3x$.

(a) $\sec 3x\, dx$ (b) $3 \sec 3x \tan 3x\, dx$ (c) $3 \sec^2 3x\, dx$

(d) $3 \tan^2 3x$ (e) None of these

**Test Form D** Name ______________ Date ____________

**Chapter 3** Class ______________ Section ____________

1. Investigate for intervals where $f(x) = \frac{1}{x^2}$ is increasing or decreasing.

2. Find all critical numbers: $f(x) = x\sqrt{2x + 1}$.

3. Find extrema: $y = \frac{2x}{(x + 4)^3}$.

4. Find the highest point on the graph of $y = \ln(\cos x)$ in the interval $\left[0, \frac{\pi}{3}\right]$.

5. Find all intervals for which the graph of the function $y = 8x^3 - 2x^4$ is concave downward.

6. Let $f(x) = x^3 - x^2 + 3$. Use the Second Derivative Test to determine which critical numbers, if any, give relative extrema.

7. Find $\lim_{x \to \infty} \frac{\sqrt{4x^2 - 1}}{x}$.

8. Use the techniques learned in this chapter to sketch the graph of $y = \frac{3}{(1 - x)^2}$.

9. Use the techniques learned in this chapter to sketch the graph of $y = x^3 - 3x + 1$.

10. Find all points of inflection for the graph of the *standard normal density function* $f(x) = \frac{1}{\sqrt{2\pi}}e^{-x^2/2}$.

11. The management of a large store has 1600 feet of fencing to fence in a rectangular storage yard using the building as one side of the yard. If the fencing is used for the remaining 3 sides, find the area of the largest possible yard.

12. Find all horizontal asymptotes: $f(x) = \frac{2}{x - 3} - \frac{x}{x + 2}$.

13. The volume of a cube is claimed to be 27 cubic inches, correct to within 0.027 cubic inch. Use differentials to estimate the propagated error in the measurement of the side of the cube.

**Test Form E** **Name** ____________________ **Date** ______________

**Chapter 3** **Class** ____________________ **Section** ______________

**A graphing calculator/utility is recommended for this test.**

**1.** Use a graphing utility to graph the functions $y_1 = \ln x^4$, $y_2 = 4 \ln x$, $y_3 = \ln x^3$, and $y_4 = 3 \ln x$. Write a short paragraph to explain why the graphs of $y_1$ and $y_2$ are not the same, but the graph of $y_3$ is the same as the graph of $y_4$.

**2.** Consider $f(x) = \dfrac{1}{(x-3)^2}$.

**a.** Sketch the graph of $f(x)$.

**b.** Calculate $f(2)$ and $f(4)$.

**c.** State why Rolle's Theorem does not apply to $f$ on the interval $[2, 4]$.

**3.** Consider $f(x) = \sqrt{x}$. Find all values, $c$, in the interval $[0, 1]$ such that the slope of the tangent line to the graph of $f$ at $c$ is parallel to the secant line through the points $(0, f(0))$ and $(1, f(1))$.

**4.** Use a graphing utility to graph the function $f(x) = 9x^4 + 4x^3 - 36x^2 - 24x$.

**a.** Adjust the viewing window and use the zoom and trace features to estimate the $x$-values of the relative extrema to one decimal place.

**b.** Use calculus to find the actual $x$-values of the relative extrema.

**5.** Use a graphing utility to graph the function $f(x) = x^3 + 3x^2 - 9x - 10$. Then use the graph to determine open intervals of increasing or decreasing.

**6.** A differentiable function $f$ has only one critical number: $x = -3$. Identify the relative extrema of $f$ at $(-3, f(-3))$ if $f'(-4) = \frac{1}{2}$ and $f'(-2) = -1$.

**7.** Use a graphing utility to graph $f(x) = \dfrac{1}{x-2} + 1$. Use the graph to determine the open intervals where the graph of the function is concave upward or concave downward.

**8.** The graph of a polynomial function, $f$, is given. On the same coordinate axes sketch $f'$ and $f''$.

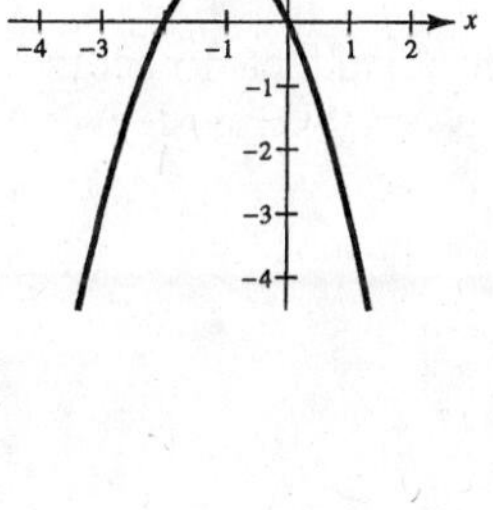

**9.** Find the limits:

(a) $\lim\limits_{x \to +\infty} \dfrac{2x}{\sqrt{x^2 + 1}}$ (b) $\lim\limits_{x \to -\infty} \dfrac{2x}{\sqrt{x^2 + 1}}$ (c) $\lim\limits_{x \to \infty} \dfrac{2x}{\sqrt{x^2 + 1}}$

**10.** Consider $f(x) = \dfrac{x^2 - 4x + 1}{x^2 + 1}$.

**a.** Find all asymptotes.

**b.** Use a graphing utility to graph $f$.

**c.** Use the graph to find the point on the graph where the curve crosses the asymptote.

**11.** Create a function whose graph has a vertical asymptote at $x = 0$ and a slant asymptote at $y = x - 1$.

**12.** An open box is to be made from a rectangular piece of cardboard, 7 inches by 3 inches, by cutting equal squares from each corner and turning up the sides.

**a.** Write the volume, $V$, as a function of the edge of the square, $x$, cut from each corner.

**b.** Use a graphing utility to graph the function, $V$. Then use the graph of the function to estimate the size of the square that should be cut from each corner and the volume of the largest such box.

**13.** Use a graphing utility to graph $f(x) = 0.2x^3 + 3.1x - 1.4$.

**a.** Use the graph to estimate (to one decimal place) the real zero of $f$.

**b.** Approximate this zero using the value found in part **a** and Newton's Method until two successive approximations differ by less than 0.001.

**14.** Consider $f(x) = x^3$.

**a.** Find an equation of the tangent line, $T$, at the point $(2, 8)$.

**b.** Graph $f$ and $T$ on the same coordinate axes using a graphing utility.

**c.** Use the graphs to estimate $f(2.1)$ and $T(2.1)$.

**d.** Calculate the actual values of $f(2.1)$ and $T(2.1)$.

**15.** The spread of flu in a certain school is given by

$$P(t) = \frac{300}{1 + e^{3.5-t}}$$

where $t$ is the number of days after students are exposed to infected students. Use this model to answer the following questions.

**a.** Estimate the initial number of students infected with the flu.

**b.** Estimate the number of days for 150 students to become infected with the flu.

**c.** What is the maximum number of students to get the flu?

**Test Form A** **Name** ________________ **Date** __________

**Chapter 4** **Class** ________________ **Section** __________

1. Evaluate the integral: $\int \sqrt[3]{t}\, dt$.

(a) $\frac{3}{4}t^{4/3} + C$ (b) $\sqrt[3]{\frac{1}{2}t^2} + C$ (c) $\frac{3}{2}t^{2/3} + C$

(d) $\frac{1}{3t^{2/3}} + C$ (e) None of these

2. Evaluate the integral: $\int 5 \sec x \tan x\, dx$.

(a) $5 \sec^3 x \tan x + C$ (b) $5 \sec x + C$ (c) $\frac{1}{5} \sec^3 x \tan x + C$

(d) $5[\sec^3 x + \sec x \tan^2 x] + C$ (e) None of these

3. Evaluate the integral: $\int \frac{x^3 + x}{x}\, dx$.

(a) $x^3 + 3x + C$ (b) $2x + C$ (c) $\frac{x^3}{3} + x + C$

(d) $\frac{2x^3 + x - 1}{x^2}$ (e) None of these

4. Evaluate the integral: $\int \frac{\sin^3 \theta}{1 - \cos^2 \theta}\, d\theta$.

(a) $-\cos \theta + C$ (b) $\cos \theta + C$ (c) $\frac{\cos \theta[3 - 3\cos^2 \theta - 2\sin^2 \theta]}{1 - \cos^2 \theta}$

(d) $\frac{1}{2} \sin^2 \theta + C$ (e) None of these

5. Find $y = f(x)$ if $f''(x) = x^2$, $f'(0) = 7$ and $f(0) = 2$.

(a) $x^2 + 9$ (b) $\frac{1}{12}x^4 + 7x + 2$ (c) $x^2 + 7x + 2$

(d) $x^4 + 84x + 24$ (e) None of these

6. Use $a(t) = -32$ feet per second squared as the acceleration due to gravity. A ball is thrown vertically upward from the ground with an initial velocity of 96 feet per second. How high will the ball go?

(a) 32 feet (b) 64 feet (c) 24 feet

(d) 144 feet (e) None of these

**7.** Let $s(n) = \sum_{i=1}^{n}\left(1 + \frac{i}{n}\right)^2\left(\frac{2}{n}\right)$. Find the limit of $s(n)$ as $n \to \infty$.

(a) $\frac{17}{12}$ (b) $\frac{10}{3}$ (c) $\frac{14}{3}$

(d) $\frac{20}{3}$ (e) None of these

**8.** Which of the following definite integrals represents the area of the shaded region?

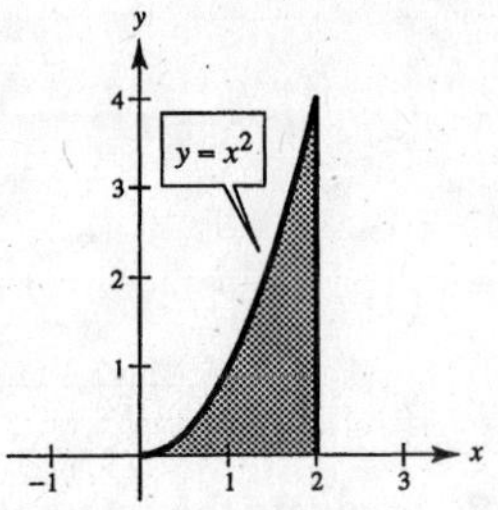


(a) $\int_0^4 x^2\, dx$ (b) $\int_0^2 x^2\, dx$

(c) $\int_1^2 x^2\, dx$ (d) $\int_0^2 x^2$

(e) None of these

**9.** Use the Fundamental Theorem of Calculus to evaluate the integral: $\int_1^4 \sqrt{x}\, dx$.

(a) 1 (b) $-\frac{14}{3}$ (c) 7

(d) $\frac{14}{3}$ (e) None of these

**10.** Find the average value of $f(x) = 2x^2 + 3$ on the interval $[0, 2]$.

(a) $\frac{22}{3}$ (b) $\frac{17}{3}$ (c) 4

(d) 27 (e) None of these

**11.** Evaluate the integral: $\int x^2(x^3 + 5)^6\, dx$.

(a) $\frac{1}{21}(x^3 + 5)^7 + C$ (b) $\frac{1}{7}(x^3 + 5)^7 + C$ (c) $\frac{x^3(x^3 + 5)^7}{21} + C$

(d) $\frac{x^3}{3}\left(\frac{x^4}{4} + 5x\right)^6 + C$ (e) None of these

**12.** Evaluate the integral: $\int_0^2 |x - 1|\, dx$.

(a) 0 (b) 1 (c) $\frac{1}{2}$

(d) 2 (e) None of these

**13.** Evaluate the integral: $\int x\sqrt{1-x}\,dx$.

(a) $-\frac{x^2}{3}(1-x)^{3/2}+C$ (b) $\frac{2-3x}{2\sqrt{1-x}}+C$ (c) $\frac{x^2}{3}(1-x)^{3/2}+C$

(d) $-\frac{2}{15}(2+3x)(1-x)^{3/2}+C$ (e) None of these

**14.** Use Simpson's Rule with $n = 4$ to approximate $\int_2^3 \frac{1}{(x-1)^2}\,dx$.

(a) 0.5004 (b) 0.5090 (c) 2.5000

(d) 1.7396 (e) None of these

**15.** Evaluate $\int \frac{e^{1/(x+1)}}{(x+1)^2}\,dx$.

(a) $\frac{e^{1/(x+1)}}{2(x+1)}+C$ (b) $\frac{e^{-x/(x+1)}}{(x+1)^2}+C$ (c) $-e^{1/(x+1)}+C$

(d) $\frac{e^{-x/(x+1)}}{(x+1)^2}$ (e) None of these

**16.** Evaluate $\int \tan 3x\,dx$.

(a) $\frac{1}{3}\ln|\sec 3x|+C$ (b) $3\sec^2 3x+C$ (c) $\frac{1}{3}\sec^2 3x$

(d) $\ln|\cos 3x|+C$ (e) None of these

**17.** Evaluate $\int \frac{x+2}{x+1}\,dx$.

(a) $\frac{x^2+4x}{x^2+2x}+C$ (b) $2x+C$ (c) $x+C$

(d) $x+\ln|x+1|+C$ (e) None of these

**18.** Evaluate $\int \frac{x+3}{x^2+9}\,dx$.

(a) $\ln|x-3|+C$ (b) $\frac{1}{3}\arctan\frac{x}{3}+C$ (c) $\frac{1}{2}\ln(x^2+9)+\arctan\frac{x}{3}+C$

(d) $\ln(x^2+9)+\frac{1}{3}\arctan\frac{x}{3}+C$ (e) None of these

**19.** Evaluate $\int \frac{dx}{\sqrt{8x + 2x - x^2}}$.

(a) $\ln\sqrt{8 + 2x - x^2}$

(b) $\arcsin\frac{x-1}{3} + C$

(c) $\sqrt{8 + 2x - x^2} + C$

(d) $\frac{1}{3}\operatorname{arcsec}\frac{x-1}{3} + C$

(e) None of these

**20.** Find $f'(x)$ for $f(x) = \cosh^2 x^2$.

(a) $-4x \cosh x^2 \sinh x^2$

(b) $4x \cosh x^2 \sinh x^2$

(c) $2x \sinh^2 x^2$

(d) $2 \cosh x^2 \sinh x^2$

(e) None of these

**Test Form B** **Name** ______________ **Date** ______________

**Chapter 4** **Class** ______________ **Section** ______________

1. Evaluate the integral: $\int \sqrt[5]{x^2}\,dx$.

(a) $\sqrt[5]{\dfrac{x^3}{3}} + C$ (b) $\dfrac{5}{7}x^{7/5} + C$ (c) $-\dfrac{1}{9x^9} + C$

(d) $-\dfrac{1}{4x^8} + C$ (e) None of these

2. Evaluate the integral: $\int 3\csc^2 x\,dx$.

(a) $\frac{1}{3}\csc^2 x + C$ (b) $6\csc^2 x \cot x + C$ (c) $-3\cot x + C$

(d) $-\frac{1}{3}\csc^3 x + C$ (e) None of these

3. Evaluate the integral: $\int \dfrac{x^4 - x^3}{x^2}\,dx$.

(a) $2x^3 - 3x^2 + C$ (b) $\dfrac{x^3}{3} - \dfrac{x^2}{2} + C$ (c) $2x - 1 + C$

(d) $\dfrac{3}{20}(4x^2 - 5x) + C$ (e) None of these

4. Evaluate the integral: $\int \dfrac{\sec^3\theta\tan\theta}{1 + \tan^2\theta}\,d\theta$.

(a) $\dfrac{1}{4}\sec^4\theta + C$ (b) $\dfrac{1}{2}\sec^2\theta + C$ (c) $\dfrac{1}{4}\sec^2\theta\tan^2\theta + C$

(d) $\sec\theta + C$ (e) None of these

5. Find $y = f(x)$ if $f''(x) = x + 2$, $f'(0) = 3$, $f(0) = -1$.

(a) $\dfrac{1}{6}x^3 + x^2 + 3x - 1$ (b) $\dfrac{x^3}{6} + 2x^2 + C$ (c) $x^3 + 6x^2 + 18x - 6$

(d) $\dfrac{1}{6}x^3 + x^2 + \dfrac{21}{2}x + \dfrac{61}{6} + C$ (e) None of these

6. Use $a(t) = -32$ feet per second squared as the acceleration due to gravity. A ball is thrown vertically upward from the ground with an initial velocity of 56 feet per second. For how many seconds will the ball be going upward?

(a) $1\frac{3}{4}$ sec (b) 24 sec (c) $\frac{7}{8}$ sec

(d) 1 sec (e) None of these

7. Let $s(n) = \sum_{i=1}^{n}\left(1 + \frac{i}{n}\right)^2\left(\frac{1}{n}\right)$. Find the limit of $s(n)$ as $n \to \infty$.

(a) $\frac{5}{3}$ (b) $\frac{7}{3}$ (c) $\frac{10}{3}$

(d) $\frac{17}{24}$ (e) None of these

8. Which of the following definite integrals represents the area of the shaded region?

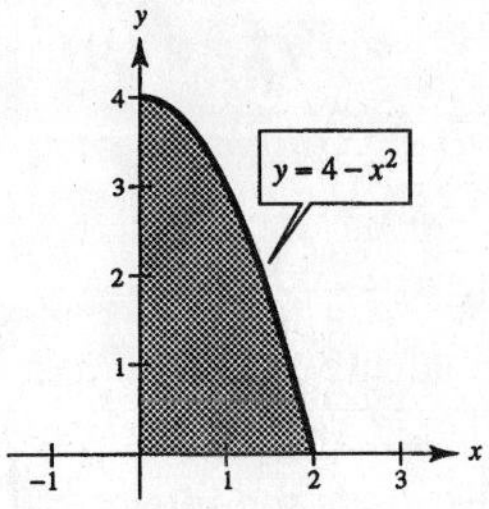


(a) $\int_0^2 (4 - x^2)$ (b) $\int_4^2 (4 - x^2)\,dx$

(c) $\int_0^2 (4 - x^2)\,dx$ (d) $\int_2^4 (4 - x^2)\,dx$

(e) None of these

9. Use the Fundamental Theorem of Calculus to evaluate the integral: $\int_1^2 \frac{1}{x^2}\,dx$.

(a) $-\frac{1}{x} + C$ (b) $-\frac{3}{4}$ (c) $\frac{1}{2}$

(d) $-\frac{3}{2}$ (e) None of these

10. Find the average value of $f(x) = 3x^2 - 2$ on the interval $[0, 2]$.

(a) 1 (b) 5 (c) 6

(d) 2 (e) None of these

11. Evaluate the integral: $\int x(x^2 - 1)^4\,dx$.

(a) $\frac{1}{10}(x^2)(x^2 - 1)^5$ (b) $\frac{1}{10}(x^2 - 1)^5 + C$ (c) $\frac{1}{5}(x^3 - x)^5 + C$

(d) $\frac{1}{5}(x^2 - 1)^5 + C$ (e) None of these

12. Evaluate the integral: $\int_{-1}^{1} |x|\,dx$.

(a) 1 (b) 0 (c) 2

(d) $-1$ (e) None of these

13. Evaluate the integral: $\int x\sqrt{x + 1}\,dx$.

(a) $\frac{3x + 2}{2\sqrt{x + 1}} + C$ (b) $\frac{2}{15}(x + 1)^{3/2}(3x - 2) + C$ (c) $\frac{1}{4}(x + 1)(x - 1) + C$

(d) $\frac{1}{3}x^2(x + 1)^{3/2} + C$ (e) None of these

**14.** Use Trapezoidal Rule with $n = 4$ to approximate $\displaystyle\int_2^3 \frac{1}{(x-1)^2}\,dx$.

(a) 0.5004 (b) 2.5000 (c) 0.5090

(d) 1.7396 (e) None of these

**15.** Evaluate $\displaystyle\int \frac{e^{\sqrt{x}}}{\sqrt{x}}\,dx$.

(a) $2e^{\sqrt{x}} + C$ (b) $\frac{1}{2}e^{\sqrt{x}} + C$ (c) $\sqrt{x}\,e^{\sqrt{x}} + C$

(d) $\sqrt{x}\,e^{\sqrt{x}+1} + C$ (e) None of these

**16.** Evaluate $\displaystyle\int x \cot x^2\,dx$.

(a) $\dfrac{x^2}{2}\sec^2 x^2 + C$ (b) $\dfrac{x^2}{4}\ln|\sin x^2| + C$ (c) $x \cot x^2 \csc x^2 + C$

(d) $\dfrac{1}{2}\ln|\sin x^2| + C$ (e) None of these

**17.** Evaluate $\displaystyle\int \frac{2x+1}{x+1}\,dx$.

(a) $2x - \ln|x+1| + C$ (b) $2x + C$ (c) $x^2 + \ln|x+1| + C$

(d) $\dfrac{2x^2+2x}{x^2+2x} + C$ (e) None of these

**18.** Evaluate $\displaystyle\int \frac{x+2}{\sqrt{4-x^2}}\,dx$.

(a) $-\dfrac{1}{2}\sqrt{4-x^2} + 2\arcsin\dfrac{x}{2} + C$ (b) $-\sqrt{4-x^2} + 2\arcsin\dfrac{x}{2} + C$

(c) $\ln|2-x| + C$ (d) $x^2 + 2x + \arcsin\dfrac{x}{2} + C$

(e) None of these

**19.** Evaluate $\displaystyle\int \frac{5}{x^2+6x+13}\,dx$.

(a) $5\ln|x^2+6x+13| + C$ (b) $5\left(\dfrac{x^3}{3} + 3x^2 + 13x\right) + C$ (c) $\dfrac{5}{2}\arctan\dfrac{x+3}{2} + C$

(d) $-\dfrac{5}{x} + \dfrac{5}{6}\ln|x| + \dfrac{5}{13}x + C$ (e) None of these

**20.** Find $f'(x)$ for $f(x) = \ln(\operatorname{sech} x)$.

(a) $\operatorname{sech} x \tanh x$ (b) $-\tanh x$ (c) $\tanh^2 x$

(d) $-\operatorname{sech} x \tanh x$ (e) None of these

**Test Form C** **Name** ____________ **Date** ____________

**Chapter 4** **Class** ____________ **Section** ____________

**A graphing calculator/utility is recommended for this test.**

**1.** Evaluate the integral: $\int (ax + b)\, dx$.

(a) $\frac{ab}{2}x^2 + C$ (b) $a + C$ (c) $\frac{a}{2}x^2 + bx + C$

(d) $\frac{a}{2}x^2 + bx$ (e) None of these

**2.** Evaluate the integral: $\int \frac{3 + 4x^{3/2}}{\sqrt{x}}\, dx$.

(a) $\frac{3}{2}\sqrt{x} + 2x^2 + C$ (b) $-\frac{3}{2}x^{-3/2} + 4 + C$ (c) $\frac{3}{2}x^{-3/2} + 2x^2 + C$

(d) $6\sqrt{x} + 2x^2 + C$ (e) None of these

**3.** The rate of growth of a particular population is given by

$$\frac{dP}{dt} = 50t^2 - 100t^{3/2},$$

where $P$ is the population size and $t$ is the time in years. The initial population is 25,000. Use a graphing utility to graph the population function. Then use the graph to estimate how many years it will take for the population to reach 50,000.

(a) 15.7 (b) 14.5 (c) 2

(d) 38.4 (e) None of these

**4.** An object has a constant acceleration of 42 feet per second squared, an initial velocity of −18 feet per second, and an initial position of 3 feet. Find the position function describing the motion of this object.

(a) $s = 42t^2 + 3$ (b) $s = 42t^2 - 18t + 3$ (c) $s = 21t^2 - 5$

(d) $s = 21t^2 - 18t + 3$ (e) None of these

**5.** Identify the sum that does not equal the others.

(a) $\sum_{n=0}^{5} (3n + 1)$ (b) $\sum_{k=1}^{6} (3k - 2)$ (c) $\sum_{j=3}^{8} (3j - 8)$

(d) $\sum_{i=1}^{6} (i + 2)$ (e) None of these

**6.** Use a graphing utility to graph $f(x) = -2x^3 - 3x^2 + 5$. Then use the upper sums to approximate the area of the region between the $x$-axis and $f$ on the interval $[0, 1]$ using 4 subintervals.

(a) 2.81 (b) 4.06 (c) 3.5

(d) 3 (e) None of these

**7.** Determine the interval(s) on which $f(x) = |x + 2|$ is integrable.

(a) $[0, 5]$ (b) $[-3, 0]$ (c) $[-1, 4]$

(d) $a$, $b$, and $c$ (e) $a$ and $c$

**8.** Consider $F(x) = \int_x^1 \sqrt{1 + t^2}\, dt$. Find $F'(x)$.

(a) $\sqrt{1 + x^2}$ (b) $\dfrac{1}{\sqrt{2}} - \dfrac{x}{\sqrt{1 + x^2}}$ (c) 1

(d) $-\sqrt{1 + x^2}$ (e) None of these

**9.** Choose the correct quantity to fill in the blank: $\int$ ____________ $dx = (ax^2 - a^2)^4 + C$.

(a) $(ax^2 - a^2)^3$ (b) $4(ax^2 - a^2)^3$ (c) $\frac{1}{5}(ax^2 - a^2)^5$

(d) $4a(ax^2 - a^2)^3$ (e) None of these

**10.** Evaluate the integral: $\int x \sec^2 x^2\, dx$.

(a) $\frac{1}{6}x^3 \sec^3 x^2 + C$ (b) $\frac{1}{2} \tan x^2 + C$ (c) $\frac{1}{2}x^2 \tan x^2 + C$

(d) $\tan x^3 + C$ (e) None of these

**11.** Use the general power rule to evaluate the integral: $\int x\sqrt{9 - 5x^2}\, dx$.

(a) $-\dfrac{1}{10}(9 - 5x^2)^{3/2} + C$ (b) $-\dfrac{1}{15}(9 - 5x^2)^{3/2} + C$ (c) $\dfrac{2}{3}(9 - 5x^2)^{3/2} + C$

(d) $-\dfrac{4}{15}(9 - 5x^2)^{3/2} + C$ (e) None of these

**12.** Evaluate the integral: $\int \dfrac{5x}{\sqrt{x + 2}}\, dx$.

(a) $\dfrac{2}{3}\sqrt{x + 2}\,(x - 4) + C$ (b) $\dfrac{5}{2}\left(x - 4 \ln\sqrt{x + 2}\right) + C$ (c) $5\left(x - 4 \ln\sqrt{x + 2}\right) + C$

(d) $\dfrac{10}{3}\sqrt{x + 2}\,(x - 4) + C$ (e) None of these

**13.** Evaluate the integral: $\int e^{(ax+b)}\,dx$.

(a) $ae^{(ax+b)} + C$

(b) $\frac{1}{a}e^{(ax+b)} + C$

(c) $e^{(ax+b)} + C$

(d) $e^{(ax^2/2+bx+c)}$

(e) None of these

**14.** Evaluate the integral: $\int \frac{t^2+1}{t+1}\,dt$.

(a) $t^2 - 2t + \ln(t+1)^4 + C$

(b) $\frac{t^2 - 2t - 1}{(t+1)^2} + C$

(c) $t + C$

(d) $\frac{1}{2}t^2 - t + \ln(t+1)^2 + C$

(e) None of these

**15.** Evaluate the indefinite integral: $\int \frac{1}{x^2 - 2x + 10}\,dx$.

(a) $\ln|x^2 - 2x + 10| + C$

(b) $\frac{1}{3}\arctan\frac{x-1}{3} + C$

(c) $\arctan\frac{3}{x-1} + C$

(d) $\frac{1}{\sqrt{11}}\arctan\frac{(x-1)}{\sqrt{11}} + C$

(e) None of these

**16.** Use a graphing utility as an aid to sketch the region bounded by the function $f(x) = \frac{1}{\sqrt{9-x^2}}$, the $x$-axis, and the lines $x = 1$ and $x = 2$. Then calculate the area of the region.

(a) $\arcsin\left(\frac{1}{3}\right)$

(b) $\frac{1}{3}\arctan\left(\frac{2}{3}\right)$

(c) $\arcsin\left(\frac{2}{3}\right) - \arcsin\left(\frac{1}{3}\right)$

(d) $\frac{1}{\sqrt{5}} - \frac{1}{\sqrt{8}}$

(e) None of these

**Test Form D** Name ______________ Date ______________

**Chapter 4** Class ______________ Section ______________

1. Evaluate the integral: $\int \sqrt{x^3}\, dx$.

2. Evaluate the integral: $\int 3 \csc x \cot x \, dx$.

3. Evaluate the integral: $\int \frac{x^3 - x^2}{x^2}\, dx$.

4. Evaluate the integral: $\int \frac{\cos^3 \theta}{2 - 2\sin^2 \theta}\, d\theta$.

5. Find the function, $y = f(x)$, if $f'(x) = 2x - 1$ and $f(1) = 3$.

6. Use $a(t) = -32$ feet per second squared as the acceleration due to gravity. An object is thrown vertically downward from the top of a 480-foot building with an initial velocity of 64 feet per second. With what velocity does the object hit the ground?

7. Let $s(n) = \sum_{i=1}^{n} \left(1 + \frac{2i}{n}\right)\left(\frac{2}{n}\right)$. Find the limit of $s(n)$ as $n \to \infty$.

8. Write the definite integral that represents the area of the region enclosed by $y = 4x - x^2$ and the $x$ axis.

9. Evaluate: $\frac{d}{dx}\int_2^x (2t^2 + 5)^2\, dt$.

10. Use the Fundamental Theorem of Calculus to evaluate $\int_{-1}^{1} (\sqrt[3]{t} - 2)\, dt$.

11. Find the average value of $f(x) = \sin x$ on the interval $\left[\frac{\pi}{4}, \frac{\pi}{2}\right]$.

12. Evaluate the integral: $\int \frac{\sec^2 x}{\sqrt{\tan x}}\, dx$.

13. Find the indefinite integral: $\int \frac{x}{\sqrt{x - 1}}\, dx$.

14. Evaluate $\int_2^{e+1} \frac{1}{x - 1}\, dx$.

**15.** Evaluate the integral: $\int \sec 2x \, dx$.

**16.** Evaluate the integral: $\int 19e^{-t/5} \, dt$.

**17.** Evaluate the integral: $\int \frac{e^x}{\sqrt{e^x + 1}} \, dx$.

**18.** Evaluate the integral: $\int \frac{5x + 16}{x^2 + 9} \, dx$.

**19.** Evaluate the integral: $\int \frac{1}{\sqrt{-3 + 4x - x^2}} \, dx$.

**20.** Find $f'(x)$ for $f(x) = \ln(\cosh 2x)$.

**Test Form E** **Name** ______________________ **Date** ____________

**Chapter 4** **Class** ______________________ **Section** ____________

**A graphing calculator/utility is recommended for this test.**

1. Evaluate the integral: $\int \frac{ax^2 + bx^3}{\sqrt{x}}\, dx.$

2. Evaluate the integral: $\int \frac{x}{(x-1)^3}\, dx.$

3. Find the function, $y = f(x)$, if $f'(x) = 2x - 1$ and $f(1) = 3$.

4. The rate of growth of a particular population is given by

$$\frac{dP}{dt} = 40t^3 - 70t^{4/3},$$

where $P$ is the population size and $t$ is the time in years. The initial population is 16,000.

   a. Determine the population function.

   b. Use a graphing utility to graph the function. Then use the graph to estimate the number of years until the population reaches 20,000. (Round your answer to one decimal place.)

   c. Use the population function to calculate the population after 8 years.

5. Use a graphing utility as an aid to sketch the region whose area is indicated by the integral:

$$\int_0^2 (-x^3 + 4x + 6)\, dx.$$

6. Determine if $f(x) = \frac{5}{2x - 3}$ is integrable on $[0, 2]$. Give a reason for your answer.

7. Use a graphing utility to graph $f(x) = \cos x - \sin x$ on the interval $[0, \pi]$. Calculate the area in the first quadrant bounded by the $x$-axis and $f$.

8. Consider $F(x) = \int_2^x \frac{1}{1 + t^4}\, dt$. Find $F'(x)$ and $F'(2)$.

9. Evaluate the integral: $\int \frac{\cos \sqrt{x}}{\sqrt{x}}\, dx.$

10. Evaluate the integral: $\int_{\pi/4}^{\pi/3} \sec^2 x\, dx.$

**11.** Evaluate the integral: $\int_0^1 x\sqrt{x^2+1}\,dx$.

**12.** Evaluate the integral: $\int \frac{1-\sin\theta}{\cos\theta}\,d\theta$.

**13.** Evaluate the integral: $\int \frac{e^{3/x}}{x^2}\,dx$.

**14.** Evaluate the integral: $\int \frac{x^2}{1+x^2}\,dx$.

**15.** Evaluate the definite integral: $\int_{-1/\sqrt{3}}^{1/\sqrt{3}} \frac{3}{\sqrt{4-9x^2}}\,dx$.

**16.** Evaluate the integral: $\int \frac{5}{(x-1)\sqrt{x^2-2x-24}}\,dx$.

**17.** Find the derivative: $h(t) = \arccos t^2$.

**Test Form A** **Name** ______________ **Date** __________

**Chapter 5** **Class** ______________ **Section** __________

**1.** Find the general solution to the first order differential equation: $2x(y + 1) - yy' = 0$.

(a) $2x^2 - y^2 \ln|y + 1| = C$ (b) $x^2 = y + \frac{1}{(y + 1)^2} + C$ (c) $\ln|y + 1| + x^2 - y = C$

(d) $x^2(y + 1)^2 - y^2 = C$ (e) None of these

**2.** Find the general solution to the first order differential equation: $y\,dx + (y - x)\,dy = 0$.

(a) $y^2 = C$ (b) $\ln|y| + x = C$ (c) $y \ln|y| + x = Cy$

(d) $y + x = \ln|x - y| + C$ (e) None of these

**3.** In 1980 the population of a town was 21,000 and in 1990 it was 20,000. Assuming the population decreases continuously at a rate proportional to the existing population, estimate the population in the year 2010.

(a) 17,619 (b) 18,000 (c) 19,048 (d) 18,141 (e) None of these

**4.** A radioactive element has a half-life of 50 days. What percentage of the original sample is left after 60 days?

(a) 43.53% (b) 49.56% (c) 37.50% (d) 25.00% (e) None of these

**5.** Find the particular solution to $y' = \sin x$ given the general solution $y = C - \cos x$ and the initial condition $y\left(\frac{\pi}{2}\right) = 1$.

(a) $-\cos x$ (b) $2 - \cos x$ (c) $-1 - \cos x$ (d) $1 - \cos x$ (e) None of these

**6.** Find the orthogonal trajectories for the family of curves $y^2 = Cx^3$.

(a) $y^3 + x^2 = K$ (b) $3y^2 + 2x^2 = K$ (c) $3ky + 2x = 0$

(d) $3y^2 = -2x^2$ (e) None of these

**7.** Find the general solution of the differential equation: $e^{2y}\,y' = x^3$.

(a) $y = \ln(x^2 + C)$ (b) $y = \frac{1}{2}\ln\left(\frac{x^4}{2}\right) + C$ (c) $y = \frac{1}{2}\ln\left(\frac{x^4}{2} + C\right)$

(d) $y = \frac{1}{2}\ln\left(\frac{x^4}{4} + C\right)$ (e) None of these

**8.** Consider the differential equation $y' = -y$ with the initial condition $y(0) = 2$. Use Euler's method with $\Delta x = 0.01$ to approximate $y(0.03)$.

(a) 2 (b) 1.458 (c) 1.940598 (d) 1.940891 (e) None of these

**9.** Find the general solution to the first-order differential equation $\dfrac{dy}{dx} + (\tan x)y = \cos x$.

(a) $y \sec x = x + C$

(b) $2y + y^2 \ln|\sec x| = \sin x + C$

(c) $\ln|y \cos x| = \sin x + C$

(d) $y \cos x = Ce^{\sin x}$

(e) None of these

**10.** Which method should be used to solve the differential equation $y' + 3xy = xy^3$?

(a) Linear

(b) Separation of variables

(c) Homogeneous

(d) Bernoulli equation

(e) None of these

**Test Form B** **Name** ____________ **Date** ____________

**Chapter 5** **Class** ____________ **Section** ____________

**1.** Find the general solution to the first order differential equation: $(4 - x)\,dy + 2y\,dx = 0$.

(a) $y = C(4 - x)^2$ (b) $4y - x^2y + xy = C$ (c) $y(4 - x^2) = C$

(d) $y^4(4 - x^2) = C$ (e) None of these

**2.** Find the general solution to the first order differential equation: $xy^2\,dy - (x^3 + y^3)\,dx = 0$.

(a) $3x^4 + 8xy^3 = C$ (b) $y^3 = \ln Cx$ (c) $y^3 - x^3 + y^3 \ln x^3 = C$

(d) $y^3 = 3x^3 \ln Cx$ (e) None of these

**3.** A certain type of bacteria increases continuously at a rate proportional to the number present. If there are 500 present at a given time and 1000 present 2 hours later, how many will there be 5 hours from the initial time given?

(a) 1750 (b) 2828 (c) 3000

(d) 2143 (e) None of these

**4.** A radioactive element has a half-life of 50 days. What percentage of the original sample is left after 85 days?

(a) 37.50% (b) 30.78% (c) 25.00%

(d) 24.06% (e) None of these

**5.** Find the particular solution to $y'' = \sin x$ given the general solution $y = -\sin x + Ax + B$ and the initial conditions $y\left(\frac{\pi}{2}\right) = 0$, $y'\left(\frac{\pi}{2}\right) = -2$.

(a) $-\sin x + 1 + \pi$ (b) $-\sin x - 2x + \pi$ (c) $-\sin x - 2x + 1 - \pi$

(d) $-\sin x + 2x + 1 + \pi$ (e) None of these

**6.** Find the solution to the initial value problem $e^{x^2} y' = xe^y$ with the initial condition $y(0) = 0$.

(a) $y = \ln\left(\frac{e^{-x^2} + 1}{2}\right)$ (b) $y = -\ln\left(\frac{e^{-x^2} - 1}{2}\right)$ (c) $y = -\ln\left(\frac{e^{-x^2} + 1}{2}\right)$

(d) $y = -\ln\left(\frac{1 - e^{-x^2}}{2}\right)$ (e) None of these

**7.** Find the general solution of the differential equation: $(x^2 + 1)\tan y \frac{dy}{dx} = x$.

(a) $y = \frac{C}{\sqrt{x^2 + 1}}$ (b) $y = C\sqrt{x^2 + 1}$ (c) $\cos y = \frac{C}{\sqrt{x^2 + 1}}$

(d) $\cos y = C\sqrt{x^2 + 1}$ (e) None of these

**8.** Consider the differential equation $y' = -xy$ with the initial condition $y(0) = 1$. Use Euler's method with $\Delta x = 0.1$ to approximate $y(0.3)$.

(a) 1 (b) 0.956 (c) $-0.198$ (d) 0.9702 (e) None of these

**9.** Find the general solution to the first-order differential equation $\frac{dy}{dx} - \frac{dy}{x} = x + 1$.

(a) $y = Cx^2 - x^3$ (b) $y = x^2 \ln x - x + Cx^2$ (c) $6xy = x^2 + C$

(d) $2y + y \ln x^4 = x^2 + C$ (e) None of these

**10.** Solve the Bernoulli equation $y' - \frac{xy}{2} = xy^5$.

(a) $y = (-2 + Ce^{-x^2})^{-1/4}$ (b) $y = (-2 + Ce^{-x})^{-1/4}$ (c) $y = (-1 + Ce^{-x^2})^{-1/2}$

(d) $y = (-2 + Ce^{-x^2})^{-1/2}$ (e) None of these

**Test Form C** **Name** ____________________ **Date** ____________

**Chapter 5** **Class** ____________________ **Section** ____________

**1.** Find the particular solution to the differential equation $y' = 3y$ given the general solution $y = Ce^{3x}$ and the initial condition $y(1) = 20$.

(a) $y = 20e^{3x-3}$ (b) $y = 20e^{3x}$ (c) $y = 20e^{x}$

(d) $y = 20e^{2x}$ (e) None of these

**2.** Find the general solution of the differential equation $xy' + 2y = 0$.

(a) $y = -2\ln|x| + C$ (b) $x^2y + 2y^2 = C$ (c) $y + 2x = C$

(d) $y = \dfrac{C}{x^2}$ (e) None of these

**3.** Find the orthogonal trajectories for the family of curves $y(x^2 + C) + 2 = 0$.

(a) $y^3 = -3\ln(Kx)$ (b) $y^2 = \ln|x| + Ky$ (c) $y = -\ln|xy^2| + K$

(d) $4y^2 + x^2 + K\ln|x| = 0$ (e) None of these

**4.** A radioactive element has a half-life of 40 days. What percentage of the original sample is left after 48 days?

(a) 49.56% (b) 43.53% (c) 25.00% (d) 37.50% (e) None of these

**5.** Determine which function is a solution to the differential equation $xy' + 2y = 0$.

(a) $e^{2x}$ (b) $\dfrac{1}{x^2}$ (c) $x^2$ (d) $2\ln\left|\dfrac{y}{x}\right|$ (e) None of these

**6.** Determine whether the function $f(x, y) = x^3 - 2x^2y + 4xy^2 - y^3$ is homogeneous, and if so, determine its degree.

(a) Homogeneous; degree 1 (b) Homogeneous; degree 2 (c) Homogeneous; degree 3

(d) Not homogeneous (e) None of these

**7.** Find the general solution of the differential equation: $y' = \dfrac{\sin x}{\cos y}$.

(a) $\sin y + \cos x = c$ (b) $\tan y = C$ (c) $\sin y = C\cos x$

(d) $\sin y - \cos x = C$ (e) None of these

**8.** Consider the differential equation $y' = y^2$ with the initial condition $y(1) = 1$. Use Euler's Method with $\Delta x = 0.1$ to approximate $y(1.3)$.

(a) 1.4285714 (b) 1.3700841 (c) 1 (d) 1.221 (e) None of these

**9.** Find the particular solution of the differential equation $y' + 2y = 4,\ y(0) = 4$.

(a) $y = 4$

(b) $y = 2 + 2e^{-2x}$

(c) $y = 2 + 2e^{2x}$

(d) $y = 2 + e^{-2x}$

(e) None of these

**10.** Solve the Bernoulli equation $y' + y = xy^3$.

(a) $y = \left(\dfrac{1}{2x + 1 + Ce^{2x}}\right)^{1/2}$

(b) $y = \left(\dfrac{2}{2x + Ce^{2x}}\right)^{1/2}$

(c) $y = \left(\dfrac{2}{2x + 1 + Ce^{2x}}\right)^{1/2}$

(d) $y = \left(\dfrac{2}{2x + 1 + Ce^{x}}\right)^{1/2}$

(e) None of these

**Test Form D** **Name** ________________ **Date** ________

**Chapter 5** **Class** ________________ **Section** ________

1. Find the general solution of the differential equation: $x \cos^2 y + \tan y \dfrac{dy}{dx} = 0.$

2. Find the general solution of the differential equation: $xy\,dx - (x^2 + y^2)\,dy = 0.$

3. A certain type of bacteria increases continuously at a rate proportional to the number present. If there are 500 present at a given time and 1000 present 2 hours later, how many hours (from the initial given time) will it take for the number of bacteria to be 2500? Round your answer to 2 decimal places.

4. Write an equation for the amount $Q$ of a radioactive substance with a half-life of 30 days, if 10 grams are present when $t = 0$.

5. Verify that the equation $y = x^2 + 2x + 2 + Ce^x$ is a solution to the differential equation $y' - y + x^2 = 0.$

6. Find the orthogonal trajectories of the family $y = Cx^2$ and sketch several members of each family.

7. Find the solution to the inital value problem: $(e^y + \cos y)y' = x,\ y(-1) = 0.$

8. Consider the differential equation $y' = \dfrac{y}{x} + 1$ with the initial condition $y(1) = 0$. Use Euler's Method with $\Delta x = 0.01$ to approximate $y(1.03)$.

9. Find the general solution to the first-order differential equation $xy' + 2y = x^2.$

10. Find the particular solution of the linear differential equation $y' + y \sin x = \sin x,\ y\left(\dfrac{\pi}{2}\right) = 2.$

**Test Form E** | **Name** ____________________ | **Date** ____________

**Chapter 5** | **Class** ____________________ | **Section** ____________

1. Solve the differential equation: $xy' = y$.

2. Find the general solution of the differential equation: $\dfrac{y'}{x} = \dfrac{e^{x^2}}{y}$.

3. Find the particular solution of the differential equation $\dfrac{dy}{dx} = 500 - y$ that satisfies the initial condition $y(0) = 7$.

4. The rate of change of $y$ with respect to $x$ is inversely proportional to the square root of $y$.

   **a.** Write a differential equation for the given statment.

   **b.** Solve the differential equation in part **a**.

5. Verify that $y^4 = 2e^{-x^2} + Ce^{-2x^2}$ is a solution to the differentaial equation $y' + xy^4 = xe^{-x^2}y^{-3}$.

6. Find the orthogonal trajectories of the family $x^2 + 4y^2 = C$ and sketch several members of each family.

7. Solve the homogenous differential equation: $(x^2 - xy + y^2)\,dx + x^2\,dy = 0$.

8. Consider the differential equation $y' = \dfrac{1}{2y}$ with the initial condition $y(1) = -1$. Use Euler's Method with $\Delta x = 0.1$ to approximate $y(1.3)$.

9. Solve, by any appropriate method, the first-order differential equation $y' + y\tan x = \tan x$.

10. Find the particular solution to the first-order linear differential equation $xy' + y = 1$ with the initial condition $y(1) = \frac{1}{5}$.

**Test Form A** **Name** ______________ **Date** ______________

**Chapter 6** **Class** ______________ **Section** ______________

**1.** Determine the area of the region bounded by the graphs of $y = x^2 - 4x$ and $y = x - 4$.

(a) $-\frac{9}{2}$ (b) $\frac{23}{6}$ (c) $\frac{9}{2}$

(d) $\frac{8}{3}$ (e) None of these

**2.** Find the volume of the solid formed by revolving the region bounded by the graphs of $y = x^3$, $y = 1$ and $x = 2$ about the $x$-axis.

(a) $\frac{127}{7}\pi$ (b) $\frac{120}{7}\pi$ (c) $\frac{240}{7}\pi$

(d) $\frac{1013}{10}\pi$ (e) None of these

**3.** Which of the following integrals represents the volume of the solid formed by revolving the region bounded by the graphs of $y = x^3$, $y = 1$ and $x = 2$ about the line $x = 2$?

(a) $2\pi \int_1^8 (2 - y)(\sqrt[3]{y} - 1)\, dy$ (b) $\pi \int_1^2 [(x^3 - 1)^2 - 1^2]\, dx$ (c) $\pi \int_1^8 \left[(\sqrt[3]{y})^2 - 1^2\right] dy$

(d) $2\pi \int_1^2 (2 - x)(x^3 - 1)\, dx$ (e) None of these

**4.** Find the volume of the solid formed by revolving the region bounded by the graphs of $y = 2x^2 + 4x$ and $y = 0$ about the $y$-axis.

(a) $\dfrac{544\pi}{15}$ (b) $4\pi$ (c) $\dfrac{16\pi}{3}$

(d) $\dfrac{16}{3}$ (e) None of these

**5.** Identify the definite integral that represents the arc length of the curve $y = \sqrt{x}$ over the interval $[0, 3]$.

(a) $\displaystyle\int_0^3 \sqrt{1 + \frac{1}{4x}}\, dx$ (b) $\displaystyle\int_0^3 \sqrt{1 + \frac{1}{2x}}\, dx$ (c) $\displaystyle\int_0^3 \sqrt{1 + x}\, dx$

(d) $\displaystyle\int_0^3 \sqrt{x}\, dx$ (e) None of these

**6.** Identify the definite integral that represents the area of the surface formed by revolving the graph of $f(x) = x^2$ on the interval $[0, \sqrt{2}]$ about the $x$ axis.

(a) $2\pi \int_0^{\sqrt{2}} x^2\sqrt{1 + x^4}\, dx$

(b) $2\pi \int_0^{\sqrt{2}} x^2\sqrt{1 + 4x^2}\, dx$

(c) $2\pi \int_0^{\sqrt{2}} x\sqrt{1 + 4x^2}\, dx$

(d) $2\pi \int_0^{2} y\sqrt{1 + \frac{1}{y}}\, dy$

(e) None of these

**7.** A force of 20 pounds stretches a spring $\frac{3}{4}$ foot on an exercise machine. Find the work done in stretching the spring 1 foot.

(a) $\frac{80}{3}$ ft-lb

(b) 15 ft-lb

(c) $\frac{80}{9}$ ft-lb

(d) $\frac{40}{3}$ ft-lb

(e) None of these

**8.** Find the $x$-coordinate of the centroid of the region bounded by the graphs of $y = \sqrt[3]{x}$, $y = 0$ and $x = 8$.

(a) $\frac{32}{7}$

(b) 4

(c) $\frac{384}{7}$

(d) $\frac{13}{3}$

(e) None of these

**Test Form B** **Name** ____________ **Date** ____________

**Chapter 6** **Class** ____________ **Section** ____________

1. Determine the area of the region bounded by the graphs of $y = -x^2 + 2x + 3$ and $y = 3$.

   (a) $\frac{4}{3}$ (b) $\frac{9}{2}$ (c) $\frac{22}{3}$

   (d) $-\frac{4}{3}$ (e) None of these

2. Find the volume of the solid formed by revolving the region bounded by the graphs of $y = x^3$, $x = 2$ and $y = 1$ about the $y$-axis.

   (a) $\frac{93}{5}\pi$ (b) $\frac{120}{7}\pi$ (c) $\frac{47}{5}\pi$

   (d) $\frac{62}{5}\pi$ (e) None of these

3. Which of the following integrals represents the volume of the solid formed by revolving the region bounded by the graphs of $y = x^3$, $y = 1$ and $x = 2$ about the line $y = 10$?

   (a) $\pi \int_1^8 (10 - y)(2 - \sqrt[3]{y})\,dy$ (b) $\pi \int_1^2 [81 - (10 - x^3)^2]\,dx$ (c) $2\pi \int_1^8 y(2 - \sqrt[3]{y})\,dy$

   (d) $\pi \int_1^2 [1 - (10 - x^3)^2]\,dx$ (e) None of these

4. Find the volume of the solid formed by revolving the region bounded by the graphs of $y = \frac{1}{2}(x - 2)^2$ and $y = 2$ about the $y$-axis.

   (a) $\dfrac{128\pi}{15}$ (b) $\dfrac{64\pi}{3}$ (c) $\dfrac{32\pi}{3}$

   (d) $\dfrac{20\pi}{3}$ (e) None of these

5. Identify the definite integral that represents the arc length of the curve $y = 1/x$ over the interval $[1, 3]$.

   (a) $\int_1^3 \sqrt{1 + (\ln x)^2}\,dx$ (b) $\int_1^3 \sqrt{1 + 1/x^2}\,dx$ (c) $\int_1^3 \sqrt{1 + 1/x^4}\,dx$

   (d) $\int_1^3 \sqrt{(1/x) + (1/x^4)}\,dx$ (e) None of these

6. Identify the definite integral that represents the area of the surface formed by revolving the graph of $f(x) = x^3$ on the interval $[0, 1]$ about the $y$-axis.

(a) $2\pi \int_0^1 x\sqrt{1 + 9x^4}\, dx$

(b) $2\pi \int_0^1 x^3\sqrt{1 + 9x^4}\, dx$

(c) $2\pi \int_0^1 x^3\sqrt{1 + 3x^2}\, dx$

(d) $2\pi \int_0^1 x\sqrt{1 + 3x^2}\, dx$

(e) None of these

7. A force of 8 pounds compresses a spring 3 inches. How much work is done on compressing the spring 6 inches?

(a) 48 in-lb

(b) 16 in-lb

(c) 24 in-lb

(d) 32 in-lb

(e) None of these

8. Find the $y$-coordinate of the centroid of the region bounded by the graphs of $y = \sqrt[3]{x}$, $y = 0$ and $x = 8$.

(a) $\frac{48}{5}$

(b) $\frac{8}{5}$

(c) 1

(d) $\frac{4}{5}$

(e) None of these

**Test Form C** **Name** ____________________ **Date** __________

**Chapter 6** **Class** ____________________ **Section** __________

**A graphing calculator/utility is recommended for this test.**

1. The integral $\int_a^b [(\sin x + 2) - e^{x^2}]\, dx$ computes the area of a region between two curves. Use a graphing utility to estimate the value of $a$.

   (a) 0 (b) 1.0 (c) $-0.6$
   (d) 3.1 (e) None of these

2. Use a graphing utility to graph the region bounded by the graphs of $y = \sqrt{x^3 - x^2}$, $y = 0$, and $x = 3$. Then use calculus to compute the volume of the solid formed by revolving this region about the $x$-axis.

   (a) $\frac{34\pi}{3}$ (b) $\frac{27\pi}{2}$ (c) $13\pi$
   (d) $\frac{\pi}{12}\left[971 - 216\sqrt{3}\right]$ (e) None of these

3. Use the integration capabilities of a graphing utility to approximate the volume of the solid formed by revolving the region bounded by the graphs of $y = \sin x$ and $y = 0$ in the interval $[0, \pi]$ about the $y$-axis. Round your answer to three decimal places.

   (a) 30.006 (b) 4.935 (c) 19.739
   (d) 3.142 (e) None of these

4. Use the integration capabilities of a graphing utility to approximate the arc length of the graph of $f(x) = \cos x$ on the interval $[0, \pi]$. Round your answer to three decimal places.

   (a) 2 (b) 3.820 (c) 3.143
   (d) 2.438 (e) None of these

5. Use the integration capabilities of a graphing utility to approximate the area of the surface formed by revolving the graph of $f(x) = \sin x$ on the interval $[0, \pi]$ about the $y$-axis. Round your answer to three decimal places.

   (a) 6.001 (b) 37.704 (c) 3.000
   (d) 14.424 (e) None of these

6. A force of 1250 pounds compresses a spring 5 inches from its natural length. Find the work done in compressing the spring 8 additional inches. [Units are in inch-pounds.]

   (a) 21,125 (b) 3250 (c) 18,000
   (d) 2000 (e) None of these

**7.** Use the Theorem of Pappus to find the volume of the solid of revolution of the torus formed by revolving the circle $x^2 + (y - 3)^2 = 4$ about the $x$-axis.

(a) $16\pi^2$
(b) $24\pi$
(c) $24\pi^2$
(d) $8\pi^2$
(e) None of these

**8.** Find the centroid of the region bounded by the graphs of $f(x) = 81 - x^2$ and $g(x) = \dfrac{81 - x^2}{4}$.

(a) $\left(\dfrac{243}{16}, 0\right)$
(b) $\left(0, \dfrac{243}{16}\right)$
(c) $\left(0, \dfrac{405}{8}\right)$
(d) $\left(0, \dfrac{81}{2}\right)$
(e) None of these

**Test Form D** **Name** ____________________ **Date** ____________

**Chapter 6** **Class** ____________________ **Section** ____________

**1.** Find the area of the region bounded by the graphs of $y = 1/x$ and $2x + 2y = 5$.

**2.** Find the volume of the solid formed by revolving the region bounded by the graphs of $y = e^x$, $y = 0$, $x = 0$, and $x = 1$ about the $x$-axis.

**3.** Use the shell method to set up the integral that represents the volume of the solid formed by revolving the region bounded by the graphs of $y = 1/x$ and $2x + 2y = 5$ about the line $y = 1/2$. (Do not evaluate the integral.)

**4.** Find the volume of the solid formed by revolving the region bounded by the graphs of $y = x^2$ and $y = 4$ about the $x$-axis.

**5.** Write the definite integral that represents the arc length of one period of the curve $y = \sin 2x$. (Do not evaluate the integral.)

**6.** Write the definite integral that represents the area of the surface formed by revolving the graph of $f(x) = \sqrt{x}$ on the interval $[0, 4]$ about the $y$-axis. (Do not evaluate the integral.)

**7.** A force of 40 pounds compresses a 12-inch spring by 3 inches. How much work is done in compressing the spring to a final length of 4 inches?

**8.** Find the centroid of the region bounded by the graphs of $y = x^3$, $y = 0$ and $x = 1$.

**Test Form E** **Name** ________________ **Date** ________

**Chapter 6** **Class** ________________ **Section** ________

**A graphing calculator/utility is recommended for this test.**

1. The integral $\int_a^b [\ln x - (x - 2)]\, dx$ computes the area of a region between two curves.

   a. Use a graphing utility to graph the curves $y = \ln x$ and $y = x - 2$.

   b. Use the graph to estimate the values of $a$ and $b$.

   c. Use the integration capabilities of a graphing utility to approximate the area of the region. Round your answer to three decimal places.

2. Use the integration capabilities of a graphing utility to approximate the volume of the solid formed by revolving the region bounded by the graphs of $y = \ln x$, $x - 2y = 1$, and the $x$-axis about the $x$-axis. Round your answer to three decimal places.

3. Consider the region bounded by the graphs of $y = \frac{1}{x}$, $y = x^2$, and $x = 2$.

   a. Sketch the region described.

   b. Find the point(s) of intersection.

   c. Calculate the volume of the solid formed when this region is revolved about the line $x = 2$.

4. A telephone wire suspended between two poles forms a catenary modeled by the equation

   $$y = 100 \cosh \frac{x}{100}, \quad -50 \le x \le 50$$

   where $x$ and $y$ are measured in feet. Approximate the length of the suspended cable if the poles are 100 feet apart.

5. Neglecting air resistance, determine the work done in propelling a 12-ton satellite to a height of 1000 miles above the surface of the earth.

6. A circular observation window on a cruise ship has a radius of 1 foot and the center of the window is 10 feet below water level. What is the fluid force on the window? (Use 64 lb/ft$^3$ as the weight-density of sea water.)

7. Consider the plane region bounded by the graphs of $f(x) = -x^2 + 2$ and $g(x) = 1$.

   a. Calculate the moment about the $x$-axis.

   b. Find the centroid of the region.

   c. Find the volume of the solid formed by revolving the region about the $x$-axis using the Theorem of Pappus.

   d. Use disc method to find the volume described in part **c.**

8. Use the integration capabilities of a graphing utility to approximate the centroid of the region bounded by the graphs $y = (\ln x)^2$ and $y = 1$. Round your answer to three decimal places.

**Test Form A** **Name** ____________________ **Date** ____________

**Chapter 7** **Class** ____________________ **Section** ____________

**1.** Evaluate the integral: $\int \frac{2x - 3}{9 + x^2}\, dx$.

(a) $\ln(9 + x^2) + C$ (b) $\frac{1}{3}\arctan\frac{x}{3} + C$ (c) $\ln(9 + x^2) - \arctan\frac{x}{3} + C$

(d) $\ln|3 + x| + C$ (e) None of these

**2.** Evaluate the integral: $\int \frac{1}{1 - \sin x}\, dx$.

(a) $x + \csc x + C$ (b) $\tan x + \sec x + C$ (c) $\ln|1 - \sin x| + C$

(d) $\frac{x}{x + \cos x} + C$ (e) None of these

**3.** Evaluate the integral: $\int x^2 \cos x\, dx$.

(a) $\frac{x^3}{3}\sin x + C$ (b) $x^2 \sin x + 2x\cos x - 2\sin x + C$

(c) $\cos\frac{x^4}{4} + C$ (d) $x^2 \sin x + x^2 \cos x + C$

(e) None of these

**4.** Evaluate the integral: $\int \arcsin t\, dt$.

(a) $\frac{1}{\sqrt{1 - t^2}} + C$ (b) $-\arccos t + C$

(c) $t \arcsin t + \sqrt{1 - t^2} + C$ (d) $t \arcsin t + \ln\sqrt{1 - t^2} + C$

(e) None of these

**5.** Evaluate the integral: $\int \sin^2 x \cos^3 x\, dx$.

(a) $\frac{1}{3}\sin^3 x - \frac{1}{5}\sin^5 x + C$ (b) $\frac{1}{12}\sin^3 x \cos^4 x + C$

(c) $\frac{1}{4}\cos^4 x - \frac{1}{6}\cos^6 x + C$ (d) $\frac{1}{2}\cos^2 x\left(\frac{1}{3}\sin^3 x - \frac{1}{5}\sin^5 x\right) + C$

(e) None of these

**6.** Evaluate the definite integral: $\int_{\pi/8}^{\pi/4} \sin^2 2\theta \, d\theta$.

(a) $\dfrac{4-\sqrt{2}}{12}$ (b) $\dfrac{4-\sqrt{2}}{24}$ (c) $\dfrac{\pi-2}{16}$

(d) $\dfrac{\pi+2}{16}$ (e) None of these

**7.** Evaluate the integral: $\int \dfrac{1}{\sqrt{9+x^2}}\, dx$.

(a) $\ln|\sec x + \tan x| + C$ (b) $\ln\left|\sqrt{x^2+9} + x\right| + C$ (c) $\dfrac{1}{3}\arctan\dfrac{x}{3} + C$

(d) $\dfrac{1}{2x}\ln(9+x^2) + C$ (e) None of these

**8.** If a trigonometric substitution in the variable $\theta$ is used to solve $\int_{5/4}^{5/2} \sqrt{25-4x^2}\, dx$, determine the lower and upper limits of integration for $\theta$.

(a) lower limit, $\arcsin\dfrac{5}{4}$; upper limit, $\arcsin\dfrac{5}{2}$ (b) lower limit, $\dfrac{\pi}{4}$; upper limit, $\pi$

(c) lower limit, $\dfrac{\pi}{6}$; upper limit, $\dfrac{\pi}{2}$ (d) lower limit, $\dfrac{\pi}{4}$; upper limit, $\arctan\dfrac{1}{2}$

(e) None of these

**9.** Evaluate the integral: $\int \dfrac{3x+4}{(x^2+4)(3-x)}\, dx$.

(a) $\dfrac{1}{2}\ln(x^2+4) + \ln|3-x| + C$ (b) $\dfrac{1}{2}\arctan\dfrac{x}{2} + \ln|3-x| + C$

(c) $\dfrac{1}{2}\arctan\dfrac{x}{2} - \ln|3-x| + C$ (d) $\ln\left|\dfrac{\sqrt{x^2+4}}{3-x}\right| + C$

(e) None of these

**10.** Find the limit: $\lim_{x\to 0} \dfrac{e^{x^2}-1}{2x^2}$.

(a) 0 (b) $\dfrac{1}{2}$ (c) $\infty$

(d) Does not exist (e) None of these

**11.** Find the limit: $\lim_{x \to 0} \left(\frac{1}{x}\right)^x$.

(a) $e$ (b) 0 (c) 1

(d) $\infty$ (e) None of these

**12.** Calculate the area under the curve $y = \frac{1}{x^2}$ above the $x$-axis on the interval $[1, \infty)$.

(a) $\infty$ (b) 1 (c) 0

(d) 2 (e) None of these

**Test Form B** **Name** ____________ **Date** ____________

**Chapter 7** **Class** ____________ **Section** ____________

**1.** Evaluate the integral: $\int \frac{2x - 3}{\sqrt{9 - x^2}}\,dx$.

(a) $\ln\sqrt{9 - x^2} + C$

(b) $-2\sqrt{9 - x^2} - 3 \arcsin \frac{x}{3} + C$

(c) $\arcsin \frac{x}{3} + C$

(d) $\ln|3 - x| + C$

(e) None of these

**2.** Evaluate the integral: $\int \frac{1}{1 - e^{-x}}\,dx$.

(a) $\ln|e^x - 1| + C$

(b) $\ln|1 - e^{-x}| + C$

(c) $e^x + C$

(d) $e^x \ln|1 - e^{-x}| + C$

(e) None of these

**3.** Evaluate the integral: $\int xe^{2x}\,dx$.

(a) $2e^{2x}(x - 2) + C$

(b) $x^2e^{x^2} + C$

(c) $\frac{e^{2x}}{4}[2x - 1] + C$

(d) $\frac{1}{2}x^2e^{2x} + C$

(e) None of these

**4.** Evaluate the integral: $\int \ln 3t\,dt$.

(a) $t \ln 3t - t + C$

(b) $\frac{1}{t} + C$

(c) $\frac{1}{3t} + C$

(d) $3t \ln 3t - 3t + C$

(e) None of these

**5.** Evaluate the integral: $\int \sin^3 x \cos^2 x\,dx$.

(a) $\frac{\cos^3 x}{3} - \frac{\cos^5 x}{5} + C$

(b) $\frac{1}{12} \sin^4 x \cos^3 x + C$

(c) $\frac{1}{4} \sin^4 x - \frac{1}{6} \sin^6 x + C$

(d) $\frac{1}{5} \cos^5 x - \frac{1}{3} \cos^3 x + C$

(e) None of these

6. Evaluate the definite integral: $\int_0^{\pi/4} \tan^3 x \, dx$.

(a) $-\frac{1}{2}$ (b) $\frac{1}{2}$ (c) $\frac{1}{2} - \ln\sqrt{2}$

(d) $\frac{1}{2} + \ln\sqrt{2}$ (e) None of these

7. Evaluate the integral: $\int \frac{1}{(16 - x^2)^{3/2}} \, dx$.

(a) $\frac{x}{16\sqrt{16 - x^2}} + C$ (b) $\frac{-1}{2\sqrt{16 - x^2}} + C$ (c) $\frac{-3}{2(16 - x^2)^{5/2}} + C$

(d) $\left(\arcsin \frac{x}{4}\right) + C$ (e) None of these

8. If a trigonometric substitution in the variable $\theta$ is used to solve $\int_{5/2}^{5/\sqrt{3}} \sqrt{4x^2 - 25} \, dx$, determine the lower and upper limits of integration for $\theta$.

(a) lower limit, 0; upper limit, $\frac{\pi}{3}$ (b) lower limit, 0; upper limit, $\frac{\pi}{6}$

(c) lower limit, $\frac{\pi}{6}$; upper limit, $\frac{\pi}{3}$ (d) lower limit, $\frac{\pi}{4}$; upper limit, $\frac{\pi}{3}$

(e) None of these

9. Evaluate the integral: $\int \frac{2}{(x + 2)^2(2 - x)} \, dx$.

(a) $\frac{1}{8}\ln|x + 2| + \frac{1}{2}\ln(x + 2)^2 - \frac{1}{8}\ln|2 - x| + C$

(b) $\frac{1}{8}\ln\left|\frac{x + 2}{2 - x}\right| - \frac{1}{2(x + 2)} + C$

(c) $-\frac{2}{(x + 2)} + \ln|2 - x| + C$

(d) $-\frac{1}{2(x + 2)} - \frac{1}{8}\ln|2 - x| + C$

(e) None of these

10. Find the limit: $\lim_{x\to\infty} \frac{x^2}{\ln x}$.

(a) 0 (b) 1 (c) $\infty$

(d) Does not exist (e) None of these

**11.** Evaluate the integral or show that it diverges: $\int_1^{\infty} e^{-x}\,dx$.

(a) $\infty$ (b) $\frac{1}{e}$ (c) 0

(d) $-\frac{1}{e}$ (e) None of these

**12.** Evaluate the integral or show that it diverges: $\int_0^3 \frac{1}{x}\,dx$.

(a) $\infty$ (b) $-\infty$ (c) 0

(d) $\ln 3$ (e) None of these

**Test Form C** Name ______________ Date ______________

**Chapter 7** Class ______________ Section ______________

**A graphing calculator/utility is recommended for this test.**

**1.** The region in the first quadrant bounded by the graph of $y = \sin x^2$, the coordinate axes, and the line $x = b, b > 0$, is revolved about the $y$-axis. Find $b$ such that the volume of the solid generated is $\frac{\pi}{2}$ cubic units.

(a) $\frac{\pi}{2}$ (b) $\sqrt{\frac{1}{3}}$ (c) $-\sqrt{\frac{\pi}{3}}$

(d) $\sqrt{\frac{\pi}{3}}$ (e) None of these

**2.** Compute the average value of $f(x) = \frac{x}{x+3}$ over the interval $[-a, a], 0 < a < 3$.

(a) $1 + \frac{3}{2a} \ln\left|\frac{3-a}{3+a}\right|$ (b) $2a + 3\ln\left|\frac{3-a}{3+a}\right|$ (c) $\frac{3}{9-a^2}$

(d) $\frac{1}{2a}$ (e) None of these

**3.** Evaluate the integral: $\int \ln x^3 \, dx$.

(a) $\frac{3}{x} + C$ (b) $\ln x^4 + C$ (c) $\frac{3}{2}(\ln x)^2 + C$

(d) $3x(\ln x - 1) + C$ (e) None of these

**4.** Find the volume of the solid formed by revolving the region bounded by the graphs of $y = \sin x$ and $y = 0$ in the interval $[0, \pi]$ about the $y$-axis.

(a) $\pi^3$ (b) $\frac{1}{2}\pi^2$ (c) $2\pi^2$

(d) $\pi$ (e) None of these

**5.** Determine which integral is obtained when the substitution $x = \sin\theta$ is made in the integral: $\int_0^1 \sqrt{1-x^2}\, dx$.

(a) $\int_0^{\pi/2} \sin\theta\cos\theta\, d\theta$ (b) $\int_0^{\pi/2} \cos^2\theta\, d\theta$ (c) $\int_0^{\pi} \cos\theta\, d\theta$

(d) $\int_0^{\pi} \cos^2\theta\, d\theta$ (e) None of these

**6.** Find the area of the region bounded by the graph of $y = 2\sqrt{x^2 + 1}$, the coordinate axes, and the line $x = 1$.

(a) $\sqrt{2} + \ln(\sqrt{2} + 1)$ (b) 4 (c) $\sqrt{2} - \ln(\sqrt{2} - 1)$

(d) $\sqrt{2} + \ln\sqrt{2}$ (e) None of these

**7.** Evaluate the integral: $\displaystyle\int \frac{x + 1}{x^2 - 5x + 6}\,dx$.

(a) $\dfrac{-x^2 - 2x + 11}{(x^2 - 5x + 6)^2} + C$ (b) $\ln\left|\dfrac{(x - 3)^4}{(x - 2)^3}\right| + C$ (c) $\ln|(x - 3)^4(x - 2)^3| + C$

(d) $\ln|(x - 3)^3(x - 2)^4| + C$ (e) None of these

**8.** Evaluate the limit: $\displaystyle\lim_{x\to 0} \frac{\sin x - x}{x^3}$.

(a) $\dfrac{1}{6}$ (b) 0 (c) $-\dfrac{1}{6}$

(d) $\infty$ (e) None of these

**9.** Evaluate the limit: $\displaystyle\lim_{x\to 0} \left(\frac{1}{x} - \frac{1}{x^2}\right)$.

(a) 0 (b) $\dfrac{1}{2}$ (c) $-1$

(d) The limit does not exist. (e) None of these

**10.** Evaluate the limit: $\displaystyle\lim_{x\to\infty} e^{-x}\ln x$.

(a) 0 (b) $\infty$ (c) 1

(d) The limit does not exist. (e) None of these

**11.** Use a graphing utility to graph the function $f(x) = \dfrac{1}{\sqrt{x}} + \dfrac{1}{\sqrt{2 - x}}$. Then calculate the area under the curve, above the $x$-axis, on the interval $[0, 2]$.

(a) $2\sqrt{2}$ (b) $4\sqrt{2}$ (c) 8

(d) 4 (e) None of these

**12.** Identify the improper integral.

(a) $\displaystyle\int_0^1 \frac{x}{3 - 2x}\,dx$ (b) $\displaystyle\int_0^2 \frac{x}{3 - 2x}\,dx$ (c) $\displaystyle\int_2^4 \frac{x}{3 - 2x}\,dx$

(d) $\displaystyle\int_0^1 \frac{1}{3 - 2x}\,dx$ (e) None of these

**Test Form D** **Name** ______________ **Date** ______________

**Chapter 7** **Class** ______________ **Section** ______________

**1.** Evaluate the integral: $\int \frac{1 + \cos x}{\sin x}\, dx$.

**2.** Evaluate the integral: $\int \frac{1}{\csc x - 1}\, dx$.

**3.** Evaluate the integral: $\int x \ln x\, dx$.

**4.** Evaluate the integral: $\int \frac{2x^2 + 4x + 22}{x^2 + 2x + 10}\, dx$.

**5.** Evaluate the integral: $\int \sin^5 t\, dt$.

**6.** Evaluate the integral: $\int \sec^3 \frac{x}{2} \tan^3 \frac{x}{2}\, dx$.

**7.** Evaluate the integral: $\int \frac{1}{\sqrt{x^2 - 4}}\, dx$.

**8.** Evaluate the definite integral: $\int_{1/2}^{1} \frac{\sqrt{1 - x^2}}{x}\, dx$.

**9.** Evaluate the integral: $\int \frac{20x^2 + x + 45}{x(4x^2 + 9)}\, dx$.

**10.** Find the limit: $\lim_{x \to 0^+} \frac{\sin 2x}{x^2}$.

**11.** Find the limit: $\lim_{x \to 0} \left(\frac{1}{x}\right)^x$.

**12.** Evaluate the integral or show that it diverges: $\int_0^{\pi} \tan x\, dx$.

**Test Form E** **Name** ____________________ **Date** ____________

**Chapter 7** **Class** ____________________ **Section** ____________

**A graphing calculator/utility is recommended for this test.**

**1.** Calculate the area of the region bounded by the graph of the function $y = \frac{1 + x^2}{(1 + x)^2}$, the coordinate axes and the line $x = 1$.

**2.** Solve the differential equation: $e^{4x} y' = x$.

**3.** Find the volume of the solid formed by revolving the region bounded by the graphs of $y = \frac{\ln x}{x}$, $x = 3$, and the $x$-axis about the line $x = 3$.

**4.** Evaluate the integral: $\int \frac{\cos^3 x}{\sqrt{\sin x}}\, dx$.

**5.** State specifically what substitution needs to be made for $x$ if this integral is to be evaluated using a trigonometric substitution: $\int \frac{x + 7}{x^2 - x - 6}\, dx$.

**6.** Calculate the average value of the function $f(x) = \frac{\sqrt{4x^2 - 1}}{x}$ on the interval $\left[\frac{1}{2}, 1\right]$.

**7.** Evaluate the integral: $\int \frac{9}{(x + 2)^2(1 - x)}\, dx$.

**8.** Evaluate the limit: $\lim_{x \to \pi/2} \left(\frac{\sin 2x}{4x^2 - \pi^2}\right)$.

**9.** Consider the limit: $\lim_{x \to 0+} (\tan x)^x$.

**a.** Use a graphing utility to estimate the limit.

**b.** Find the limit analytically.

**10.** Find the differentiable functions $f$ and $g$ that satisfy the condition that $\lim_{x \to \infty} \frac{f(x)}{g(x)} = 7$ while $\lim_{x \to \infty} f(x) = \infty$ and $\lim_{x \to \infty} g(x) = \infty$.

**11.** Evaluate each of the integrals:

**a.** $\int_0^1 \frac{1}{\sqrt{x}}\, dx.$

**b.** $\int_1^\infty \frac{1}{\sqrt{x}}\, dx.$

**c.** Use parts **a** and **b** to evaluate the integral: $\int_0^\infty \frac{1}{\sqrt{x}}\, dx.$

**12.** Consider the region under the curve $y = \frac{1}{x}$, above the $x$-axis, on the interval $1 \leq x < \infty$.

**a.** Sketch the region described.

**b.** Calculate the area of the region.

**c.** Calculate the volume of the solid obtained by revolving the region about the $x$-axis.

**Test Form A** **Name** ______________ **Date** __________

**Chapter 8** **Class** ______________ **Section** __________

1. Find the fourth term of the sequence $\left\{\frac{(-1)^{n+1}2^n}{3n-1}\right\}$, $n = 1, 2, 3, \ldots$.

(a) $\frac{16}{11}$ (b) $-\frac{16}{11}$ (c) $-\frac{8}{11}$

(d) $-\frac{16}{13}$ (e) None of these

2. Determine if the following sequence converges or diverges: $\left\{\frac{2n-1}{3n^2+1}\right\}$, $n = 1, 2, 3, \ldots$.
If the sequence converges, find its limit.

(a) Converges to $\frac{2}{3}$ (b) Converges to 0 (c) Converges to $-\frac{1}{3}$

(d) Diverges (e) None of these

3. Determine if the following sequence converges or diverges: $\left\{\frac{n!}{(n-2)!}\right\}$, $n = 2, 3, 4, \ldots$.
If the sequence converges, find its limit.

(a) Converges to 1 (b) Converges to 0 (c) Converges to $-2$

(d) Diverges (e) None of these

4. Find the sum: $\sum_{n=0}^{\infty} 3\left(\frac{1}{2}\right)^n$.

(a) $\frac{3}{2}$ (b) 3 (c) 6

(d) $\frac{50}{9}$ (e) None of these

5. Determine which of the following series converges.

(a) $\sum_{n=1}^{\infty} \frac{1}{n}$ (b) $\sum_{n=0}^{\infty} 3\left(\frac{4}{3}\right)^n$ (c) $\sum_{n=0}^{\infty} \frac{(n+1)!}{2^n}$

(d) $\sum_{n=1}^{\infty} \frac{1}{n^{3/2}}$ (e) None of these

6. Determine which series diverges.

(a) $\sum_{n=0}^{\infty} \frac{1}{2^n}$ (b) $\sum_{n=1}^{\infty} (4 + (-1)^n)$ (c) $\sum_{n=0}^{\infty} \frac{(-1)^n}{(n+1)!}$

(d) $\sum_{n=1}^{\infty} \frac{1}{n^2}$ (e) None of these

**7.** Determine which test can be used to prove the divergence of the series $\sum_{n=1}^{\infty} \frac{n+1}{3n+1}$.

(a) Geometric Series Test (b) $p$-Series Test (c) Ratio Test

(d) $n$th-Term Test for Divergence (e) None of these

**8.** Determine which test can be used to show that the series $\sum_{n=1}^{\infty} \frac{n}{8n^3 + 6n^2 - 7}$ converges.

(a) Geometric Series Test (b) $p$-Series Test (c) Ratio Test

(d) Limit Comparison Test (e) None of these

**9.** A ball is dropped from a height of 24 feet. Each time it drops $h$ feet, it rebounds $(2/3)h$ feet. Find the total distance traveled by the ball.

(a) 72 feet (b) 144 feet (c) 120 feet

(d) 84 feet (e) None of these

**10.** Investigate the series $\sum_{n=1}^{\infty} \frac{1 \cdot 3 \cdot 5 \cdot \cdot \cdot (2n-1)}{n!}$ for convergence or divergence.

(a) Diverges by Ratio Test (b) Converges by $n$th-Term Test for Divergence

(c) Diverges by Root Test (d) Converges by Integral Test

(e) None of these

**11.** Investigate the series $\sum_{n=2}^{\infty} \frac{1}{(\ln n)^n}$ for convergence or divergence.

(a) Converges by Ratio Test (b) Diverges by $n$th-Term Test for Divergence

(c) Converges by Root Test (d) Diverges by Integral Test

(e) None of these

**12.** Determine which of the following series is a telescopic series.

(a) $\sum_{n=0}^{\infty} \frac{1}{2^n}$ (b) $\sum_{n=1}^{\infty} \frac{1}{n(n+2)}$ (c) $\sum_{n=1}^{\infty} \frac{1}{n^2}$

(d) $\sum_{n=1}^{\infty} \frac{1}{\sqrt{n}}$ (e) None of these

**13.** Determine if $\sum_{n=1}^{\infty} \frac{\cos n\pi}{n^2}$ is convergent or divergent. If convergent, classify the series as absolutely convergent or conditionally convergent.

(a) Divergent (b) Conditionally convergent

(c) Absolutely convergent (d) None of these

**14.** Determine which of the following tests could be used to show that the harmonic series diverges.

(a) Geometric Series Test (b) Ratio Test (c) Telescopic Series Test

(d) Integral Test (e) None of these

**15.** Find the number of terms necessary to approximate the sum of the series with an error of less than 0.001:

$$\sum_{n=1}^{\infty} \frac{1}{n^{3/2}}.$$

(a) $10\sqrt{2}$ (b) 4,000,000 (c) 1000

(d) 4000 (e) None of these

**16.** Find the third term of the Taylor polynomial, centered at $\frac{\pi}{2}$, for $f(x) = \cos x$.

(a) $-\frac{1}{5!}\left(x - \frac{\pi}{2}\right)^5$ (b) $\frac{1}{3!}x^3$ (c) $\frac{1}{3!}\left(x - \frac{\pi}{2}\right)^3$

(d) $-\left(x - \frac{\pi}{2}\right)^5$ (e) None of these

**17.** Find the radius of convergence of the power series $\sum_{n=1}^{\infty} \frac{(2x)^n}{n}$.

(a) $\frac{1}{2}$ (b) 2 (c) $\infty$

(d) 0 (e) None of these

**18.** Determine the interval of convergence of the series $\sum_{n=1}^{\infty} \frac{(x-2)^n}{n3^n}$.

(a) $[-1, 5)$ (b) $[-5, 1)$ (c) $[-3, 3)$

(d) $\left[\frac{5}{3}, \frac{7}{3}\right)$ (e) None of these

**19.** Use the series $\frac{1}{1-x} = \sum_{n=0}^{\infty} x^n = 1 + x + x^2 + x^3 + \cdot \cdot \cdot$, to find a power series expansion for the function $f(x) = \frac{1}{1-x^2}$.

(a) $1 + x^2 + x^3 + x^4 + \cdot \cdot \cdot$ (b) $x + x^2 + x^3 + x^4 + \cdot \cdot \cdot$

(c) $1 + x^2 + x^4 + x^6 + \cdot \cdot \cdot$ (d) $x^2 + x^3 + x^4 + x^5 + \cdot \cdot \cdot$

(e) None of these

**20.** Investigate $\sum_{n=1}^{\infty} \left(\frac{n+1}{n}\right)^n$ for convergence or divergence.

(a) Converges by Root Test (b) Diverges by Root Test

(c) Converges by Ratio Test (d) Diverges by *n*th-Term Test for Divergence

(e) None of these

**Test Form B** **Name** ________________ **Date** ____________

**Chapter 8** **Class** ________________ **Section** ____________

**1.** Find the fourth term of the sequence $\left\{\frac{(-1)^n(2^n+1)}{n!}\right\}$, $n = 1, 2, 3, \ldots$.

(a) $\frac{17}{4}$ (b) $\frac{-17}{4}$ (c) $\frac{9}{24}$

(d) $\frac{17}{24}$ (e) None of these

**2.** Determine if the following sequence converges or diverges: $\left\{\frac{n!}{(n-2)!}\right\}$, $n = 2, 3, 4, \ldots$.

If the sequence converges, find its limit.

(a) Converges to 2 (b) Converges to 0 (c) Converges to 4

(d) Diverges (e) None of these

**3.** Determine if the following sequence converges or diverges: $\left\{\frac{2n-1}{3n+1}\right\}$, $n = 1, 2, 3, \ldots$.

If the sequence converges, find its limit.

(a) Converges to $\frac{2}{3}$ (b) Converges to 0 (c) Converges to $-\frac{1}{3}$

(d) Diverges (e) None of these

**4.** Find the sum: $\sum_{n=0}^{\infty} 2\left(-\frac{1}{2}\right)^n = 2 - 1 + \frac{1}{2} - \frac{1}{4} + \cdots$.

(a) $\frac{4}{3}$ (b) $-1$ (c) 0

(d) 4 (e) None of these

**5.** Determine which series converges.

(a) $\sum_{n=1}^{\infty} (4 + (-1)^n)$ (b) $\sum_{n=1}^{\infty} \frac{2^n}{(n+1)!}$ (c) $\sum_{n=0}^{\infty} 5\left(\frac{3}{2}\right)^n$

(d) $\sum_{n=1}^{\infty} \frac{1}{\sqrt{n}}$ (e) None of these

**6.** Determine which series diverges.

(a) $\sum_{n=0}^{\infty} \frac{n!}{3n! - 1}$ (b) $\sum_{n=1}^{\infty} \frac{1}{n^6}$ (c) $\sum_{n=0}^{\infty} 5\left(\frac{1}{10}\right)^n$

(d) $\sum_{n=0}^{\infty} \frac{n}{2^n}$ (e) None of these

**7.** Determine which test can be used to prove the divergence of the series $\sum_{n=1}^{\infty} \frac{1}{\sqrt{n}}$.

(a) Geometric Series Test (b) $p$-Series Test (c) Ratio Test

(d) $n$th-Term Test for Divergence (e) None of these

**8.** A force is applied to a particle, which moves in a straight line, in such a way that after each second the particle moves only one-half the distance that it moved in the preceding second. If the particle moved 20 cm in the first second, how far will it move altogether?

(a) 30 cm (b) 10 cm (c) 40 cm

(d) 45 cm (e) None of these

**9.** Identify the type of series: $\sum_{n=1}^{\infty} \frac{3}{(2n-1)(2n+1)}$.

(a) Geometric Series (b) $p$-Series (c) Telescopic Series

(d) Harmonic Series (e) None of these

**10.** Investigate $\sum_{n=1}^{\infty} \frac{(-1)^{n+1}}{n}$ for convergence or divergence.

(a) Diverges by Integral Test (b) Converges by Alternating Series Test

(c) Diverges by $n$th-Term Test for Divergence (d) Converges by Ratio Test

(e) None of these

**11.** Determine which test would be appropriate to show that the series $\sum_{n=1}^{\infty} \frac{n!}{1 \cdot 3 \cdot 5 \cdots (2n-1)}$ converges.

(a) Geometric Series Test (b) $p$-Series Test

(c) Ratio Test (d) Limit Comparison Test

(e) None of these

**12.** Investigate the series $\sum_{n=1}^{\infty} \frac{\sqrt{n}}{4n^3 - 6n^2 + 5}$ for convergence or divergence.

(a) Converges by $n$th-Term Test (b) Diverges by $n$th-Term Test for Divergence

(c) Converges by Root Test (d) Converges by Limit Comparison Test

(e) None of these

**13.** Determine whether the series $\sum_{n=1}^{\infty} \frac{(-1)^n}{\sqrt{n}}$ is convergent or divergent. If convergent, classify the series as absolutely convergent or conditionally convergent.

(a) Divergent (b) Conditionally convergent

(c) Absolutely convergent (d) None of these

**14.** Determine which of the following tests can be used to show that $\sum_{n=1}^{\infty}\left(\frac{2n-1}{3n+5}\right)^n$ converges.

(a) Root Test
(b) Ratio Test
(c) Geometric Series Test
(d) $p$-Series Test
(e) None of these

**15.** Find the number of terms necessary to approximate the sum of the series with an error of less than 0.001: $\sum_{n=1}^{\infty}\frac{(-1)^n}{n}$.

(a) $\frac{1}{999}$
(b) 1000
(c) 999
(d) $-999$
(e) None of these

**16.** Find the third term of the Maclaurin polynomial for the function $f(x) = \sin x$.

(a) $\frac{1}{3!}x^3$
(b) $\frac{1}{5!}x^5$
(c) $-\cos x$
(d) $\frac{1}{3!}\left(x - \frac{\pi}{2}\right)^3$
(e) None of these

**17.** Find the radius of convergence of the power series $\sum_{n=0}^{\infty}\left(\frac{x}{2}\right)^n$.

(a) $\frac{1}{2}$
(b) 2
(c) $\infty$
(d) 0
(e) None of these

**18.** Determine the interval of convergence of the series $\sum_{n=1}^{\infty}\frac{(x+2)^n}{n3^n}$.

(a) $[-1, 5)$
(b) $[-5, 1)$
(c) $[-3, 3)$
(d) $\left[\frac{5}{3}, \frac{7}{3}\right)$
(e) None of these

**19.** Use the series $\frac{1}{1-x} = 1 + x + x^2 + x^3 + \cdots$, to find a power series expansion for the function $f(x) = \ln|1 - x|$.

(a) $C + x + \frac{x^2}{2} + \frac{x^3}{3} + \frac{x^4}{4} + \cdots$
(b) $C + \ln|1 + x + x^2 + x^3 + \cdots|$
(c) $C + 1 + 2x + 3x^2 + \cdots$
(d) $C - x - \frac{x^2}{2} - \frac{x^3}{3} - \frac{x^4}{4} - \cdots$
(e) None of these

**20.** Investigate the series $\sum_{n=1}^{\infty}\frac{1}{1+e^{-n}}$ for convergence or divergence.

(a) Converges by Root Test
(b) Converges by $n$th-Term Test for Divergence
(c) Converges by Ratio Test
(d) Diverges by Geometric Series Test
(e) None of these

**Test Form C** **Name** ______________ **Date** ______________

**Chapter 8** **Class** ______________ **Section** ______________

**A graphing calculator/utility is recommended for this test.**

**1.** Simplify: $\dfrac{(n+3)!}{(n+1)!}$.

(a) $n+3$ (b) $(n+3)(n+2)$ (c) $\dfrac{n+3}{n+1}$

(d) $(n+3)(n+1)$ (e) None of these

**2.** Let $\{a_n\} = \left\{n + \dfrac{1}{n}\right\}$. Use a graphing utility to graph the corresponding function, $f(x) = x + \dfrac{1}{x}$.

Then use the graph of $f$ as an aid in determining which of the following statements is true about $\{a_n\}$.

(a) The sequence is unbounded. (b) The sequence is bounded but not convergent.

(c) The sequence is monotonic and bounded. (d) The sequence is unbounded but convergent.

(e) None of these

**3.** Choose the series that diverges by the $n$th-Term Test for Divergence.

(a) $\displaystyle\sum_{n=0}^{\infty} \frac{14}{n}$ (b) $\displaystyle\sum_{n=0}^{\infty} \frac{n-6}{n}$ (c) $\displaystyle\sum_{n=0}^{\infty} \frac{100n^{14}}{4^n}$

(d) $\displaystyle\sum_{n=0}^{\infty} \frac{n-6}{n!}$ (e) None of these

**4.** Determine whether the following series is convergent or divergent. If it is convergent, find its sum.

$$\sum_{n=0}^{\infty} \left(\frac{e}{\pi}\right)^n.$$

(a) Converges to $\dfrac{\pi}{\pi - e}$ (b) Converges to $\dfrac{e}{e - \pi}$ (c) Converges to $\dfrac{1}{\pi - e}$

(d) Diverges (e) None of these

**5.** Find the sum: $\displaystyle\sum_{n=0}^{\infty} \left(\frac{1}{3^n} - \frac{1}{4^n}\right)$.

(a) $-2$ (b) $\dfrac{1}{6}$ (c) $-1$

(d) 0 (e) None of these

**6.** Determine which of the following tests can be used to show that the harmonic series diverges.

(a) Geometric Series Test (b) $n$th-Term Test for Divergence (c) Telescopic Series Test

(d) Integral Test (e) None of these

**7.** Use the fact that $\sum_{n=1}^{\infty} \frac{1}{n^2} = \frac{\pi^2}{6}$ to find $\sum_{n=3}^{\infty} \frac{1}{n^2}$.

(a) $\frac{\pi^2}{6}$ (b) $\frac{\pi^2 - 12}{6}$ (c) $\frac{\pi^2 - 6}{6}$

(d) $\frac{2\pi^2 - 15}{12}$ (e) None of these

**8.** Apply the Integral Test to the series: $\sum_{n=1}^{\infty} \frac{1}{n^2 + 1}$.

(a) Converges to $\frac{\pi}{4}$ (b) Diverges (c) Converges

(d) The test is inconclusive (e) None of these

**9.** Determine the convergence or divergence of the series using the Limit Comparison Test: $\sum_{n=1}^{\infty} \frac{1}{\sqrt{n(n+1)}}$. Name the series used in the comparison.

(a) Converges using $\sum_{n=1}^{\infty} \frac{1}{n^2}$ (b) Converges using $\sum_{n=1}^{\infty} \frac{1}{n(n+1)}$

(c) Diverges using $\sum_{n=1}^{\infty} \frac{1}{\sqrt{n}}$ (d) Diverges using $\sum_{n=1}^{\infty} \frac{1}{n}$

(e) None of these

**10.** Determine the convergence or divergence of the series $\sum_{n=2}^{\infty} \frac{n}{(n+1)\ln n}$ by comparing it with the series $\sum_{n=1}^{\infty} \frac{1}{n}$ and $\sum_{n=1}^{\infty} \frac{1}{n^2}$. Use the Direct Comparison Test and a graphing utility to graph the corresponding functions.

(a) Converges by comparison with $\sum_{n=2}^{\infty} \frac{1}{n}$. (b) Diverges by comparison with $\sum_{n=2}^{\infty} \frac{1}{n}$.

(c) Converges by comparison with $\sum_{n=2}^{\infty} \frac{1}{n^2}$. (d) Diverges by comparison with $\sum_{n=2}^{\infty} \frac{1}{n^2}$.

(e) None of these

**11.** Find the number of terms necessary to approximate the sum of the series with an error less than 0.001:

$$\sum_{n=0}^{\infty} \left(-\frac{e}{\pi}\right)^n.$$

(a) 4 (b) 46 (c) 5

(d) 47 (e) None of these

12. If the series $\sum_{n=1}^{\infty} a_n$ is a conditionally convergent, determine which of the following series must diverge.

(a) $\sum_{n=1}^{\infty} a_n^2$ (b) $\sum_{n=1}^{\infty} |a_n|$ (c) $\sum_{n=1}^{\infty} (-1)^{2n} a_n$

(d) $\sum_{n=1}^{\infty} (-a_n)$ (e) None of these

13. Investigate the convergence or divergence of the series: $\sum_{n=1}^{\infty} \frac{n!}{e^n}$.

(a) Converges by Ratio Test (b) Converges by Geometric Series Test

(c) Diverges by Ratio Test (d) Diverges by Geometric Series Test

(e) None of these

14. Use the third Maclaurin polynomial to approximate the value of $e^{0.2}$. Round your answer to four decimal places.

(a) 1.2213 (b) 1.2214 (c) 1.2227

(d) 2.6667 (e) None of these

15. Let $f$ be the function $f(x) = \sum_{n=1}^{\infty} \frac{(-1)^n(x-2)^n}{n}$. Find the interval of convergence of $\int f(x)\,dx$.

(a) $(1, 3)$ (b) $(1, 3[$ (c) $)1, 3)$

(d) $)1, 3[$ (e) None of these

16. Find a power series, centered at 2, for the function $f(x) = \frac{1}{3x-2}$.

(a) $\sum_{n=0}^{\infty} -\frac{(3x)^n}{2^{n+1}}$ (b) $\sum_{n=0}^{\infty} -\frac{3^n(x-2)^n}{2^{n+1}}$ (c) $\sum_{n=0}^{\infty} \frac{(-1)^n 3^n (x-2)^n}{4^{n+1}}$

(d) $\sum_{n=0}^{\infty} -\frac{3^n(x-2)^n}{4^{n+1}}$ (e) None of these

**Test Form D** Name ______________ Date ______________

**Chapter 8** Class ______________ Section ______________

1. Find the fourth term of the sequence: $\left\{\dfrac{(-1)^n 3^{n+1}}{2n-1}\right\}$, $n = 1, 2, 3, \ldots$.

2. Determine if the following sequence converges or diverges: $\left\{\dfrac{2^n}{1+n}\right\}$, $n = 1, 2, 3, \ldots$.
If the sequence converges, find its limit.

3. Determine if the following sequence converges or diverges: $\left\{\left(\dfrac{n+1}{n}\right)^n\right\}$, $n = 1, 2, 3, \ldots$.
If the sequence converges, find its limit.

4. Find the sum: $\displaystyle\sum_{n=0}^{\infty} 3\left(-\frac{1}{2}\right)^n$.

5. Find the sum: $\displaystyle\sum_{n=1}^{\infty} \frac{1}{n^2+4n+3}$.

For each of the series in 6 through 14, determine convergence or divergence and state the test used.

6. $\displaystyle\sum_{n=1}^{\infty} \frac{1}{\sqrt[3]{n}}$

7. $\displaystyle\sum_{n=1}^{\infty} \frac{n}{\sqrt{n^3+2n}}$

8. $\displaystyle\sum_{n=0}^{\infty} \frac{1}{3^n}$

9. $\displaystyle\sum_{n=1}^{\infty} \frac{(-1)^n}{n^2}$

10. $\displaystyle\sum_{n=1}^{\infty} \left(\frac{1}{n} - \frac{1}{n+1}\right)$

11. $\displaystyle\sum_{n=1}^{\infty} \frac{5}{2n-1}$

12. $\displaystyle\sum_{n=1}^{\infty} \frac{2 \cdot 4 \cdot 6 \cdot \cdot \cdot 2n}{n!}$

13. $\displaystyle\sum_{n=1}^{\infty} \left(\frac{2n}{5n-1}\right)^n$

14. $\displaystyle\sum_{n=1}^{\infty} \frac{1}{1+e^{-n}}$

15. Find the number of terms necessary to approximate the sum of the series with an error of less than 0.001: $\displaystyle\sum_{n=1}^{\infty} \frac{1}{n^2}$.

16. Find the radius of convergence of the series $\displaystyle\sum_{n=1}^{\infty} \frac{2 \cdot 5 \cdot 8 \cdot \cdot \cdot (3n-1)}{3 \cdot 7 \cdot 11 \cdot \cdot \cdot (4n-1)} x^n$.

**17.** For what values of $x$ does the series $\sum_{n=0}^{\infty} \frac{2^n (x-4)^n}{n}$ converge?

**18.** Using the power series for $\sin x = x - \frac{x^3}{3!} + \frac{x^5}{5!} - \frac{x^7}{7!} + \frac{x^9}{9!} - \cdots$, find the power series for $\cos \sqrt{x}$.

**19.** Find the fourth term of the Taylor Series, centered at $x = 1$, for $f(x) = \ln x$.

**20.** Using the power series $e^x = 1 + x + \frac{x^2}{2!} + \frac{x^3}{3!} + \cdots$, approximate (by using four terms) the value of $\int_0^{0.4} e^{\sqrt{x}}\, dx$. Round your answer to three decimal places.

**Test Form E** **Name** ______________ **Date** ______________

**Chapter 8** **Class** ______________ **Section** ______________

**A graphing calculator/utility is recommended for this test.**

1. Determine if the following sequence converges or diverges: $\{(\pi - e)^n\}$. If it converges, find its limit.

2. Consider the sequence whose $n$th terms is $a_n = \dfrac{\sin n}{n}$. Use the graph of the corresponding function, $f(x) = \dfrac{\sin x}{x}$, as an aid in answering each of the following:

   **a.** Determine whether $\{a_n\}$ is monotonic.

   **b.** Determine whether $\{a_n\}$ is bounded.

   **c.** Determine whether $\{a_n\}$ converges.

3. Determine whether the following series converges or diverges and state the test used: $\displaystyle\sum_{n=0}^{\infty} \frac{2n!}{3n! + 5}$.

4. Consider the fraction $\frac{7}{33}$.

   **a.** Write the fraction as an infinite geometric series.

   **b.** Determine the value of the first term, $a$, and the common ratio, $r$ of the series.

   **c.** Write the series from part **a** using summation notation.

5. Use sigma notation to write a formula for the sum of the infinite geometric series:

$$1 + \frac{x}{3} + \frac{x^2}{9} + \frac{x^3}{27} + \frac{x^4}{81} + \cdots.$$

6. Consider the series: $\displaystyle\sum_{n=1}^{\infty} n^{-3}$.

   **a.** Use the Integral Test to show that the series converges.

   **b.** Approximate the sum of the series using four terms.

   **c.** Calculate the maximum error in the approximation in part **b.** Round your answer to the nearest thousandth.

   **d.** Give an upper bound for the sum of this series using the information obtained in parts **b** and **c.**

7. Determine the convergence or divergence of the series using the Limit Comparison Test: $\displaystyle\sum_{n=1}^{\infty} \frac{1}{\sqrt[3]{n^3 + 2n}}$.

   Name the series used in the comparison.

**8.** Consider the series: $\sum_{n=2}^{\infty} \frac{1}{\ln n}$.

**a.** Use the harmonic series as a comparison series. Name this series and state whether it converges or diverges.

**b.** Compare the given series with the series named in part **a** by graphing the corresponding functions and applying the Direct Comparison Test. State and justify your conclusion.

**9.** Show that the series converges by applying the Alternating Series Test: $\sum_{n=2}^{\infty} \frac{(-1)^n}{n(\ln n)}$.

**10.** Consider the series: $\sum_{n=1}^{\infty} \frac{(-1)^n}{2^n}$.

**a.** Show that the series is absolutely convergent.

**b.** Calculate the sum of the first six terms. Round your answer to three decimal places.

**c.** Find the number of terms necessary to approximate the sum of the series with an error less than 0.001.

**11.** Investigate the convergence or divergence of the series and state the test used: $\sum_{n=1}^{\infty} \frac{3^{n-1}}{n2^n}$.

**12.** Let $f$ be the function $f(x) = \arctan x$.

**a.** Find the third Maclaurin polynomial for $f$.

**b.** Approximate the value of $\arctan(-0.3)$ using the polynomial found in part **a.**

**13.** Let $f$ be the function $f(x) = \sum_{n=0}^{\infty} (-1)^n \left(\frac{x}{3}\right)^n$. Find the series and the interval of convergence for $\int f(x)\, dx$.

**14.** Use the power series $e^x = 1 + x + \frac{x^2}{2!} + \frac{x^3}{3!} + \frac{x^4}{4!} + \cdots = \sum_{n=0}^{\infty} \frac{x^n}{n!}$ to find a power series for $f(x) = e^{-2x}$. Write the first four terms and the general term of the power series.

**15.** Use the trigonometric identity $\cos^2 \theta = \frac{1 + \cos 2\theta}{2}$ and the power series $\cos x = \sum_{n=0}^{\infty} \frac{(-1)^n x^{2n}}{(2n)!}$ to find a power series for the function $f(x) = \cos^2 x$.

**Test Form A** **Name** ______________ **Date** ______________

**Chapter 9** **Class** ______________ **Section** ______________

1. Find the focus of the parabola given by the equation $y = -\frac{1}{4}x^2 + 2x - 5$.

(a) $(4, -2)$ (b) $(3, -1)$ (c) $(4, 0)$

(d) $(3, 5)$ (e) None of these

2. Suppose the equation for an ellipse is given by $\dfrac{(x-3)^2}{16} + \dfrac{(y+2)^2}{b^2} = 1$. Find $b$ so that the eccentricity is 0.75.

(a) 3 (b) 6 (c) $\sqrt{7}$

(d) 5 (e) None of these

3. Which of the following is a graph of the equation $\dfrac{(y+2)^2}{9} - \dfrac{(x-3)^2}{16} = 1$?

(a)

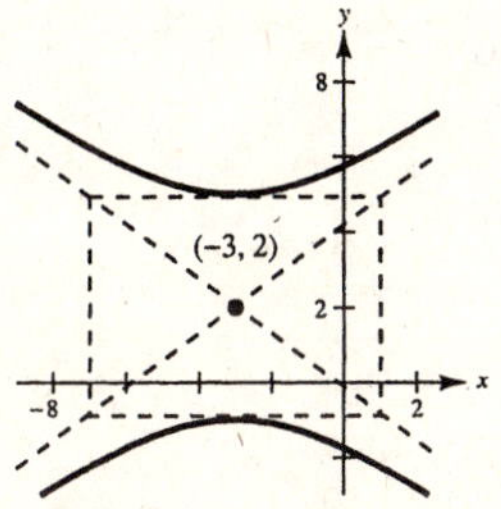


(b)

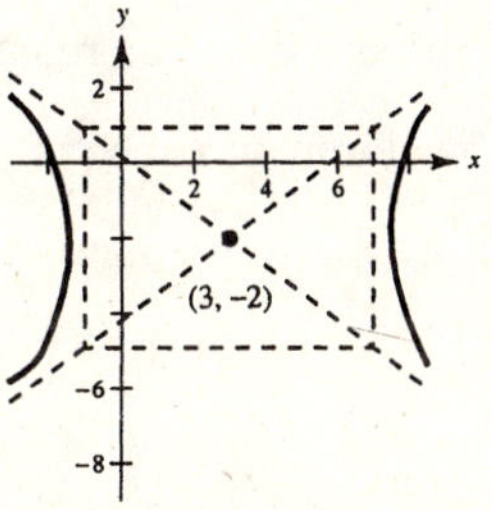


(c)

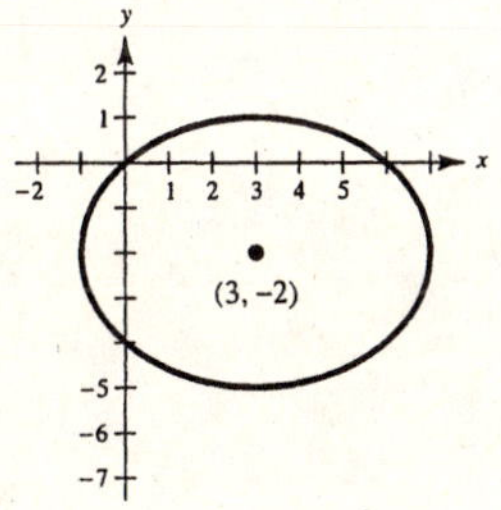


(d)

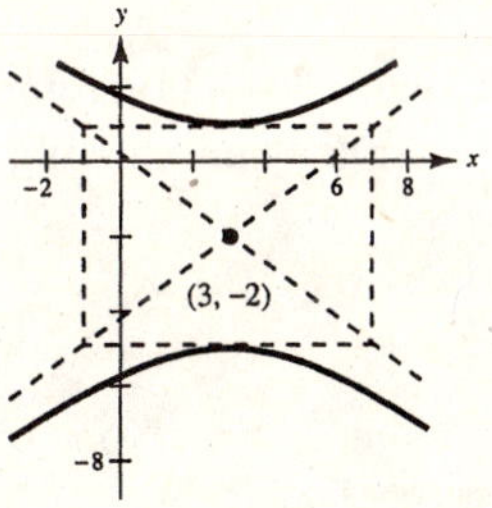


(e) None of these

**4.** Identify the curve represented by the parametric equations $x = 2 - t$ and $y = t^2$.

(a)

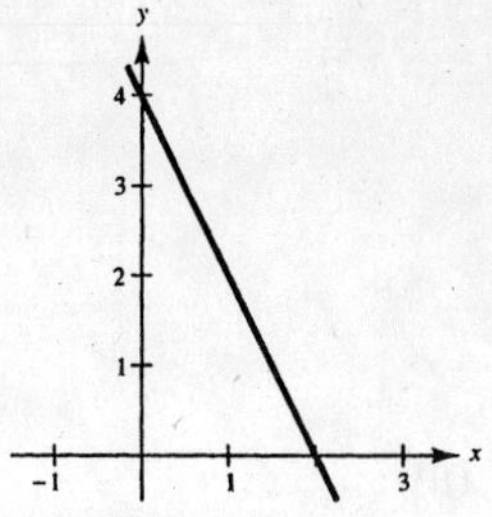


(b)

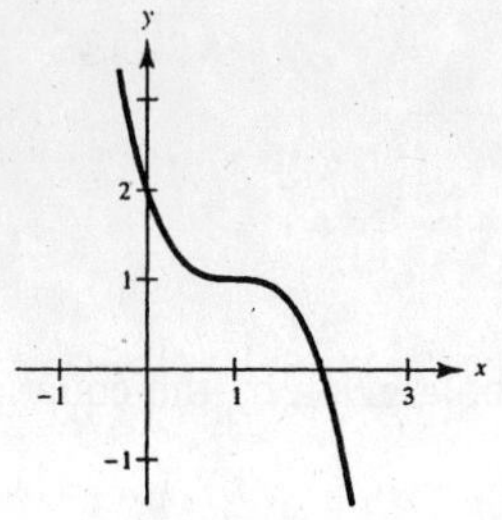


(c)

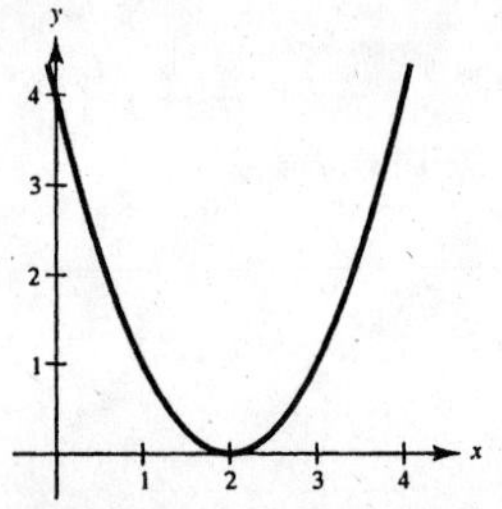


(d)

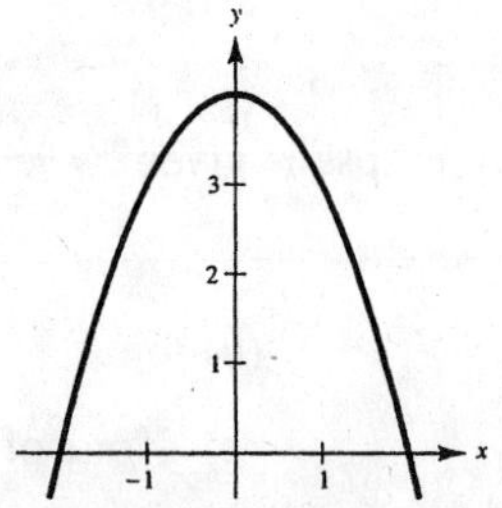


(e) None of these

**5.** Find the corresponding rectangular equation represented by the parametric equations $x = t - 2$ and $y = \dfrac{1}{t-1}$ by eliminating the parameter.

(a) $x + y = -3$  (b) $y = \dfrac{1}{x+1}$  (c) $y = \dfrac{1}{x-3}$

(d) $y = \dfrac{1}{x-1}$  (e) None of these

**6.** Find the corresponding rectangular equation represented by the parametric equations $x = 1 + \sec\theta$ and $y = 2 + \tan\theta$ by eliminating the parameter.

(a) $x^2 + y^2 - 2x - 4y + 4 = 0$  (b) $x^2 - y^2 - 2x + 4y - 2 = 0$  (c) $x^2 - y^2 - 2x + 4y - 4 = 0$

(d) $y = 2 + \dfrac{x+1}{x-1}$  (e) None of these

**7.** Find the parametric equation for $y$ if $x = t + 1$ for the line passing through the points $(2, 1)$ and $(-2, 3)$.

(a) $y = t$  (b) $y = -\frac{1}{2}t$  (c) $y = -\frac{1}{2}t - \frac{1}{2}$

(d) $y = -\frac{1}{2}t + \frac{3}{2}$  (e) None of these

**8.** Find $\dfrac{dy}{dx}$ for the curve given by $x = t^2$ and $y = \sqrt{t-1}$.

(a) $\dfrac{1}{4t\sqrt{t-1}}$  (b) $\dfrac{1}{2\sqrt{t-1}}$  (c) $\dfrac{t}{\sqrt{t-1}}$

(d) $2t$  (e) None of these

**9.** Find the equation of the tangent line for the curve given by $x = 2t$ and $y = t^2 + 5$ at the point where $t = 1$.

(a) $y = 2x + 2$ (b) $y = tx - 2t + 6$ (c) $y = x + 4$

(d) $y = x - 4$ (e) None of these

**10.** Find $\frac{d^2y}{dx^2}$ for the curve given by $x = 2\cos\theta$ and $y = \sin\theta$.

(a) $-\frac{1}{4}\csc^3\theta$ (b) $\frac{1}{2}\csc^2\theta$ (c) $-2\sec^2\theta$

(d) $\frac{1}{2}\cot\theta\csc\theta$ (e) None of these

**11.** Find the arc length of the curve given by $x = t^2$ and $y = 2t^2 + 1$, $1 < t < 3$.

(a) $16\sqrt{5}$ (b) 40 (c) 24

(d) $8\sqrt{5}$ (e) None of these

**12.** Identify the ordered pair that is not a polar representation for the point sketched.

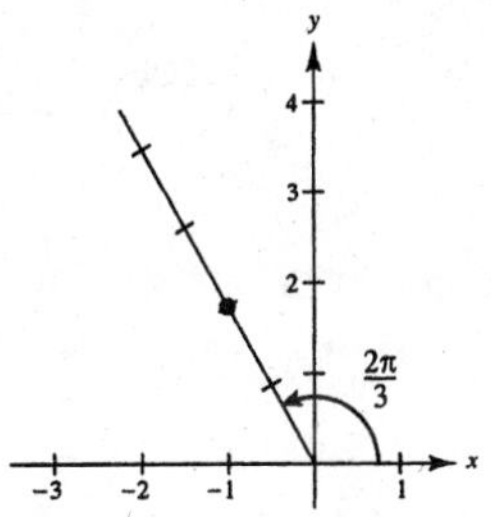


(a) $\left(2, \frac{2\pi}{3}\right)$ (b) $\left(2, -\frac{\pi}{3}\right)$

(c) $\left(-2, -\frac{\pi}{3}\right)$ (d) $\left(2, -\frac{4\pi}{3}\right)$

(e) None of these

**13.** Find the corresponding rectangular coordinates for the polar point $\left(4, -\frac{\pi}{6}\right)$.

(a) $(2\sqrt{3}, -2)$ (b) $(2\sqrt{3}, 2)$ (c) $(2, -2\sqrt{3})$

(d) $(2, 2\sqrt{3})$ (e) None of these

**14.** Convert the rectangular equation $x^2 + y^2 - 2y = 0$ to polar form.

(a) $r = 2\cos\theta$ (b) $r = \frac{1}{2}\csc\theta$ (c) $r = 2\sin\theta$

(d) $r = -2\sin\theta$ (e) None of these

**15.** Convert the polar equation $r = \frac{2}{2\sin\theta - 3\cos\theta}$ to rectangular form.

(a) $r = \frac{2}{2y - 3x}$ (b) $2y - 3x = 2$ (c) $2x - 3y = 2$

(d) $x^2 + y^2 = \frac{2}{2y - 3x}$ (e) None of these

**16.** Find the values of $\theta$ at which there are horizontal tangent lines on the graph of $r = 1 + \sin\theta$.

(a) $-\frac{\pi}{6}, -\frac{\pi}{3}$ (b) $0, \pi$ (c) $\frac{\pi}{4}, \frac{3\pi}{4}$

(d) $\frac{\pi}{2}, \frac{7\pi}{6}, \frac{11\pi}{6}$ (e) None of these

**17.** Find the equation for the graph to the right:

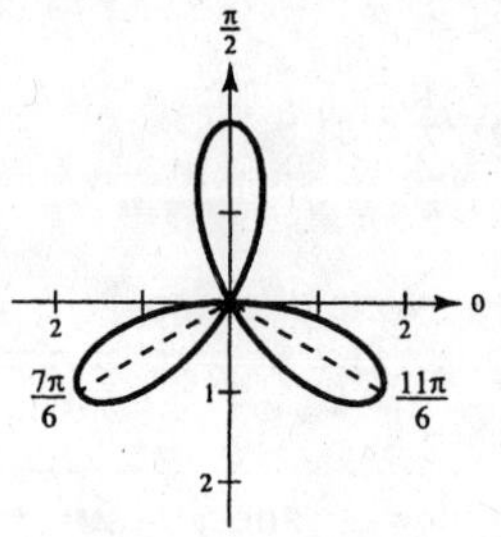


(a) $r = 1 + \sin 3\theta$

(b) $r = 2\cos 3\theta$

(c) $r = -2\sin 3\theta$

(d) $r = 2\sin 6\theta$

(e) None of these

**18.** Calculate the area inside the cardioid $r = 1 + \cos\theta$.

(a) $3\pi$ (b) $\frac{3\pi}{4}$ (c) $\frac{3\pi}{2}$

(d) $\frac{\pi}{2}$ (e) None of these

**19.** Find the values of $\theta$ at the points of intersection: $r^2 = \cos\theta$, $r = \cos\theta$.

(a) 0 (b) $0, \frac{\pi}{2}, \frac{3\pi}{2}$ (c) $0, \pi$

(d) $\frac{\pi}{2}, \frac{3\pi}{2}$ (e) None of these

**20.** Calculate the distance around the graph of the polar curve $r = 3\sin\theta$.

(a) $6\pi$ (b) $\frac{3}{2}\pi$ (c) 16

(d) $3\pi$ (e) None of these

**Test Form B** **Name** ________________ **Date** ________

**Chapter 9** **Class** ________________ **Section** ________

1. Find the equation of the directrix of the parabola given by the equation $y = -\frac{1}{4}x^2 = 2x - 5$.

(a) $y = -2$ (b) $x = 3$ (c) $y = 0$

(d) $x = 5$ (e) None of these

2. Suppose the equation for an ellipse is given by $\frac{(x-3)^2}{a^2} + \frac{(y+2)^2}{7} = 1$. Find $a$ so that the eccentricity is 0.75.

(a) 5 (b) 6 (c) 3

(d) 4 (e) None of these

3. Find the standard form of the equation $x^2 - 3y^2 - 4x + 6y + 10 = 0$.

(a) $\frac{(y+1)^2}{3} - \frac{(x-2)^2}{9} = 1$ (b) $\frac{(x-2)^2}{9} - \frac{(y+1)^2}{3} = 1$ (c) $\frac{(y-1)^2}{3} - \frac{(x-2)^2}{9} = 1$

(d) $(x-2)^2 - 3(y-1)^2 = -9$ (e) None of these

4. Identify the curve represented by the parametric equations $x = t^2$ and $y = 2 + t$.

(a)
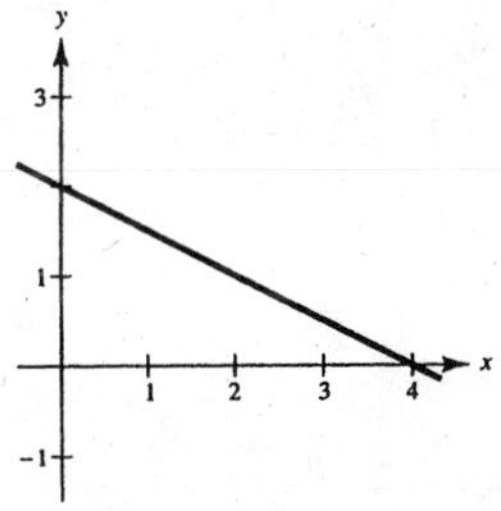

(b)
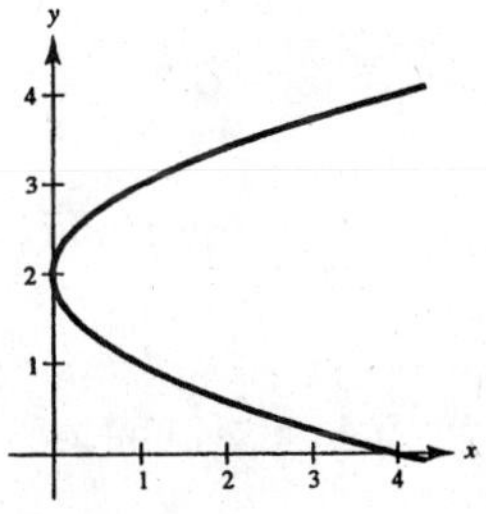

(c)
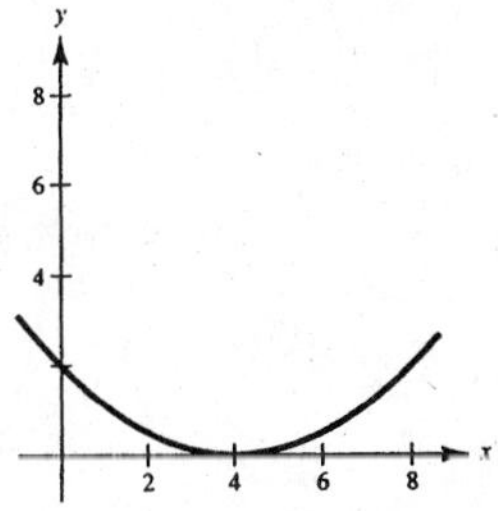

(d)
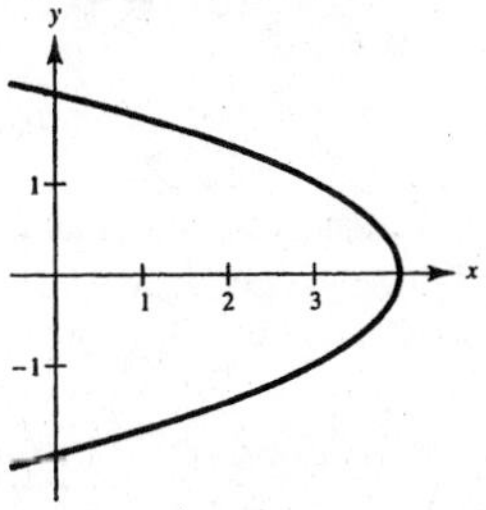

(e) None of these

**5.** Find the corresponding rectangular equation for the curve represented by the parametric equations $x = t^2 + 2$ and $y = t^2 - 1$ by eliminating the parameter.

(a) $x + y = 1$
(b) $y = x + 1$
(c) $x = y + 1$
(d) $y = x - 3$
(e) None of these

**6.** Find the corresponding rectangular equation for the curve represented by the parametric equations $x = 3 + 2\cos\theta$ and $y = 1 + \sin\theta$ by eliminating the parameter.

(a) $x^2 + 4y^2 - 6x - 8y + 9 = 0$
(b) $x^2 - 4y^2 - 6x + 8y + 1 = 0$
(c) $\dfrac{x^2 - 9}{4} + \dfrac{y^2 - 1}{1} = 1$
(d) $x = 2y + 1$
(e) None of these

**7.** Find the parametric equation for $y$ if $x = t - 1$ for the line passing through the points $(2, 1)$ and $(-2, 3)$.

(a) $y = t$
(b) $y = -\frac{1}{2}t$
(c) $y = -\frac{1}{2}t + \frac{5}{2}$
(d) $y = -\frac{1}{2}t + \frac{3}{2}$
(e) None of these

**8.** Find $\dfrac{dy}{dx}$ for the curve given by $x = \sqrt{t}$ and $y = (t - 1)^3$.

(a) $3(t - 1)^2$
(b) $\dfrac{1}{6\sqrt{t}(t - 1)^2}$
(c) $\dfrac{6(t - 1)^2}{\sqrt{t}}$
(d) $6\sqrt{t}(t - 1)^2$
(e) None of these

**9.** Find the equation of the tangent line for the curve given by $x = 3t - 1$ and $y = t^2$ at the point where $t = 1$.

(a) $2x - 3y - 1 = 0$
(b) $3y = 2x + 1$
(c) $y - 1 = \frac{2}{3}t(x - 2)$
(d) $y = 2x - 3$
(e) None of these

**10.** Find $\dfrac{d^2y}{dx^2}$ for the curve given by $x = \dfrac{1}{2}t^2$ and $y = t^2 + t$.

(a) $-\dfrac{1}{t^2}$
(b) $-\dfrac{1}{t^3}$
(c) $\dfrac{2t + 1}{t}$
(d) $2$
(e) None of these

**11.** Find the arc length for the curve given by $x = t^2$ and $y = 2t^2 - 1$, $1 < t < 4$.

(a) $30\sqrt{5}$
(b) $75$
(c) $15\sqrt{5}$
(d) $5\sqrt{5}$
(e) None of these

**12.** Find the corresponding rectangular coordinates for the polar point $\left(-4, \frac{\pi}{6}\right)$.

(a) $(-2\sqrt{3}, -2)$ (b) $(-2, -2\sqrt{3})$ (c) $(2\sqrt{3}, 2)$

(d) $(2, 2\sqrt{3})$ (e) None of these

**13.** Identify the ordered pair that is not a polar representation for the point sketched.

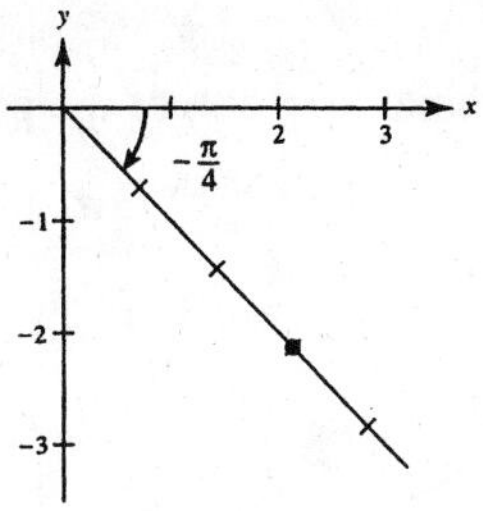


(a) $\left(3, -\frac{\pi}{4}\right)$ (b) $\left(-3, \frac{3\pi}{4}\right)$

(c) $\left(3, \frac{\pi}{4}\right)$ (d) $\left(3, \frac{7\pi}{4}\right)$

(e) None of these

**14.** Convert the rectangular equation $x^2 = 3y$ to polar form.

(a) $r = 3 \sin \theta - \cos^2 \theta$ (b) $r = 3 \sec \theta \tan \theta$ (c) $r = 3 \csc \theta \cot \theta$

(d) $r = 3 \cos \theta - \sin^2 \theta$ (e) None of these

**15.** Convert the polar equation $r = 2 \sin \theta$ to rectangular form.

(a) $x^2 + y^2 - 2y = 0$ (b) $x^2 - 3y^2 = 0$ (c) $x^2 + y^2 - 2x = 0$

(d) $x - y = 0$ (e) None of these

**16.** Find the value(s) of $\theta$ at which there are vertical tangent lines on the graph of $r = 1 + \sin \theta$.

(a) $\frac{\pi}{2}$ (b) $0, \pi$ (c) $\frac{\pi}{6}, \frac{5\pi}{6}$

(d) $\frac{7\pi}{6}, \frac{11\pi}{6}$ (e) None of these

**17.** Find the equation for the graph to the right:

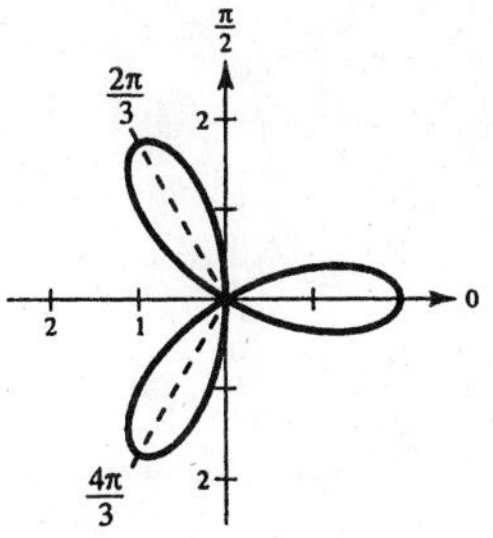


(a) $r = 2 \sin 3\theta$

(b) $r = 2 \cos 3\theta$

(c) $r = 2 \sin 6\theta$

(d) $r = 2 \cos 6\theta$

(e) None of these

**18.** Calculate the area of common interior of $r = 2 \sin \theta$ and $r = 2 \cos \theta$.

(a) $\pi$ (b) $\frac{\pi}{2} - 1$ (c) $\frac{\pi}{4}$

(d) $\frac{\pi}{2} - 2$ (e) None of these

**19.** Find the value(s) of $\theta$ at the points of intersection: $r = 3 + \sin\theta$, $r = 3 - \sin\theta$.

(a) $0, \pi$ (b) $\frac{\pi}{2}, \pi$ (c) $\frac{\pi}{2}, \frac{3\pi}{2}$

(d) $0, \frac{\pi}{2}$ (e) None of these

**20.** Calculate the distance around the graph of the polar curve $r = -3\cos\theta$.

(a) $\frac{3}{2}\pi$ (b) $3\pi$ (c) $6\pi$

(d) 16 (e) None of these

**Test Form C** **Name** ________________ **Date** ____________

**Chapter 9** **Class** ________________ **Section** ____________

**A graphing calculator/utility is recommended for this test.**

**1.** Find an equation of the parabola with focus at (3, 4) and directrix $x = 1$.

(a) $y^2 - 4x - 8y + 24 = 0$ (b) $y^2 - 8z - 6y + 41 = 0$ (c) $x^2 - 6x - 6y + 33 = 0$

(d) $x^2 - 6x - 4y + 25 = 0$ (e) None of these

**2.** Find the center of the ellipse: $4x^2 + 9y^2 - 32x + 18y + 37 = 0$.

(a) $(16, -9)$ (b) $(-8, 2)$ (c) $(4, -1)$

(d) $(2, 3)$ (e) None of these

**3.** Find the vertices of the hyperbola satisfying the equation $\left(\frac{x}{4}\right)^2 - \left(\frac{y}{3}\right)^2 = 1$.

(a) $(0, 0)$ (b) $y = \pm\frac{3}{4}x$ (c) $(0, -3), (0, 3)$

(d) $(-4, 0), (4, 0)$ (e) None of these

**4.** Use a graphing utility to graph the curve given by the parametric equations $x = 3 \sin t$ and $y = 3 \cos t$. Indicate the orientation of the graph.

(a)

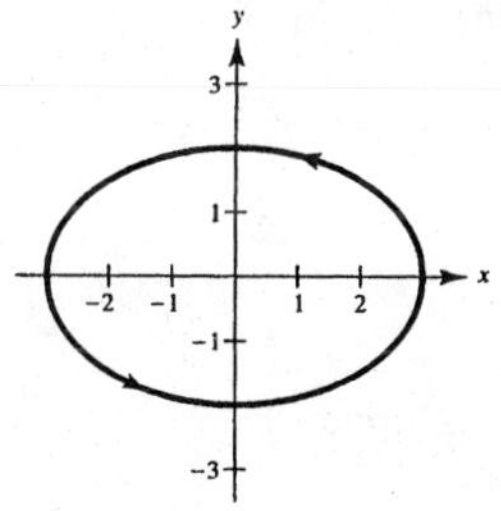


(b)

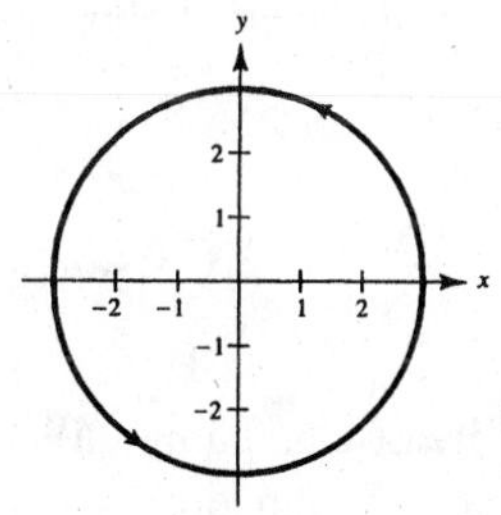


(c)

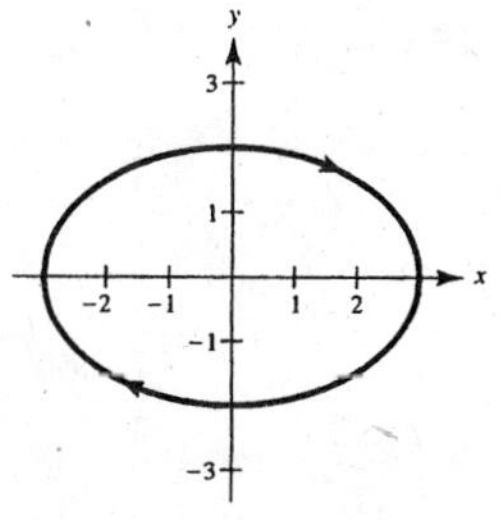


(d)

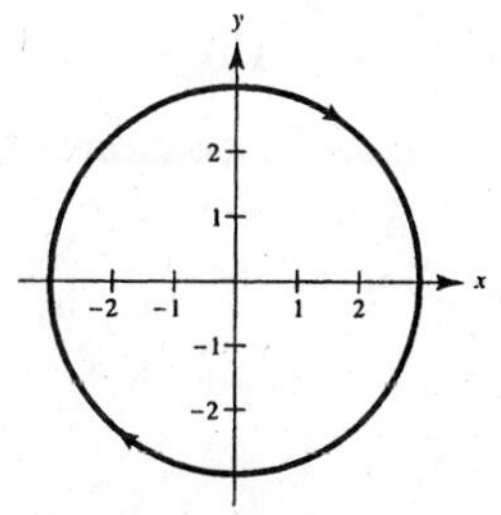


(e) None of these

**5.** Use a graphing utility to graph the curve given by the parametric equations $x = t^2 - 1$ and $y = 1 - t^2$, $t \geq 0$. Indicate the direction of the curve.

(a)

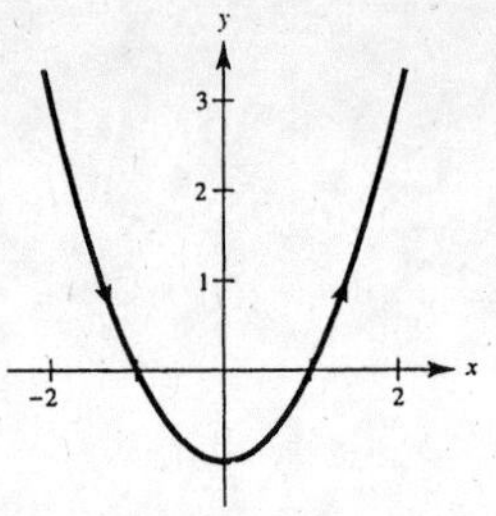


(b)

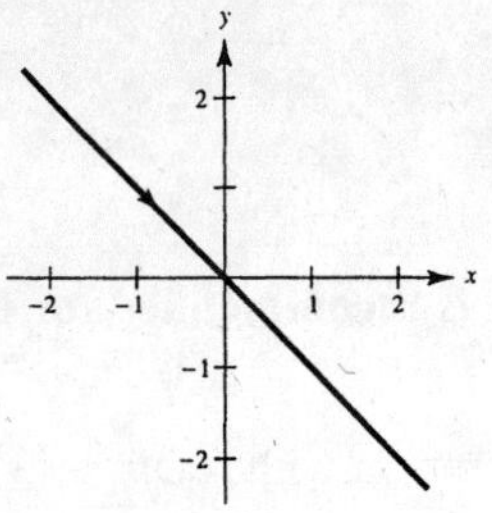


(c)

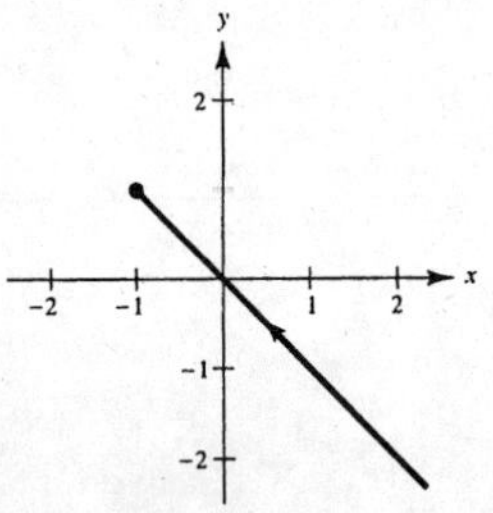


(d)

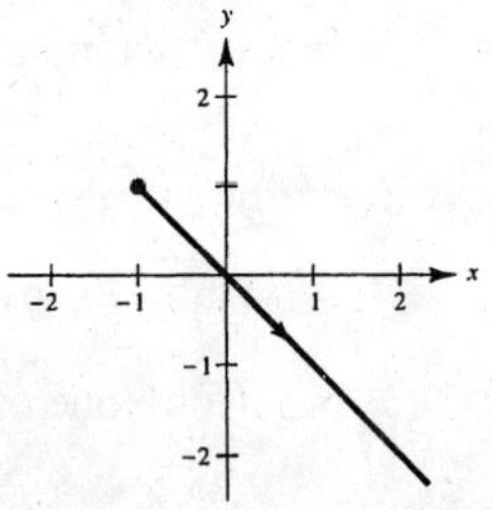


(e) None of these

**6.** Consider the curve given by the parametric equations $x = 4\cos^3\theta$ and $y = 4\sin^3\theta$. Use the graph of the curve to determine those points at which the curve is not smooth.

(a) $(4, 0), (0, 4), (-4, 0), (0, -4)$

(b) $(1, 0), (0, 1), (-1, 0), (0, 1)$

(c) $(4, 0)$ and $(-4, 0)$

(d) The curve is smooth everywhere.

(e) None of these

**7.** Find the length of one arch of the cycloid given by the parametric equations $x = a(\theta - \sin\theta)$ and $y = a(1 - \cos\theta)$.

(a) $8\pi$

(b) $a\pi$

(c) $8a$

(d) $2\pi a$

(e) None of these

**8.** Use the integration capability of a graphing utility to approximate the circumference of the curve given by the parametric equations $x = 2 + 4\cos\theta$ and $y = -1 + \sin\theta$. Round your answer to three decimal places.

(a) 8.579

(b) 17.157

(c) 17.158

(d) 53.407

(e) None of these

**9.** Find all points (if any) of horizontal tangency to the curve given by $x = \cos\theta$ and $y = 2\sin 2\theta$.

(a) $\dfrac{\pi}{4}$

(b) $\left(\dfrac{1}{\sqrt{2}}, 2\right)$

(c) $\left(\dfrac{1}{\sqrt{2}}, 2\right), \left(\dfrac{1}{\sqrt{2}}, -2\right), \left(-\dfrac{1}{\sqrt{2}}, 2\right), \left(-\dfrac{1}{\sqrt{2}}, -2\right)$

(d) There are none.

(e) None of these

**10.** Find a polar coordinate equation for $x^2 + y^2 + 2x + 6y = 0$.

(a) $r = \sqrt{10}$ (b) $r = 2 \sec \theta + 6 \tan \theta$ (c) $r = -2 \sin \theta - 6 \cos \theta$

(d) $r = -2 \cos \theta - 6 \sin \theta$ (e) None of these

**11.** Use a graphing utility to graph the polar equation $r^2 = 4 \cos \theta$. Then, from the graph, estimate the value of $\theta$ between 0 and $\frac{\pi}{2}$ where the curve has a horizontal tangent line.

(a) 5.34 (b) $\frac{\pi}{4}$ (c) 0.94

(d) 2 (e) None of these

**12.** Calculate the area enclosed by the graph of $r = 3 \cos 3\theta$.

(a) $\pi$ (b) $\frac{9\pi}{8}$ (c) $\frac{9\pi}{4}$

(d) $\frac{9\pi}{2}$ (e) None of these

**13.** Use the graphs of the curves given by $r = 3 \tan \theta \sin \theta$ and $r = 3 \cos \theta$ to estimate the points of intersection.

(a) $(1.5, 1.5)$ and $(1.5, -1.5)$ (b) $(0, 0), (1.5, 1.5), (1.5, -1.5)$

(c) $(0, 0)$ and $(1.5, 1.5)$ (d) $(0, 0), (1.5, 1.5)$ and $(-1.5, 1.5)$

(e) None of these

**14.** Identify the graph of the polar equation $r = 2 \sec \theta$.

(a)

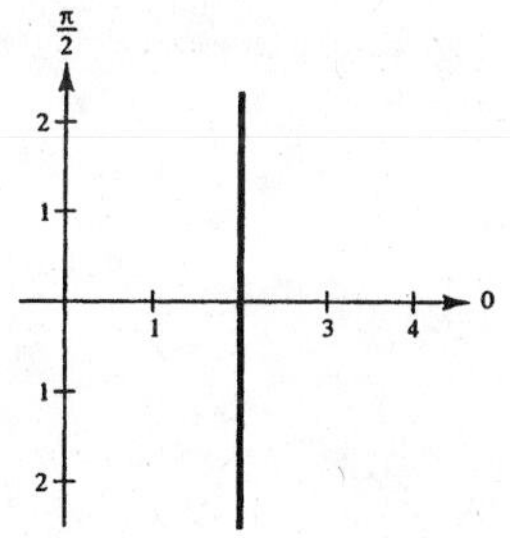


(b)

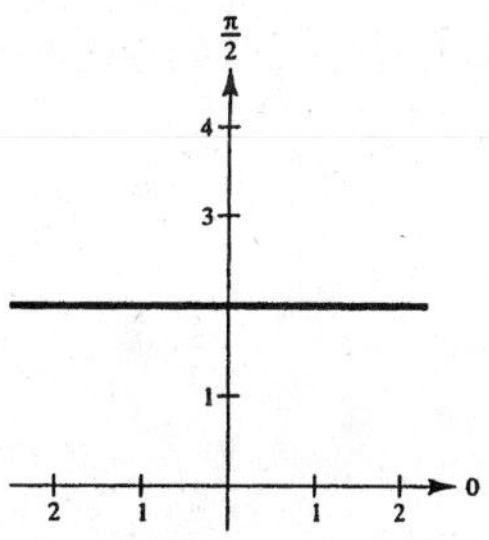


(c)

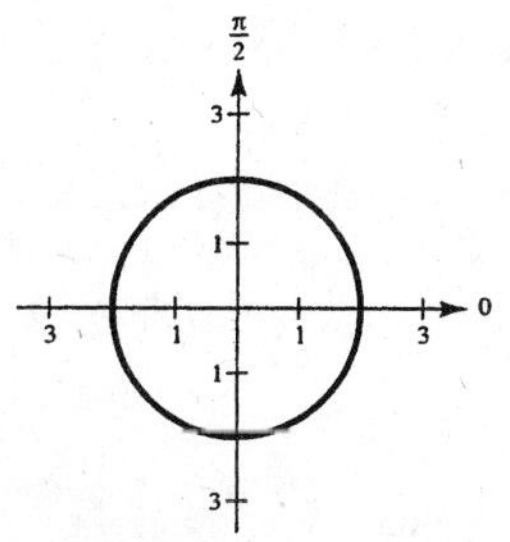


(d)

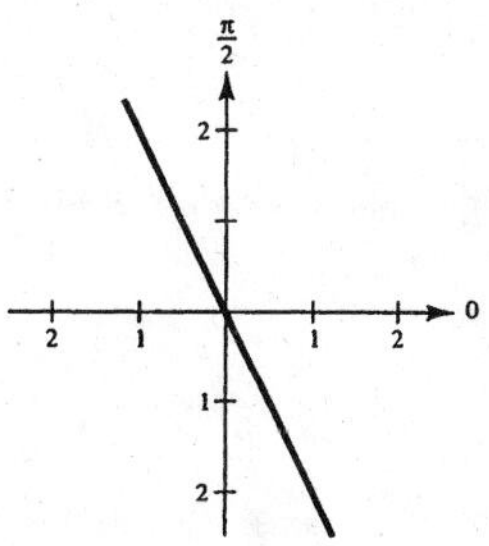


(e) None of these

**15.** Use a graphing utility as an aid in identifying the graph of $r = \dfrac{5}{2 - 2\sin\theta}$.

(a)

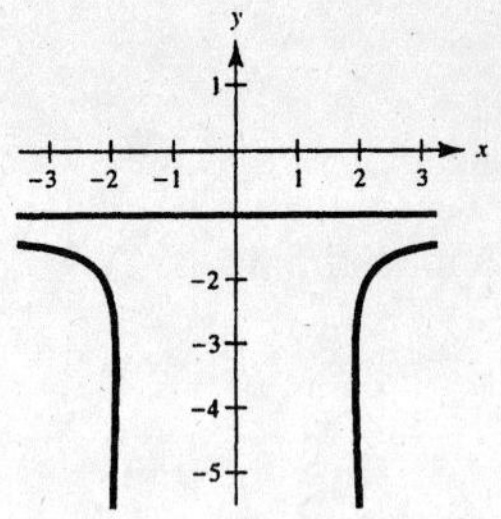

(b)

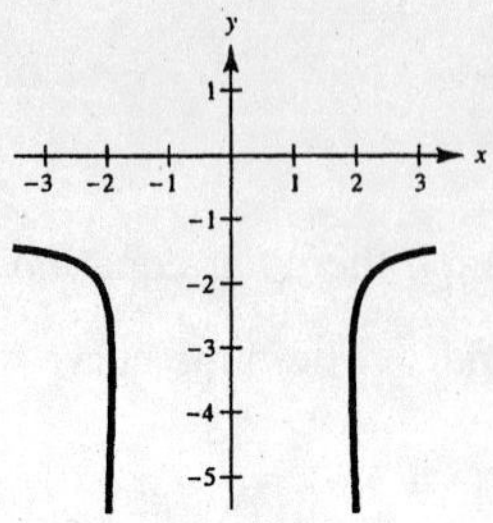

(c)

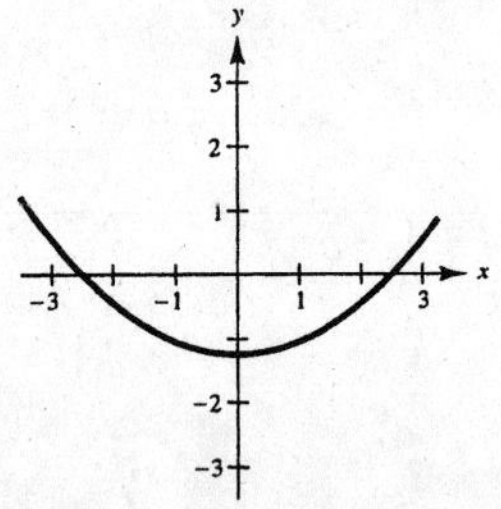

(d)

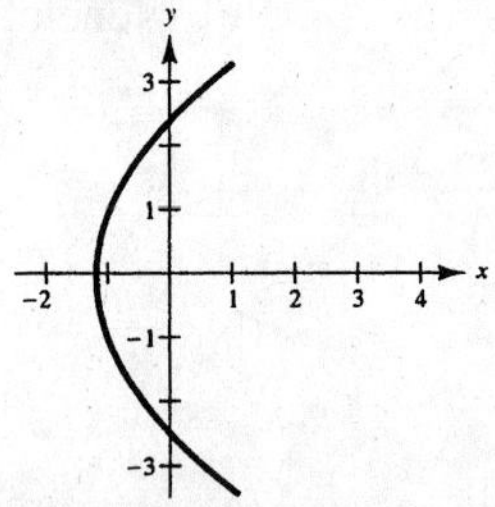

(e) None of these

**Test Form D** **Name** ______________________ **Date** ____________

**Chapter 9** **Class** ______________________ **Section** ____________

1. Find the standard form of the equation for the parabola with vertex $(4, 0)$ and passing through the point $(1, 2)$.

2. Write the equation in standard form and sketch its graph: $4y^2 - x^2 - 8y - 4x = 16$.

3. Write an equation for the tangent line to the ellipse $x^2 + 4y^2 = 8$ at the point $(-2, 1)$.

4. Sketch the curve represented by the parametric equations $x = 4 \sin \theta$ and $y = 3 \cos \theta$.

5. Find the corresponding rectangular equation by eliminating the parameter: $x = t^2 - 1, y = t + 2$.

6. Find the corresponding rectangular equation by eliminating the parameter: $x = 2 + \sec \theta, y = 1 + \tan \theta$.

7. Find the parametric equation for $y$ if $x = e^t$ for the line passing through the points $(2, 1)$ and $(-2, 3)$.

8. Find $\dfrac{dy}{dx}$ for the curve given by $x = 2 \cos \theta$ and $y = 2 + \sin \theta$.

9. Find the equation of the tangent line for the curve represented by $x = \sqrt{t}$ and $y = \frac{1}{2}t^2$ at the point where $t = 4$.

10. Find $\dfrac{d^2y}{dx^2}$ for the curve given by $x = t^3 + 2$ and $y = t^2 + t$.

11. Calculate the length of the arc of the curve given by $x = \frac{8}{3}t^{3/2}$ and $y = 2t - t^2$ between $t = 1$ and $t = 3$.

12. Sketch the polar point: $\left(-3, \dfrac{\pi}{3}\right)$.

13. Find the corresponding rectangular coordinates for the polar point $\left(-2, \dfrac{7\pi}{6}\right)$.

14. Convert the rectangular equation $2y - 3x = 2$ to polar form.

**15.** Convert the polar equation $r = 3 \cos \theta$ to rectangular form.

**16.** Find the value(s) of $\theta$ that give relative extrema of the function $r = 1 + 2 \sin \theta$.

**17.** Sketch a graph of $r = 3 \sin 2\theta$.

**18.** Calculate the area inside one petal of $r = 2 \cos 3\theta$.

**19.** Find the value(s) of $\theta$ at the points of intersection of $r = 5 \sin \theta$ and $r = 2 + \sin \theta$.

**20.** Calculate the area outside $r = \sin \theta$ and inside $r = 2 \sin \theta$.

**Test Form E** **Name** ______________ **Date** ______________

**Chapter 9** **Class** ______________ **Section** ______________

**A graphing calculator/utility is recommended for this test.**

**1.** Consider the parabola given by the equation $y = 4x - x^2$.

**a.** Find the equations for the tangent lines to the graph of the parabola at the points where $x = 1$ and $x = 3$.

**b.** Sketch a graph of the parabola and the tangent lines found in part **a.**

**c.** Calculate the area of the region enclosed by the parabola and the tangent lines.

**2.** Olympic Stadium in Montreal, Canada, is constructed in the shape of an ellipse with major and minor axes of 480 and 280 meters, respectively. Find an equation (in general form) of this ellipse.

**3.** Let $C$ be the curve represented by the parametric equations $x = \sqrt{t}$ and $y = 3t - 1$.

**a.** Find the corresponding rectangular equation for $C$ by eliminating the parameter.

**b.** Sketch the curve represented by these parametric equations. Indicate the orientation of the curve.

**c.** Find the domain of the rectangular equation.

**4.** Consider the curve represented by the parametric equation $x = 2 \cos t$ and $y = 1 + \sin t$.

**a.** Find the corresponding rectangular equation.

**b.** Use a graphing utility to graph the curve. Sketch the curve and indicate the orientation.

**5.** Let $x = 2(\theta - \sin \theta)$ and $y = 2(1 - \cos \theta)$.

**a.** Use a graphing utility to graph the cycloid given by the parametric equations.

**b.** Find the points at which the curve is not smooth.

**6.** Consider the curve given by the parametric equations $x = t^3 + 2t - 1$ and $y = t^2 - 5 + 5,\ t \geq 0$.

**a.** Use a graphing utility to graph the curve and sketch the curve.

**b.** Write the integral to calculate the arc length of the curve where $1 \leq t \leq 5$.

**c.** Use the integration capability of a graphing utility to approximate the integral found in part **b.** Round your answer to three decimal places.

**7.** Find relative extrema for the curve given by the parametric equations $x = t^3 + 2t - 1$ and $y = t^2 - t + 5,\ t \geq 0$.

**8.** For the curve given by $x = t^2 + t - 2$ and $y = t^3 - 2t - 1$, find the slope and concavity at the point (4, 3).

**9.** Let $r = \cos\theta + \sin\theta$.

**a.** Show that the given polar equation is the equation of a circle by writing the equation in standard form in rectangular coordinates.

**b.** Name the center and radius of the circle.

**10.** Find an equation of the tangent line at $\theta = \dfrac{\pi}{4}$ for the curve given by the polar equation $r = \sin 3\theta$.

**11.** Use a graphing utility to graph the polar equation $r = 4\cos 5\theta$. Determine the value of $\theta$ for the line of symmetry for the petal that lies completely in the first quadrant.

**12.** Sketch the curve and calculate the area enclosed by $r = 2\sin 3\theta$.

**13.** Use the graphs of the curves given by $r = \cos\theta$ and $r = 1 + \sin\theta$ to find the values of $\theta$ at the points of intersection.

**Test Form A** **Name** ______________ **Date** ____________

**Chapter 10** **Class** ______________ **Section** ____________

**1.** Find a unit vector in the direction of **v** if **v** is the vector from $P(1, 2, -3)$ to $Q(4, -1, 0)$.

(a) $3\mathbf{i} - 3\mathbf{j} + 3\mathbf{k}$ (b) $\mathbf{i} - \mathbf{j} + \mathbf{k}$ (c) $-\frac{1}{\sqrt{3}}\mathbf{i} + \frac{1}{\sqrt{3}}\mathbf{j} - \frac{1}{\sqrt{3}}\mathbf{k}$

(d) $\frac{1}{\sqrt{3}}\mathbf{i} - \frac{1}{\sqrt{3}}\mathbf{j} + \frac{1}{\sqrt{3}}\mathbf{k}$ (e) None of these

**2.** Calculate $\mathbf{j} \times \mathbf{i}$.

(a) $-\mathbf{j}$ (b) $-\mathbf{k}$ (c) $\mathbf{k}$

(d) $\mathbf{i}$ (e) None of these

In problems 3 through 8, let $\mathbf{u} = 3\mathbf{i} - \mathbf{j} - 2\mathbf{k}$, $\mathbf{v} = -2\mathbf{i} - 3\mathbf{j} + 2\mathbf{k}$ and $\mathbf{w} = \mathbf{i} + 2\mathbf{k}$.

**3.** Calculate $\mathbf{u} \cdot \mathbf{v}$.

(a) $-7$ (b) $-13$ (c) $\langle -8, -2, -11 \rangle$

(d) $1$ (e) None of these

**4.** Calculate $\mathbf{u} \times \mathbf{v}$.

(a) $\langle 4, 10, -11 \rangle$ (b) $\langle -8, -2, -11 \rangle$ (c) $-7$

(d) $\langle -4, -10, 11 \rangle$ (e) None of these

**5.** Calculate $\mathbf{u} \cdot (\mathbf{v} \times \mathbf{w})$.

(a) $-30$ (b) $-18\mathbf{i} - 6\mathbf{j} - 6\mathbf{k}$ (c) $-16$

(d) $-24$ (e) None of these

**6.** Calculate $\text{proj}_{\mathbf{v}}\mathbf{u}$.

(a) $-\frac{7}{17}$ (b) $\langle \frac{14}{17}, \frac{21}{17}, -\frac{14}{17} \rangle$ (c) $\sqrt{17}$

(d) $\langle 14, 21, -14 \rangle$ (e) None of these

**7.** Calculate $\cos\theta$ where $\theta$ is the angle between **u** and **v**.

(a) $\frac{-7}{2\sqrt{17}}$ (b) $\frac{-7}{\sqrt{238}}$ (c) $\frac{-13}{\sqrt{238}}$

(d) $\frac{-13}{2\sqrt{17}}$ (e) None of these

**8.** Which of the following vectors is orthogonal to both **u** and **v**?

(a) **k** (b) $-8\mathbf{i} + 2\mathbf{j} - 11\mathbf{k}$ (c) $\mathbf{v} \times \mathbf{u}$

(d) **w** (e) None of these

**9.** Which of the following is an orthogonal pair of vectors?

(a) $2\mathbf{i} - \mathbf{j}, \mathbf{i} + \mathbf{k}$ (b) $\mathbf{i} - \mathbf{j} + 2\mathbf{k}, -\mathbf{i} - \mathbf{j} - \mathbf{k}$ (c) $3\mathbf{i} - 2\mathbf{k}, 2\mathbf{j} - \mathbf{k}$

(d) $5\mathbf{i} + \mathbf{j} + \mathbf{k}, -\mathbf{i} + 2\mathbf{j} + 3\mathbf{k}$ (e) None of these

**10.** Which of the following is a set of parametric equations for the line through the points $(-2, 0, 3)$ and $(4, 3, 3)$?

(a) $x = -2 + 2t$, $y = t$, $z = 3 + t$

(b) $x = -2 + 2t$, $y = 3t$, $z = 3$

(c) $x = -2 + 2t$, $y = t$, $z = 3$

(d) $x = -2 + 4t$, $y = 3t$, $z = 3 + 3t$

(e) None of these

**11.** Write an equation of the plane that contains the line given by $\dfrac{x}{1} = \dfrac{y-1}{3} = \dfrac{z+1}{2}$ and is perpendicular to the line given by $\dfrac{x-1}{-17} = \dfrac{y+5}{1} = \dfrac{z-3}{7}$.

(a) $17x - y - 7z - 6 = 0$ (b) $x + 3y + 2z + 8 = 0$ (c) $x + 3y + 2z - 1 = 0$

(d) $17x - y - 7z + 1 = 0$ (e) None of these

**12.** Which of the following is a sketch of the plane given by $2x + y + 3z = 6$?

(a)

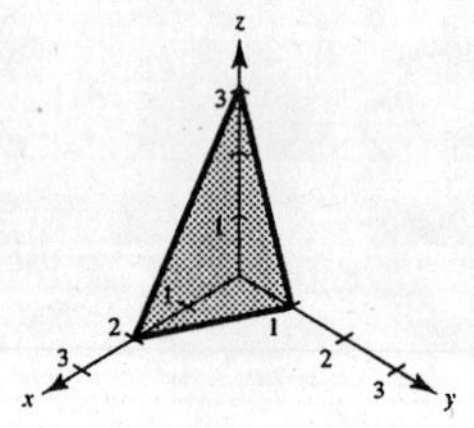

(b)

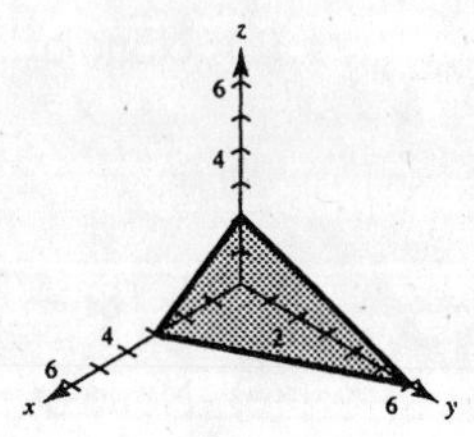

(c)

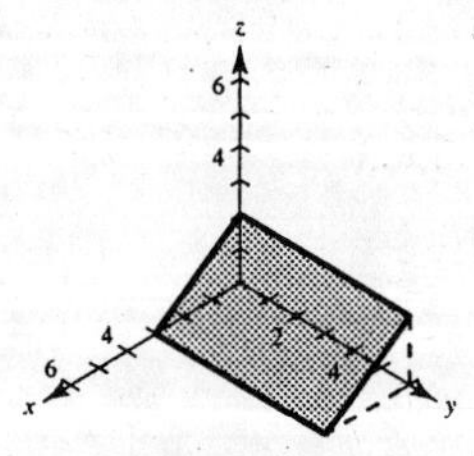

(d)

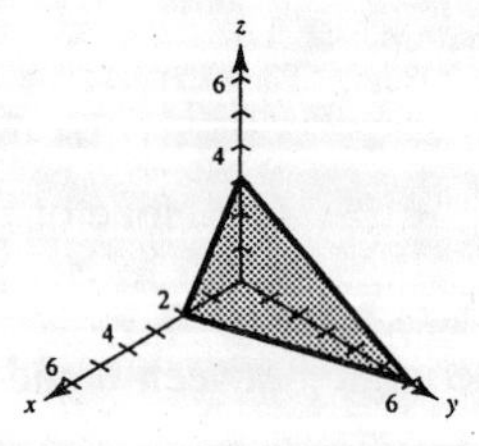

(e) None of these

**13.** Calculate the distance from the point $(-1, 2, 3)$ to the line given by $x = 2 - t$, $y = 1$, and $z = 1 + t$.

(a) $\dfrac{5}{\sqrt{2}}$ (b) $\sqrt{\dfrac{3}{14}}$ (c) $\dfrac{5}{\sqrt{14}}$

(d) $\sqrt{\dfrac{3}{2}}$ (e) None of these

**14.** Find the center and the radius of the sphere given by $x^2 + y^2 + z^2 + 2x - 2y + 6z + 7 = 0$.

(a) Center: $(1, 1, 3)$
Radius: 2

(b) Center: $(-1, 1, -3)$
Radius: $\sqrt{37}$

(c) Center: $(-1, 1, -3)$
Radius: 2

(d) Center: $(1, -1, 3)$
Radius: $\sqrt{37}$

(e) None of these

**15.** Identify the quadric surface given by $z = \dfrac{x^2}{4} + \dfrac{y^2}{16}$.

(a) Elliptic cone (b) Elliptic paraboloid (c) Hyperbolic paraboloid

(d) Hyperboloid of one sheet (e) Hyperboloid of two sheets

**16.** Identify the quadric surface sketched at the right.

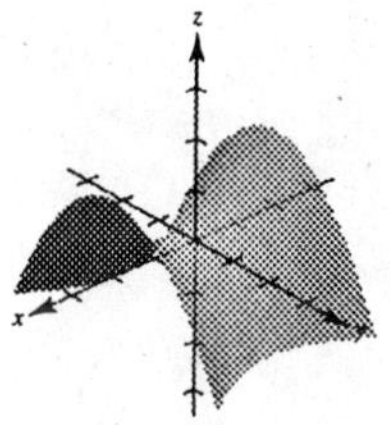


(a) Hyperboloid of two sheets (b) Elliptic paraboloid

(c) Hyperboloid of one sheet (d) Elliptic cone

(e) None of these

**17.** Find the equation of the surface of revolution if the generating curve, $y = 2x + 1$, is revolved about the $y$-axis.

(a) $4x^2 - y^2 + 4z^2 + 2y - 1 = 0$ (b) $3x^2 - z^2 + 4x + 1 = 0$

(c) $4x^2 - y^2 - z^2 + 4x + 1 = 0$ (d) $3x^2 - y^2 - z^2 + 4x + 1 = 0$

(e) None of these

**18.** Express the cylindrical point $\left(4, \dfrac{\pi}{2}, 3\right)$ in spherical coordinates.

(a) $\left(5, \dfrac{\pi}{2}, 60°\right)$ (b) $\left(5, \dfrac{\pi}{2}, \dfrac{3}{5}\right)$ (c) $\left(5, \dfrac{\pi}{2}, 53\right)$

(d) $\left(5, \dfrac{\pi}{2}, 0.93\right)$ (e) None of these

**19.** Find an equation in spherical coordinates for the surface given by the rectangular equation $x^2 - y^2 = z$.

(a) $\rho = \cot\phi \csc\phi$

(b) $r^2 = z \sec 2\theta$

(c) $\rho = \cot\phi \csc\phi \sec 2\theta$

(d) $\rho = \tan\phi \csc\phi \sec 2\theta$

(e) None of these

**20.** Find the work done in moving a particle from $P(3, -1, 0)$ to $Q(2, 3, 1)$ if the magnitude and direction of the force are given by $\mathbf{v} = \langle 5, 6, -2 \rangle$.

(a) 5

(b) 35

(c) 17

(d) 13

(e) None of these

**Test Form B** **Name** ______________ **Date** ______________

**Chapter 10** **Class** ______________ **Section** ______________

**1.** Find a unit vector in the direction of **v** if **v** is the vector from $P(2, -1, 3)$ to $Q(1, 0, -4)$.

(a) $\mathbf{i} - \mathbf{j} + 7\mathbf{k}$

(b) $-\mathbf{i} + \mathbf{j} - \mathbf{k}$

(c) $-\frac{1}{\sqrt{51}}\mathbf{i} + \frac{1}{\sqrt{51}}\mathbf{j} - \frac{7}{\sqrt{51}}\mathbf{k}$

(d) $\frac{1}{\sqrt{51}}\mathbf{i} - \frac{1}{\sqrt{51}}\mathbf{j} + \frac{1}{\sqrt{51}}\mathbf{k}$

(e) None of these

**2.** Calculate $\mathbf{j} \times \mathbf{k}$.

(a) $-\mathbf{i}$

(b) $-\mathbf{k}$

(c) $-\mathbf{j}$

(d) $\mathbf{i}$

(e) None of these

In problems 3 through 8, let $\mathbf{u} = \mathbf{i} + 2\mathbf{j} - \mathbf{k}$, $\mathbf{v} = 3\mathbf{i} - \mathbf{j} + 4\mathbf{k}$ and $\mathbf{w} = 2\mathbf{j} - \mathbf{k}$.

**3.** Calculate $\mathbf{u} \cdot \mathbf{v}$.

(a) $\langle 7, -7, -7 \rangle$

(b) $-3$

(c) $\langle -1, 1, 1 \rangle$

(d) 24

(e) None of these

**4.** Calculate $\mathbf{u} \times \mathbf{v}$.

(a) $\langle 7, -7, -7 \rangle$

(b) $-3$

(c) $\langle -1, 1, 1 \rangle$

(d) 24

(e) None of these

**5.** Calculate $(\mathbf{u} \times \mathbf{v}) \cdot \mathbf{w}$.

(a) $-4$

(b) 0

(c) $-7$

(d) $\mathbf{i} - 8\mathbf{j}$

(e) None of these

**6.** Calculate the component of **u** in the direction of **v**.

(a) $\frac{24}{\sqrt{26}}$

(b) $-\frac{3}{\sqrt{26}}$

(c) $\frac{9}{\sqrt{26}}$

(d) $-\frac{9}{26}\mathbf{i} + \frac{3}{26}\mathbf{j} - \frac{6}{13}\mathbf{k}$

(e) None of these

**7.** Calculate $\cos\theta$ where $\theta$ is the angle between **u** and **v**.

(a) $\frac{9}{2\sqrt{26}}$

(b) $\frac{9}{2\sqrt{39}}$

(c) $\frac{-3}{2\sqrt{26}}$

(d) $\frac{-3}{2\sqrt{39}}$

(e) None of these

**8.** Which of the following vectors is orthogonal to both **u** and **w**?

(a) **j**
(b) **w** × **u**
(c) **v**
(d) 4**j** + **k**
(e) None of these

**9.** Which of the following statements is true about the vectors $\mathbf{u} = \frac{1}{2}\mathbf{i} + \frac{1}{3}\mathbf{j} - \frac{1}{4}\mathbf{k}$ and $\mathbf{v} = -2\mathbf{i} - \frac{4}{3}\mathbf{j} + \mathbf{k}$?

(a) **u** and **v** are orthogonal.
(b) **u** and **v** are parallel.
(c) **u** is a unit vector of **v**.
(d) The angle between **u** and **v** is $\frac{\pi}{4}$.
(e) None of these

**10.** Which of the following is a set of parametric equations for the line through the points $(-3, 2, 0)$ and $(4, 3, 3)$?

(a) $x = 4 + 7t$, $y = 3 + t$, $z = 3 + 3t$
(b) $x = 4 - 3t$, $y = 3 + 2t$, $z = 3$
(c) $x = -3 + t$, $y = 2 + 5t$, $z = 3t$
(d) $x = -3 + 4t$, $y = 2 + 3t$, $z = 3t$
(e) None of these

**11.** Find an equation of the plane that passes through the points $(2, 1, -4)$, $(-3, 1, 3)$ and $(-2, -1, 0)$.

(a) $2x - 4y + 5z + 10 = 0$
(b) $7x + 6y + 5z + 20 = 0$
(c) $7x - 4y + 5z + 10 = 0$
(d) $7x - 4y + 5z + 2 = 0$
(e) None of these

**12.** If the equation of a cylinder is given by $x^2 - z = 0$, then

(a) The rulings are parallel to the $x$-axis.
(b) The graph is a parabola.
(c) The rulings are parallel to the $z$-axis.
(d) The generating curve is a parabola.
(e) None of these

**13.** Calculate the distance from the point $(1, -2, 3)$ to the plane given by $(x - 3) + 2(y + 1) - 4z = 0$.

(a) $\sqrt{14}$
(b) $\dfrac{14}{\sqrt{21}}$
(c) $\dfrac{16}{\sqrt{21}}$
(d) $\sqrt{65}$
(e) None of these

**14.** Find the center and radius of the sphere given by the equation $x^2 + y^2 + z^2 - 4x - 2y + 2z = 10$.

(a) Center: $(-2, -1, 1)$ Radius: 4
(b) Center: $(2, 1, -1)$ Radius: 4
(c) Center: $(-2, -1, 1)$ Radius: $\sqrt{10}$
(d) Center: $(2, 1, -1)$ Radius: $\sqrt{10}$
(e) None of these

**15.** Identify the quadric surface given by $z^2 = \frac{x^2}{4} + \frac{y^2}{16}$.

(a) Elliptic cone (b) Elliptic paraboloid (c) Hyperbolic paraboloid

(d) Hyperboloid of one sheet (e) Hyperboloid of two sheets

**16.** Identify the quadric surface sketched at the right.

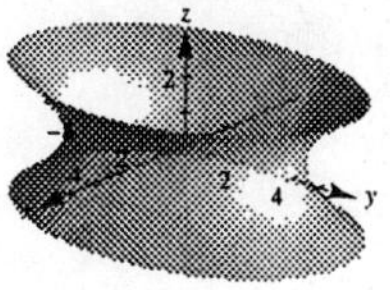


(a) Hyperboloid of two sheets (b) Elliptic paraboloid

(c) Hyperboloid of one sheet (d) Elliptic cone

(e) None of these

**17.** Find the equation of revolution if the generating curve, $y - 2x = 1$, is revolved about the $x$-axis.

(a) $4x^2 - y^2 + 4z^2 + 2y - 1 = 0$ (b) $3x^2 - z^2 + 4x + 1 = 0$

(c) $4x^2 - y^2 - z^2 + 4x + 1 = 0$ (d) $3x^2 - y^2 - z^2 + 4x + 1 = 0$

(e) None of these

**18.** Express the spherical point $\left(3, \frac{3\pi}{4}, \frac{\pi}{3}\right)$ in rectangular coordinates.

(a) $\left(\frac{-3}{\sqrt{2}}, \frac{3}{\sqrt{2}}, \frac{3\sqrt{3}}{2}\right)$ (b) $\left(\frac{-3\sqrt{3}}{2\sqrt{2}}, \frac{3\sqrt{3}}{2\sqrt{2}}, \frac{3}{2}\right)$ (c) $\left(\frac{3}{2\sqrt{2}}, \frac{3\sqrt{3}}{2\sqrt{2}}, \frac{-3}{\sqrt{2}}\right)$

(d) $\left(\frac{3}{2}, \frac{3\sqrt{3}}{2}, \frac{-3}{\sqrt{2}}\right)$ (e) None of these

**19.** Find an equation in cylindrical coordinates for the surface given by the rectangular equation $x^2 - y^2 = z$.

(a) $r^2 = z$ (b) $r^2 = z \sec 2\theta$ (c) $\rho = \cot \phi \csc \phi \sec 2\theta$

(d) $\rho = \cot \phi \csc \phi$ (e) None of these

**20.** Find the work done in moving a particle from $P(0, 2, 1)$ to $Q(3, -2, 2)$ if the magnitude and direction of the force are given by $\mathbf{v} = \langle 4, -2, 3 \rangle$.

(a) 15 (b) 21 (c) 168

(d) 23 (e) None of these

**Test Form C** **Name** ________________ **Date** ____________

**Chapter 10** **Class** ________________ **Section** ____________

**1.** A vector **v** has initial point $(-1, -3)$ and terminal point $(2, 1)$. Find the unit vector in the direction of **v**.

(a) $\frac{1}{\sqrt{5}}\mathbf{i} - \frac{2}{\sqrt{5}}\mathbf{j}$ (b) $\frac{3}{5}\mathbf{i} + \frac{4}{5}\mathbf{j}$ (c) $\frac{4}{5}\mathbf{i} + \frac{3}{5}\mathbf{j}$

(d) $\frac{1}{\sqrt{5}}(\mathbf{i} + 2\mathbf{j})$ (e) None of these

**2.** Find a vector with magnitude $\sqrt{10}$ in the direction of $\mathbf{v} = \langle -1, 1 \rangle$.

(a) $\left\langle -\frac{1}{\sqrt{2}}, \frac{1}{\sqrt{2}} \right\rangle$ (b) $\langle -\sqrt{10}, \sqrt{10} \rangle$ (c) $\langle -\sqrt{5}, \sqrt{5} \rangle$

(d) $\langle -\sqrt{2}, \sqrt{2} \rangle$ (e) None of these

**3.** Forces with magnitudes of 200 and 300 pounds act on a machine part at angles of 60° and −45°, respectively, with the positive $x$-axis. Find the direction of the resultant forces.

(a) −7.1° (b) 7.1° (c) 82.8°

(d) 7.5° (e) None of these

**4.** Find the standard equation for the sphere that has points $(4, -3, 5)$ and $(-6, 1, -1)$ as endpoints of a diameter.

(a) $(x + 1)^2 + (y + 1)^2 + (z - 2)^2 = 38$ (b) $(x - 1)^2 + (y - 1)^2 + (z + 2)^2 = 36$

(c) $(x + 1)^2 + (y + 1)^2 + (z - 2)^2 = 36$ (d) $(x - 5)^2 + (y + 2)^2 + (z - 3)^2 = 6$

(e) None of these

**5.** Determine which vector is parallel to the vector $\mathbf{v} = \langle 2, -3, -1 \rangle$.

(a) $\langle 4, 6, -2 \rangle$ (b) $\langle -\frac{2}{3}, 1, \frac{1}{3} \rangle$ (c) $\langle 1, -\frac{3}{2}, \frac{1}{2} \rangle$

(d) $\langle 6, -9, 3 \rangle$ (e) None of these

**6.** Find the unit vector in the direction of the vector $\mathbf{v} = 2\mathbf{i} + \mathbf{j} + 2\mathbf{k}$.

(a) $\frac{2}{\sqrt{5}}\mathbf{i} + \frac{1}{\sqrt{5}}\mathbf{j} + \frac{3}{\sqrt{5}}\mathbf{k}$ (b) $\frac{2}{9}\mathbf{i} + \frac{1}{9}\mathbf{j} + \frac{2}{9}\mathbf{k}$ (c) $\mathbf{i} + \mathbf{j} + \mathbf{k}$

(d) $\frac{2}{3}\mathbf{i} + \frac{1}{3}\mathbf{j} + \frac{2}{3}\mathbf{k}$ (e) None of these

7. Find $\mathbf{u} \cdot \mathbf{v}$ if $\|\mathbf{u}\| = 40$, $\|\mathbf{v}\| = 15$, and the angle between vectors **u** and **v** is $\dfrac{2\pi}{3}$.

(a) $-300\sqrt{3}$ (b) $\dfrac{900}{\pi}$ (c) 300

(d) $-300$ (e) None of these

8. Find the vector component of **u** orthogonal to **v** for $\mathbf{u} = \mathbf{i} + 2\mathbf{j}$ and $\mathbf{v} = -4\mathbf{i} + 4\mathbf{j}$.

(a) $-\frac{1}{2}\mathbf{i} + \frac{1}{2}\mathbf{j}$ (b) $\frac{3}{2}\mathbf{i} + \frac{3}{2}\mathbf{j}$ (c) $-\frac{1}{\sqrt{2}}\mathbf{i} + \frac{1}{\sqrt{2}}\mathbf{j}$

(d) $\frac{1}{2}\mathbf{i} + \frac{5}{2}\mathbf{j}$ (e) None of these

9. Calculate the angle that vector $\mathbf{v} = 3\mathbf{i} - 5\mathbf{j} + \mathbf{k}$ makes with the positive $y$-axis.

(a) 59.5° (b) 32.3° (c) 147.7°

(d) 80.3° (e) None of these

10. $\mathbf{v} \cdot \mathbf{v} =$ ____________

(a) $\|\mathbf{v}\|$ (b) $\dfrac{\mathbf{v}}{\|\mathbf{v}\|}$ (c) $\|\mathbf{v}\|^2$

(d) $\text{proj}_{\mathbf{u}}\mathbf{v}$ (e) None of these

11. Calculate $\mathbf{u} \times \mathbf{v}$ for $\mathbf{u} = 3\mathbf{i} + 4\mathbf{j} - \mathbf{k}$ and $\mathbf{v} = 2\mathbf{i} - \mathbf{j} + 3\mathbf{k}$.

(a) $-11\mathbf{i} + 11\mathbf{j} + 11\mathbf{k}$ (b) $13\mathbf{i} - 7\mathbf{j} + 5\mathbf{k}$ (c) $11\mathbf{i} + 11\mathbf{j} - 11\mathbf{k}$

(d) $11\mathbf{i} - 11\mathbf{j} - 11\mathbf{k}$ (e) None of these

12. Find the area of the parallelogram having vectors $\mathbf{v}_1 = -\mathbf{i} + 2\mathbf{j} + 2\mathbf{k}$ and $\mathbf{v}_2 = 3\mathbf{i} - 2\mathbf{j} + \mathbf{k}$ as adjacent sides.

(a) 3 (b) 10 (c) $\sqrt{101}$

(d) $\sqrt{69}$ (e) None of these

13. Find an equation of the plane determined by the points $(1, 2, -3)$, $(2, 3, 1)$, and $(0, -2, -1)$.

(a) $18\mathbf{i} - 6\mathbf{j} - 3\mathbf{k} = 0$ (b) $6x - 2y - z = 0$ (c) $6x - 2y - z = 5$

(d) $14x + 6y + 3z = 17$ (e) None of these

14. Find the point of intersection of the line given by the parametric equations $x = 3 + 2t$, $y = 7 + 8t$, and $z = -2 + t$ with the $yz$-plane.

(a) $\left(0, -5, -\frac{7}{2}\right)$ (b) $\left(0, \frac{37}{3}, -\frac{4}{3}\right)$ (c) $(3, 7, -2)$

(d) $(-7, -33, -7)$ (e) None of these

**15.** Find a generating curve for the surface of revolution given by $2x^2 + y^2 + z^2 = 3x$.

(a) $y = \sqrt{3x - 2x^2}$ (b) $z = \sqrt{3x - 2x^2}$ (c) $x = y^2 + z^2$

(d) Both (a) and (b) (e) None of these

**16.** Write the equation in standard form and identify the quadric surface given by $4x^2 - 9y^2 - 36z = 0$.

(a) $36z = 4x^2 - 9y^2$, Cone

(b) $z = \frac{x^2}{9} - \frac{y^2}{4}$, Hyperbolic paraboloid

(c) $z = \frac{x^2}{9} - \frac{y^2}{4}$, Hyperboloid of one sheet

(d) $x^2 = \frac{9}{4}y^2 + 9z$, Hyperbolic paraboloid

(e) None of these

**17.** Express the spherical coordinate point $\left(6, -\frac{\pi}{6}, \frac{\pi}{3}\right)$ in cylindrical coordinates.

(a) $\left(\frac{3}{2}, -\frac{3\sqrt{3}}{2}, 3\right)$ (b) $\left(-\frac{3\sqrt{3}}{2}, \frac{3}{2}, 3\right)$ (c) $\left(3\sqrt{3}, 3, \frac{\pi}{3}\right)$

(d) $\left(3\sqrt{3}, -\frac{\pi}{6}, 3\right)$ (e) None of these

**18.** Find a rectangular equation for the graph represented by the spherical equation $\rho = 2 \cos \phi$.

(a) $x^2 + y^2 + z^2 = 2x$ (b) $x^2 + y^2 + z^2 = 2z$ (c) $y^2 + z^2 - x^2 = 0$

(d) $z = 2x^2$ (e) None of these

**Test Form D** **Name** ______________ **Date** ________

**Chapter 10** **Class** ______________ **Section** ________

1. Find a unit vector in the direction of **v** if **v** is the vector from $P(2, 1, -3)$ to $Q(-1, 0, 4)$.

2. Calculate $\mathbf{i} \times \mathbf{k}$.

In problems 3 through 8, let $\mathbf{u} = -\mathbf{i} + 2\mathbf{j} - \mathbf{k}$, $\mathbf{v} = 2\mathbf{i} + \mathbf{j} + \mathbf{k}$ and $\mathbf{w} = 3\mathbf{i} - \mathbf{k}$.

3. Calculate $\mathbf{u} \cdot \mathbf{v}$.

4. Calculate $\mathbf{u} \times \mathbf{v}$.

5. Calculate $\mathbf{u} \cdot (\mathbf{v} \times \mathbf{w})$.

6. Calculate $\mathbf{proj}_{\mathbf{v}}\mathbf{u}$.

7. Calculate $\theta$ where $\theta$ is the angle between **u** and **v**.

8. Calculate a vector perpendicular to both **u** and **w**.

9. Let $L_1 : x = 2 - 4t, y = -1 + 5t, z = 7t$ and $L_2 : x = 2 + t, y = 5 - 2t, z = 1 + 2t$. Show that the vectors $\mathbf{v}_1$ and $\mathbf{v}_2$ parallel to lines $L_1$ and $L_2$, respectively, are orthogonal.

10. Find parametric equations for the line through the point $(4, -1, 3)$ and parallel to the line given by $\frac{x}{1} = \frac{y-2}{2} = \frac{z+1}{5}$.

11. Write an equation of the plane that is determined by the lines given by $\frac{x-6}{-6} = \frac{y+1}{2} = \frac{z}{-3}$ and $\frac{x-4}{2} = \frac{y-3}{1} = \frac{z+1}{1}$.

12. Sketch the plane given by $3x + 4y + 2z = 12$.

13. Find the numbers $x$ and $y$ such that the point $(x, y, 1)$ lies on the line passing through the points $(2, 5, 7)$ and $(0, 3, 2)$.

14. Write the equation of the sphere, $4x^2 + 4y^2 + 4z^2 - 4x + 16y - 8z + 9 = 0$, into standard form and find the center and radius.

15. Identify and sketch the quadric surface given by $\frac{x^2}{1} + \frac{y^2}{4} - z = 0$.

**16.** Sketch the surface represented by the equation $y = x^2$.

**17.** Find the equation of the surface of revolution if the generating curve, $2x + 3z = 1$, is revolved about the $x$-axis.

**18.** Convert the point $(3, -3, 7)$ from rectangular to cylindrical coordinates.

**19.** Find an equation in rectangular coordinates for the spherical coordinate equation $\rho = 9 \csc \phi \csc \theta$ and identify the surface.

**20.** Find the area of the parallelogram having vectors $\mathbf{u} = \mathbf{i} - 2\mathbf{j} + 6\mathbf{k}$ and $\mathbf{v} = -5\mathbf{i} + 2\mathbf{k}$ as adjacent sides.

**Test Form E** **Name** ______________ **Date** ______________

**Chapter 10** **Class** ______________ **Section** ______________

**1.** A vector **v** has initial point $(2, -1)$ and terminal point $(-2, 3)$. Find the unit vector in the direction of **v**.

**2.** Find a vector with magnitude 3 in the direction of the vector $\mathbf{v} = \langle 1, 2 \rangle$.

**3.** Three forces with magnitude of 50, 20, and 40 pounds act on an object at angles of 60°, 30°, and $-90°$, respectively, with the positive $x$-axis. Find the direction and magnitude of the resultant forces.

**4.** Find the standard equation for the sphere that has points $(-2, 5, -1)$ and $(3, 7, -3)$ as endpoints of a diameter.

**5.** Determine whether each vector is parallel to the vector $\mathbf{v} = -4\mathbf{i} - \mathbf{j} + 5\mathbf{k}$. If it is, find $c$ such that $\mathbf{v} = c\mathbf{u}$.

**a.** $\mathbf{u} = 8\mathbf{i} + 4\mathbf{j} + 10\mathbf{k}$

**b.** $\mathbf{u} = -8\mathbf{i} - 4\mathbf{j} + 10\mathbf{k}$

**c.** $\mathbf{u} = 3\mathbf{i} + \frac{3}{4}\mathbf{j} - \frac{15}{4}\mathbf{k}$

**d.** $\mathbf{u} = -2\mathbf{i} + \frac{1}{2}\mathbf{j} + \frac{5}{2}\mathbf{k}$

**6.** A vector **v** has initial point $(2, 1, 3)$ and terminal point $(-4, 2, -1)$.

**a.** Write **v** in the component form.

**b.** Write **v** as a linear combination of the standard unit vectors.

**c.** Find the magnitude of **v**.

**d.** Find the unit vector in the direction of **v**.

**e.** Find the unit vector in the direction opposite that of **v**.

**7.** Find $\mathbf{u} \cdot \mathbf{v}$ if $\|\mathbf{u}\| = 7$, $\|\mathbf{v}\| = 12$, and the angle between **u** and **v** is $\frac{\pi}{4}$.

**8.** Let $\mathbf{u} = -\mathbf{i} + 2\mathbf{j} + 2\mathbf{k}$ and $\mathbf{v} = 3\mathbf{i} - 2\mathbf{j} + \mathbf{k}$.

(a) Find the projection of **u** onto **v**.

(b) Find the vector component of **u** orthogonal to **v**.

**9.** Find direction cosines for the vector with initial point $(2, -1, 3)$ and terminal point $(4, 3, -5)$.

**10.** Determine a scalar, $k$, so that the vectors $\mathbf{u} = 3\mathbf{i} + 2\mathbf{j}$ and $\mathbf{v} = 2\mathbf{i} + k\mathbf{j}$ are orthogonal.

**11.** Let $\mathbf{u} = -\mathbf{i} + 2\mathbf{j} + 2\mathbf{k}$ and $\mathbf{v} = 3\mathbf{i} - 2\mathbf{j} + \mathbf{k}$.

**a.** Find $\mathbf{u} \times \mathbf{v}$.

**b.** Show that the vector $\mathbf{u} \times \mathbf{v}$ is orthogonal to $\mathbf{v}$.

**12.** Calculate the area of the triangle with vertices $(3, -1, 2)$, $(2, 1, 5)$, and $(1, -2, -2)$.

**13.** Find parametric equations for the line perpendicular to the lines given by $\dfrac{x-2}{-4} = \dfrac{y+3}{-7} = \dfrac{z+1}{3}$ and $\dfrac{x-3}{7} = \dfrac{y+2}{-2} = \dfrac{z-8}{3}$ passing through the point $(2, 4, -1)$.

**14.** Find the point of intersection of the line $\dfrac{x+2}{2} = \dfrac{y-3}{3} = \dfrac{z-1}{1}$ and the plane $2x + 3y + z - 3 = 0$.

**15.** Sketch the surface represented by $y = x^2 + z^2$.

**16.** Identify each of the quadric surfaces:

**a.** $3x^2 + 3y^2 + 3z^2 - 2x + 3y - 11 = 0$

**b.** $2x^2 + 5z^2 = y^2$

**c.** $2x^2 - y^2 - z^2 = 1$

**17.** Consider the surface whose rectangular equation is $x^2 + y^2 - z^2 = 1$.

**a.** Find an equation for the surface in cylindrical coordinates.

**b.** Find an equation for the surface in spherical coordinates.

**18.** Sketch the surface represented by the spherical equation $\rho = 2 \csc \phi$.

**Test Form A** **Name** ______________ **Date** ______________

**Chapter 11** **Class** ______________ **Section** ______________

**1.** Find the domain of the vector-valued function $\mathbf{r}(t) = t^2\mathbf{i} + \frac{1}{t}\mathbf{j} + \sqrt{1-t}\,\mathbf{k}$.

(a) $(-\infty, \infty)$ (b) $(-\infty, 1]$ (c) $(-\infty, 1)$

(d) $(-\infty, 0) \cup (0, 1]$ (e) None of these

**2.** Find the velocity vector of a particle that moves along the curve $C$ given by $\mathbf{r}(t) = \cos t\mathbf{i} + \sin t\mathbf{j} - 16t^2\mathbf{k}$.

(a) $-\sin t\mathbf{i} + \cos t\mathbf{j} - 32t\mathbf{k}$ (b) $\sin t\mathbf{i} - \cos t\mathbf{j} - 32t\mathbf{k}$ (c) $-\cos t\mathbf{i} - \sin t\mathbf{j} - 32\mathbf{k}$

(d) $\cos t\mathbf{i} + \sin t\mathbf{j} - 32\mathbf{k}$ (e) None of these

**3.** Find the velocity vector for an object having $a(t) = e^t\mathbf{j} - 32\mathbf{k}$, if $\mathbf{v}(0) = 3\mathbf{i} + 2\mathbf{j} + \mathbf{k}$.

(a) $e^t\mathbf{j} - 32t\mathbf{k} + C$ (b) $C\mathbf{i} + e^t\mathbf{j} - 32t\mathbf{k}$ (c) $3\mathbf{i} + (e^t + 1)\mathbf{j} + (-32t + 1)\mathbf{k}$

(d) $3\mathbf{i} + (e^t + 2)\mathbf{j} + (-32t + 1)\mathbf{k}$ (e) None of these

**4.** Let $\mathbf{T}(t) = \frac{1}{\sqrt{9t^2+5}}(3t\mathbf{i} + \mathbf{j} - 2\mathbf{k})$. Find $\frac{d\mathbf{T}}{dt}$.

(a) $\frac{1}{\sqrt{9t^2+5}}(3\mathbf{i})$ (b) $\frac{-2t}{(9t^2+5)^{3/2}}(3\mathbf{i})$ (c) $\frac{3}{(9t^2+5)^{3/2}}(5\mathbf{i} - 3t\mathbf{j} + 6t\mathbf{k})$

(d) $\frac{-2t}{(9t^2+5)^{3/2}}(3t\mathbf{i} + \mathbf{j} - 2\mathbf{k})$ (e) None of these

**5.** Let $\mathbf{r}(t) = 2\cos t\mathbf{i} + 2\sin t\mathbf{j} + 3t\mathbf{k}$. Calculate the unit tangent vector.

(a) $-\frac{2}{\sqrt{13}}\sin t\mathbf{i} + \frac{2}{\sqrt{13}}\cos t\mathbf{j} + \frac{3}{\sqrt{13}}\mathbf{k}$ (b) $\frac{2}{\sqrt{13}}\cos t\mathbf{i} + \frac{2}{\sqrt{13}}\sin t\mathbf{j} + \frac{3t}{\sqrt{13}}\mathbf{k}$

(c) $-2\sin t\mathbf{i} + 2\cos t\mathbf{j} + 3\mathbf{k}$ (d) $-\sin t\mathbf{i} + \cos t\mathbf{j} + \mathbf{k}$

(e) None of these

**6.** Find the normal component of acceleration for the curve given by $\mathbf{r}(t) = t\mathbf{i} + \frac{\sqrt{6}}{2}t^2\mathbf{j} + t^3\mathbf{k}$.

(a) $\sqrt{6 + 72t^2}$ (b) $\sqrt{6}$ (c) $6t$

(d) $2\sqrt{3\sqrt{6}t}$ (e) None of these

**7.** Calculate the curvature of the curve given by $\mathbf{r}(t) = 2t\mathbf{i} + t^2\mathbf{j} + \ln t\mathbf{k}$.

(a) $\dfrac{2t}{(2t^2 + 1)^2}$ (b) 2 (c) $\dfrac{2}{t^5(2t^2 + 1)^2}$

(d) $\dfrac{4}{81}$ (e) None of these

**8.** Find the arc length of the curve $C$ given by $\mathbf{r}(t) = t\mathbf{i} + \dfrac{\sqrt{6}}{2}t^2\mathbf{j} + t^3\mathbf{k}, \quad -1 \le t \le 1$.

(a) 8 (b) 4 (c) 0

(d) 2 (e) None of these

**9.** Identify the plane curve given by the vector-valued function $\mathbf{r}(t) = 3t\mathbf{i} + (t + 1)\mathbf{j}$.

(a) Ellipse (b) Hyperbola (c) Line

(d) Parabola (e) None of these

**10.** Find the limit: $\displaystyle\lim_{t \to 1}\left(\frac{t^2 + 1}{t - 1}\mathbf{i} + \frac{t^2 + 7t - 8}{t - 1}\mathbf{j} + \frac{t^2 - 1}{t + 1}\mathbf{k}\right)$.

(a) $9\mathbf{j}$ (b) $\mathbf{i}$ (c) $2\mathbf{i} + 9\mathbf{j}$

(d) Does not exist (e) None of these

**Test Form B** **Name** ______________ **Date** ______________

**Chapter 11** **Class** ______________ **Section** ______________

1. Find the domain of the vector-valued function $\mathbf{r}(t) = 3t\mathbf{i} - \sqrt{1 - t^2}\mathbf{j} + 4\mathbf{k}$.

(a) $(-\infty, \infty)$ (b) $(-\infty, -1] \cup [1, \infty)$ (c) $-1 \le t \le 1$

(d) $-1 < t < 1$ (e) None of these

2. Find the acceleration vector of a particle that moves along the curve $C$ given by $\mathbf{r}(t) = (\ln t)\mathbf{i} + \frac{1}{t}\mathbf{j} + t^3\mathbf{k}$.

(a) $\frac{1}{t}\mathbf{i} - \frac{1}{t^2}\mathbf{j} + 3t^2\mathbf{k}$ (b) $-\frac{1}{t^2}\mathbf{i} + \frac{2}{t^3}\mathbf{j} + 6t\mathbf{k}$ (c) $\frac{1}{t}\mathbf{i} + (\ln t)\mathbf{j} + 3t^2\mathbf{k}$

(d) $-t\mathbf{i} + \mathbf{j} + 6t^4\,\mathbf{k}$ (e) None of these

3. Find the velocity vector for an object having $\mathbf{a}(t) = e^t\mathbf{j} - 32\mathbf{k}$, if $\mathbf{v}(0) = 3\mathbf{i} - 2\mathbf{j} + \mathbf{k}$.

(a) $e^t\mathbf{j} - 32\mathbf{k} + C$ (b) $3\mathbf{i} + (e^t - 2)\mathbf{j} + (-32t + 1)\mathbf{k}$

(c) $3\mathbf{i} + \left(\frac{e^{t+1}}{t+1} - 3\right)\mathbf{j} + (-32t + 1)\mathbf{k}$ (d) $3\mathbf{i} + (e^t - 3)\mathbf{j} + (-32t + 1)\mathbf{k}$

(e) None of these

4. Let $\mathbf{T}(t) = \frac{1}{\sqrt{10 + 4t^2}}(3\mathbf{i} - \mathbf{j} + 2t\mathbf{k})$. Find $\frac{d\mathbf{T}}{dt}$.

(a) $\frac{1}{\sqrt{10 + 4t^2}}(2\mathbf{k})$ (b) $\frac{4}{(10 + 4t^2)^{3/2}}(-3t\mathbf{i} + t\mathbf{j} + 5\mathbf{k})$

(c) $\frac{-4t}{(10 + 4t^2)^{3/2}}(2\mathbf{k})$ (d) $\frac{-4t}{(10 + 4t^2)^{3/2}}(-3t\mathbf{i} + t\mathbf{j} + 5\mathbf{k})$

(e) None of these

5. Let $\mathbf{r}(t) = 3\cos t\mathbf{i} + 3\sin t\mathbf{j} + 2t\mathbf{k}$. Calculate the principal unit normal vector.

(a) $-\cos t\mathbf{i} - \sin t\mathbf{j}$ (b) $-\frac{3}{\sqrt{13}}\sin t\mathbf{i} + \frac{3}{\sqrt{13}}\cos t\mathbf{j} + \frac{2}{\sqrt{13}}\mathbf{k}$

(c) $\frac{1}{\sqrt{18 + 4t^2}}(3\cos t\mathbf{i} + 3\sin t\mathbf{j} + 2t\mathbf{k})$ (d) $-\frac{1}{\sqrt{2}}\cos t\mathbf{i} - \frac{1}{\sqrt{2}}\sin t\mathbf{j}$

(e) None of these

6. Find the tangential component of acceleration for the curve given by $\mathbf{r}(t) = t\mathbf{i} + \frac{\sqrt{6}}{2}t^2\mathbf{j} + t^3\mathbf{k}$.

(a) $6t$ (b) $\sqrt{6 + 72t^2}$ (c) $\sqrt{6}$

(d) $2\sqrt{3\sqrt{6}t}$ (e) None of these

7. Calculate the curvature of the curve given by $\mathbf{r}(t) = \frac{1}{3}t^3\mathbf{i} + \frac{\sqrt{2}}{2}t^2\mathbf{j} + t\mathbf{k}$ at the point where $t = 2$.

(a) $\frac{\sqrt{2}}{(t^2 + 1)^2}$

(b) $\frac{4}{5}$

(c) $\frac{\sqrt{2}}{25}$

(d) $\frac{2\sqrt{13}}{125}$

(e) None of these

8. Find the arc length of the curve given by $\mathbf{r}(t) = 2t\mathbf{i} + t^2\mathbf{j} + (\ln t)\mathbf{k},\ 1 \le t \le 2$.

(a) 3

(b) $\frac{1}{3}$

(c) $3 + \ln 2$

(d) $\frac{3}{2}$

(e) None of these

9. Identify the plane curve given by the vector-valued function $\mathbf{r}(t) = 3t^2\mathbf{i} + (1 - t)\mathbf{j}$.

(a) Parabola

(b) Hyperbola

(c) Line

(d) Circle

(e) None of these

10. Find the limit: $\lim_{t \to 0} \left( \cos t\mathbf{i} + \frac{2 \sin t}{t}\mathbf{j} + t^3\mathbf{k} \right)$.

(a) 0

(b) 3

(c) $\mathbf{i}$

(d) $\mathbf{i} + 2\mathbf{j}$

(e) None of these

**Test Form C** **Name** ______________ **Date** __________

**Chapter 11** **Class** ______________ **Section** __________

**A graphing calculator/utility is recommended for this test.**

**1.** Use a graphing utility to graph the vector-valued function $\mathbf{r}(t) = \sqrt{8 - t^3}\,\mathbf{i} + t^2\mathbf{j}$ and give the orientation of the curve.

(a)

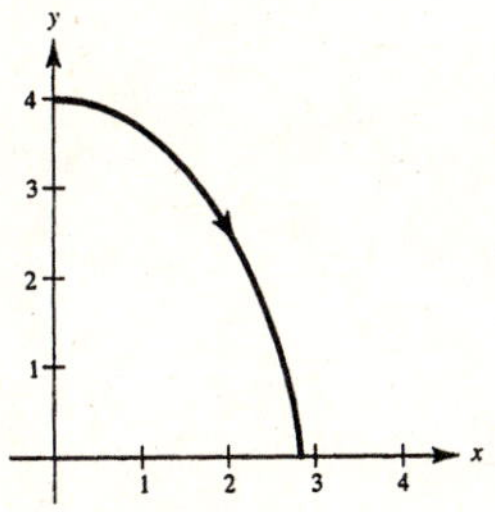


(b)

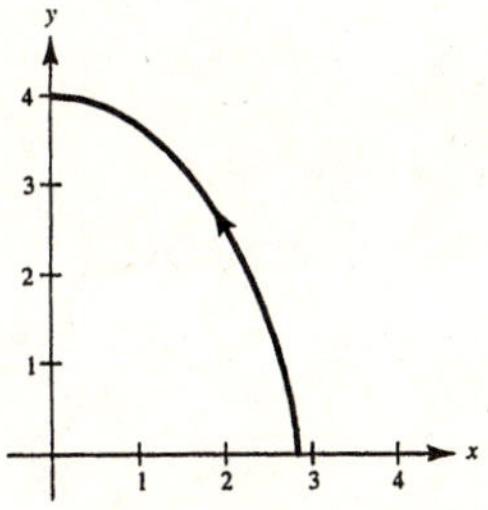


(c)

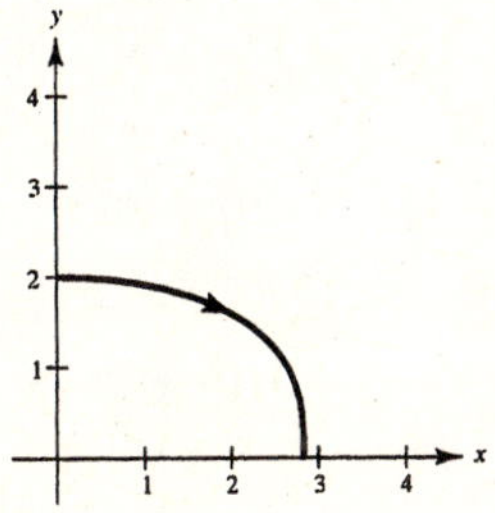


(d)

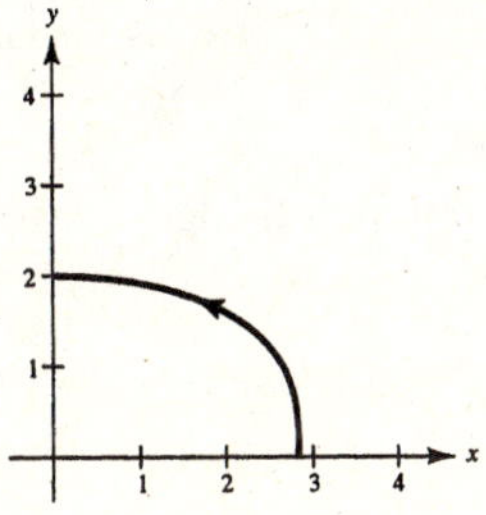


(e) None of these

**2.** Evaluate the limit: $\lim_{t \to 3} \left( \frac{t-3}{t^2+9}\mathbf{i} + \frac{t+3}{t^2+9}\mathbf{j} + \frac{t-3}{t^2-9}\mathbf{k} \right)$.

(a) $-\frac{1}{6}\mathbf{i} + \frac{1}{3}\mathbf{j} + \frac{1}{6}\mathbf{k}$ (b) $\frac{1}{3}\mathbf{j}$ (c) 0

(d) $\frac{1}{3}\mathbf{j} + \frac{1}{6}\mathbf{k}$ (e) None of these

**3.** Calculate $\mathbf{r}'(t)$ for the vector-valued function $\mathbf{r}(t) = \frac{1}{\sqrt{10 + 4t^2}}(3\mathbf{i} - \mathbf{j} + 2t\mathbf{k})$.

(a) $\frac{4}{(10 + 4t^2)^{3/2}}(-3t\mathbf{i} + t\mathbf{j} + 5\mathbf{k})$ (b) $\frac{1}{\sqrt{10 + 4t^2}}(2\mathbf{k})$

(c) $\frac{4t}{(10 + 4t^2)^{3/2}}(3\mathbf{i} - \mathbf{j} + 2t\mathbf{k})$ (d) $\frac{4t}{(10 + 4t^2)^{3/2}}(2\mathbf{k})$

(e) None of these

4. Find the points on the plane curve represented by $\mathbf{r}(t) = 2\cos^3 t\mathbf{i} + 2\sin^3 t\mathbf{j}$ at which the curve is not smooth.

(a) $(2, 0), (0, 2), (-2, 0), (0, -2)$
(b) $(1, 0), (0, 1), (-1, 0), (0, -1)$
(c) $(2, 0)$ and $(-2, 0)$
(d) $\frac{n\pi}{2}, n = \pm 1, \pm 2, \pm 3, \ldots$
(e) None of these

5. Find the speed of an object having the position vector $\mathbf{r}(t) = \sin t\mathbf{i} + \sqrt{2}\cos t\mathbf{j} + \sin t\mathbf{k}$.

(a) $\cos t\mathbf{i} - \sqrt{2}\sin t\mathbf{j} + \cos t\mathbf{k}$
(b) $2$
(c) $-\sin t\mathbf{i} - \sqrt{2}\cos t\mathbf{j} - \sin t\mathbf{k}$
(d) $\sqrt{2}$
(e) None of these

6. A golf ball is hit off the ground at an angle of $\pi/6$ and travels 400 feet. How long was it in the air?

(a) $\frac{25}{\sqrt{3}}$ sec
(b) $\frac{50}{\sqrt{3}}$ sec
(c) $\frac{10}{3^{1/4}}$ sec
(d) $\frac{5}{3^{1/4}}$ sec
(e) None of these

7. Find a symmetric representation of the tangent line to the space curve given by $\mathbf{r}(t) = t\mathbf{i} + t^2\mathbf{j} + t^3\mathbf{k}$ at the point $(2, 4, 8)$.

(a) $\frac{x-2}{1} = \frac{y-4}{2} = \frac{z-8}{3}$
(b) $\frac{x-2}{1} = \frac{y-4}{4} = \frac{z-8}{12}$
(c) $\frac{x-2}{2} = \frac{y-4}{4} = \frac{z-8}{8}$
(d) $x = 2 + t,\ y = 4 + 4t,\ z = 8 + 12t$
(e) None of these

8. Find the principal unit normal vector for the curve represented by $\mathbf{r}(t) = t^2\mathbf{i} + t\mathbf{j}$ when $t = -1$.

(a) $\frac{1}{\sqrt{5}}(\mathbf{i} - 2\mathbf{j})$
(b) $\frac{1}{\sqrt{5}}(\mathbf{i} + 2\mathbf{j})$
(c) $\frac{1}{\sqrt{5}}(2\mathbf{i} + \mathbf{j})$
(d) $\frac{1}{\sqrt{5}}(2\mathbf{i} - \mathbf{j})$
(e) None of these

**9.** Find the curvature of the plane curve given by $\mathbf{r}(t) = 3\cos t\mathbf{i} + 3\sin t\mathbf{j}$ at the point $(\sqrt{2}, \sqrt{7})$.

(a) $\frac{1}{9}$ (b) 3 (c) $\frac{1}{2\sqrt{3}}$

(d) $\frac{1}{3}$ (e) None of these

**10.** Use the integration capabilities of a graphing utility to approximate the arc length of the space curve given by $\mathbf{r}(t) = t^3\mathbf{i} + t^2\mathbf{j} + t\mathbf{k}$ from the point $(1, 1, 1)$ to the point $(27, 9, 3)$. Round your answer to three decimal places.

(a) 19,699.420 (b) 27.389 (c) 733.409

(d) 8.257 (e) None of these

**Test Form D** **Name** ________________ **Date** ____________

**Chapter 11** **Class** ________________ **Section** ____________

1. Find the domain of the vector-valued function $\mathbf{r}(t) = 2t\mathbf{i} + t^2\mathbf{j} + \ln t\mathbf{k}$.

2. Find the limit: $\lim_{t \to 1} \left[ \frac{\sin(t - 1)}{t - 1}\mathbf{i} + \frac{t + 3}{t - 2}\mathbf{j} + \cos \pi t\mathbf{k} \right]$.

3. Let $\mathbf{r}(t) = t \sin t\mathbf{i} + t \cos t\mathbf{j} + t\mathbf{k}$. Find $\mathbf{r}'(\pi)$.

4. Calculate the speed of the object having the position vector $\mathbf{r}(t) = \sin t\mathbf{i} + \cos t\mathbf{j} - 16t^2\mathbf{k}$ at the point where $t = 1$.

5. Let $\mathbf{r}(t) = 2(1 - \cos t)\mathbf{i} + 2(t - \sin t)\mathbf{j}$. Find values of $t$ for which $\mathbf{v}(t) = 0$.

6. Calculate $\mathbf{r}(t)$ if $\mathbf{a}(t) = -32\mathbf{k}$, $\mathbf{v}(0) = 3\mathbf{i} - 2\mathbf{j} + \mathbf{k}$ and $\mathbf{r}(0) = 5\mathbf{j} + 2\mathbf{k}$.

7. Find the arc length of the curve $C$ given by $\mathbf{r}(t) = 3t\mathbf{i} + \sqrt{3}t^2\mathbf{j} + \frac{2}{3}t^3\mathbf{k}, 0 \leq t \leq 1$.

8. Show that the helix given by $\mathbf{r}(t) = \cos t\mathbf{i} + \sin t\mathbf{j} + t\mathbf{k}$ has constant curvature.

9. Let $\mathbf{r}(t) = e^{-t}(\mathbf{i} + \mathbf{j} + \mathbf{k})$. Calculate the unit tangent vector.

10. Find the normal component of acceleration for the curve given by $\mathbf{r}(t) = 3t\mathbf{i} + t^3\mathbf{j} + t\mathbf{k}$.

**Test Form E** **Name** ____________________ **Date** ____________

**Chapter 11** **Class** ____________________ **Section** ____________

**A graphing calculator/utility is recommended for this test.**

**1.** Use a graphing utility to graph the vector-valued function $\mathbf{r}(t) = e^t\mathbf{i} + (t + 1)\mathbf{j}$ and indicate the orientation of the curve.

**2.** Evaluate the limit: $\lim_{x\to 0}\left[\left(\frac{e^{2t}}{t^2 - 1}\right)\mathbf{i} + \left(\frac{1 - \cos t}{t}\right)\mathbf{j} + \sqrt{4 - t^2}\,\mathbf{k}.\right]$

**3.** Calculate $\mathbf{r}'(t)$ for $\mathbf{r}(t) = \frac{1}{\sqrt{9t^2 + 5}}\langle 3t, 1, -2\rangle$.

**4.** Find the open intervals on which the plane curve given by $\mathbf{r}(t) = 2(t - \sin t)\mathbf{i} + 2(1 - \cos t)\mathbf{j}$ is smooth.

**5.** Find the velocity, speed and acceleration of the vector-valued function $\mathbf{r}(t) = \cos t\mathbf{i} + \sin t\mathbf{j} - 16t^2\mathbf{k}$ at $t = \frac{\pi}{4}$.

**6.** A projectile is fired from ground level at an angle of elevation of 30° with an initial velocity of 75 feet per second.

**a.** Write the vector-valued function for the path of the projectile.

**b.** Use a graphing utility to graph the path of the projectile.

**c.** Use the graph to approximate the maximum height and the range of the projectile.

**7.** Consider the space curve represented by the vector-valued function $\mathbf{r}(t) = t\mathbf{i} + t^3\mathbf{j} + 3t\mathbf{k}$.

**a.** Find the unit tangent vector at $t = 2$.

**b.** Find a set of parametric equations for the tangent line at the point corresponding to $t = 2$.

**8.** Let $\mathbf{r}(t) = t\mathbf{i} + t^2\mathbf{j}$ represent a plane curve.

**a.** Find $\mathbf{T}(t)$, $\mathbf{T}(1)$, and $\mathbf{N}(1)$.

**b.** Sketch the graph of the plane curve represented by $\mathbf{r}(t)$; at the point corresponding to $t = 1$, sketch the vectors $\mathbf{T}(1)$ and $\mathbf{N}(1)$.

**9.** Find the curvature at the point where $t = 1$ of the curve given by $\mathbf{r}(t) = t\mathbf{i} + \frac{1}{2}t^2\mathbf{j} + \frac{1}{3}t^3\mathbf{k}$.

**10.** Use the integration capabilities of a graphing utility to approximate the arc length of the space curve given by $\mathbf{r}(t) = t\mathbf{i} + t^3\mathbf{j} + 3t\mathbf{k}$ from the point (1, 1, 3) to the point (2, 8, 6). Round your answer to three decimal places.

**Test Form A** **Name** ______________ **Date** ______________

**Chapter 12** **Class** ______________ **Section** ______________

**1.** Find $f_x(x, y)$ for $f(x, y) = e^{xy}(\cos x \sin y)$.

(a) $-ye^{xy}(\sin x \sin y)$ (b) $e^{xy}(\sin y)(y \cos x - \sin x)$ (c) $e^{xy}(\sin x \sin y)$

(d) $e^{xy}(\sin y)(\cos x - \sin x)$ (e) None of these

**2.** Find $\dfrac{\partial z}{\partial y}$ for $x^2 - (y + z)^2 = 0$.

(a) $-2y - 2z$ (b) $2x - 2y - 2z$ (c) $\dfrac{x}{y + z}$

(d) $-1$ (e) None of these

**3.** Find $f_{xy}(x, y)$ for $f(x, y) = \dfrac{4x^2}{y} + \dfrac{y^2}{2x}$.

(a) $-\dfrac{8x}{y^2} - \dfrac{y}{x^2}$ (b) $8x - y$ (c) $8x + y$

(d) $\dfrac{16x^3y - 8x^4 + 2xy^3 - y^4}{2x^2y^2}$ (e) None of these

**4.** The volume of a right circular cone is given by $V = \dfrac{\pi r^2 h}{3}$. The radius of the base and the height are measured to be 12 inches and 16 inches respectively. The possible error in each measurement is 0.03 inch. Use differentials to approximate the maximum possible error in the computed volume.

(a) $5.28\ \pi$ in.$^3$ (b) $0.24\ \pi$ in.$^3$ (c) $0.27$ in.$^3$

(d) $0.000027\ \pi$ in.$^3$ (e) None of these

**5.** Let $f(x, y, z) = z^2e^{xy}$. Find $\nabla f$ at the point $(-1, 0, 3)$.

(a) $9\mathbf{i} - 9\mathbf{j} + 6\mathbf{k}$ (b) $9\mathbf{i} + 9\mathbf{j} + 6\mathbf{k}$ (c) $-9\mathbf{j} + 6\mathbf{k}$

(d) $yz^2e^{xy}\mathbf{i} + xz^2e^{xy}\mathbf{j} + 2ze^{xy}\mathbf{k}$ (e) None of these

**6.** Calculate the directional derivative of $f(x, y, z) = z^2e^{xy}$ at the point $(-1, 0, 3)$ in the direction of the vector $\mathbf{v} = 3\mathbf{i} + \mathbf{j} - 5\mathbf{k}$.

(a) $-\dfrac{66\sqrt{35}}{35}$ (b) $-\dfrac{39\sqrt{35}}{35}$ (c) $-39$

(d) $-9\mathbf{j} + 6\mathbf{k}$ (e) None of these

**7.** For the function $f(x, y) = x^2e^y$, find the maximum value of the directional derivative at the point $(-2, 0)$.

(a) $-4\mathbf{i} + 4\mathbf{j}$ (b) $0$ (c) $4\sqrt{2}$

(d) $8$ (e) None of these

**8.** Find an equation for the tangent plane to the surface given by $f(x, y) = \frac{1}{7}(4x^2 - 2y^2)$ at the point $(-2, -1, 2)$.

(a) $16x - 4y + 28 = 0$

(b) $16x - 4y - 7z - 42 = 0$

(c) $16x - 4y + 7z = 0$

(d) $16x - 4y + 7z + 14 = 0$

(e) None of these

**9.** Find the minimum value of the function $f(x, y) = x^2 + y^2 - xy - 4$.

(a) 0

(b) 2

(c) $-12$

(d) $-4$

(e) None of these

**10.** Let $f(x, y, z) = 4x^2 + y^2 + 5z^2$. Use Lagrange Multipliers to find the point on the plane $2x + 3y + 4z = 12$ at which $f(x, y, z)$ has its least value.

(a) $(1, 2, 1)$

(b) $\left(\frac{5}{11}, \frac{30}{11}, \frac{8}{11}\right)$

(c) $\frac{120}{11}$

(d) $\left(\frac{4}{7}, -\frac{2}{7}, \frac{82}{7}\right)$

(e) None of these

**Test Form B** **Name** ______________ **Date** __________

**Chapter 12** **Class** ______________ **Section** __________

**1.** Find $f_y(x, y)$ for $f(x, y) = e^{xy}(\cos x \sin y)$.

(a) $xe^{xy}(\cos x \cos y)$ (b) $e^{xy}(\cos x \cos y)$ (c) $e^{xy}(\cos x)(\cos y + x \sin y)$

(d) $e^{xy}(\cos x)(\cos y + \sin y)$ (e) None of these

**2.** Find $\frac{\partial z}{\partial x}$ for $x^2 - (y + z)^2 = 0$.

(a) $-2y - 2x$ (b) $2x - 2y - 2z$ (c) $\frac{x}{y + z}$

(d) $-1$ (e) None of these

**3.** Find $f_{xx}(x, y)$ for $f(x, y) = \frac{4x^2}{y} + \frac{y^2}{2x}$.

(a) $4x + \frac{x}{2}$ (b) $\frac{8}{y} + \frac{y^2}{x^3}$ (c) $\frac{8}{y}$

(d) $\frac{8x}{y} + \frac{y^2}{2}$ (e) None of these

**4.** Let $f(x, y) = \sqrt{x^2 + y^2}$. Use a total differential to approximate the change in $f(x, y)$ as $(x, y)$ varies from the point $(3, 4)$ to the point $(3.04, 3.98)$.

(a) 0.4 (b) 0.04 (c) 0.08

(d) 0.008 (e) None of these

**5.** Let $f(x, y) = x^2e^y$. Find $\nabla f$ at the point $(-2, 0)$.

(a) $-4\mathbf{i} + 4\mathbf{j}$ (b) 4 (c) $-4\mathbf{i} - 4\mathbf{j}$

(d) $(\nabla x)^2e^{\nabla y}$ (e) None of these

**6.** Calculate the directional derivative of $f(x, y) = x^2e^y$ at the point $(-2, 0)$ in the direction of the vector $\mathbf{v} = 3\mathbf{i} + \mathbf{j}$.

(a) $-8$ (b) $-12\mathbf{i} + 4\mathbf{j}$ (c) $\frac{8\sqrt{10}}{5}$

(d) $-\frac{4\sqrt{10}}{5}$ (e) None of these

**7.** For the function $f(x, y, z) = z^2e^{xy}$, find the maximum value of the directional derivative at the point $(-1, 0, 3)$.

(a) 9 (b) $3\sqrt{13}$ (c) 0

(d) $-9\mathbf{j} + 6\mathbf{k}$ (e) None of these

**8.** Find an equation for the tangent plane to the surface given by $f(x, y) = x^2y$ at the point $(2, -1, -4)$.

(a) $4x - 4y + z = 0$ (b) $4x - 4y + z - 8 = 0$ (c) $4x - 4y + z + 16 = 0$

(d) $4x - 4y - z - 8 = 0$ (e) None of these

**9.** Find the saddle point for $f(x, y) = x^2 - y^2 - 2x - 6y - 3$.

(a) $(1, -3, 0)$ (b) $(-1, 3, 5)$ (c) $(-1, 3, 0)$

(d) $(1, -3, 5)$ (e) None of these

**10.** Use Lagrange Multipliers to maximize $f(x, y, z) = 4x^2 + y^2 + z^2$ with the constraint that $2x - y + z = 4$.

(a) $\frac{16}{3}$ (b) $\left(\frac{2}{3}, -\frac{4}{3}, \frac{4}{3}\right)$ (c) 36

(d) $\frac{16}{9}$ (e) None of these

**Test Form C** **Name** ______________ **Date** ______________

**Chapter 12** **Class** ______________ **Section** ______________

1. Describe the domain of $f(x, y, z) = \sqrt{1 - x^2 - \frac{y^2}{9} - \frac{z^2}{4}}$.

(a) $\left\{(x, y, z): x^2 + \frac{y^2}{9} + \frac{z^2}{4} \geq 1\right\}$

(b) $\left\{(x, y, z): x^2 + \frac{y^2}{9} + \frac{z^2}{4} < 1\right\}$

(c) $\left\{(x, y, z): x^2 + \frac{y^2}{9} + \frac{z^2}{4} \leq 1\right\}$

(d) All real numbers

(e) None of thcsc

2. Find $\lim_{\Delta x \to 0} \frac{f(x + \Delta x, y) - f(x, y)}{\Delta x}$ for $f(x, y) = x^2 + y^2 - 2x$.

(a) $2x + \Delta x - 2$

(b) $2x - 2$

(c) $2x + 2y - 2$

(d) 1

(e) None of these

3. Discuss the continuity of the function $f(x, y) = \sqrt{x - y}$.

(a) $f$ is continuous at each point in the $xy$-plane where $y \leq x$.

(b) $f$ is continuous at each point in the $xy$-plane.

(c) $f$ is continuous at each point in the $xy$-plane where $y \geq x$.

(d) $f$ is continuous at each point in the $xy$-plane where $y = x$.

(e) None of these

4. Find $f_z\left(3, \sqrt{11}, -4\right)$ for $f(x, y, z) = \sqrt{x^2 + y^2 + z^2}$.

(a) $\frac{1}{2}$

(b) $-\frac{1}{9}$

(c) $-\frac{2}{3}$

(d) $\frac{1}{12}$

(e) None of these

5. The height and radius of a right circular cylinder are approximately 10 inches and 3 inches, respectively. The maximum error in the measurement of the height is $\pm 0.02$ inches and in the measurement of the radius is $\pm 0.01$ inches. Use the total differential to estimate the relative error in the calculated volume of the cylinder.

(a) $0.78\pi$

(b) 76%

(c) 0.87%

(d) 0.3%

(e) None of these

6. At a certain instant, the height of a right circular cone is 30 inches and is increasing at a rate of 2 inches per second. At the same instant, the radius of the base is 20 inches and is increasing at the rate of 1 inch per second. At what rate is the volume increasing at that instant? $\left(V = \frac{1}{3}\pi r^2 h\right)$

(a) $\frac{2000}{3}\pi$ in.$^3$/sec

(b) $\frac{80}{3}\pi$ in.$^3$/sec

(c) $400\pi$ in.$^3$/sec

(d) $4000\pi$ in.$^3$/sec

(e) None of these

**7.** Find the first partial derivative of $z$ with respect to $x$: $x^3y + y^3z + z^3x = 11$.

(a) $-\dfrac{y^3 + 3xz^2}{3x^2y + z^3}$ (b) $y^3 + 3xz^2$ (c) $-\dfrac{3x^2y + z^3}{y^3 + 3xz^2}$

(d) $3x^2y + z^3$ (e) None of these

**8.** Let $f(x, y) = 3x^2 - 3xy - y^2$. Find the maximum value of the directional derivative of $f$ at the point $(1, 1)$.

(a) 4 (b) $\sqrt{10}$ (c) $-1$

(d) $\sqrt{34}$ (e) None of these

**9.** Find a unit vector normal to the surface given by $f(x, y) = x^2 - 3xy + y^2$ at the point where $x = y = 1$.

(a) $\mathbf{i} + \mathbf{j} + \mathbf{k}$ (b) $-\mathbf{i} - \mathbf{j}$ (c) $\dfrac{1}{\sqrt{3}}\mathbf{i} + \dfrac{1}{\sqrt{3}}\mathbf{j} + \dfrac{1}{\sqrt{3}}\mathbf{k}$

(d) $-\dfrac{1}{\sqrt{2}}\mathbf{i} - \dfrac{1}{\sqrt{2}}\mathbf{j}$ (e) None of these

**10.** Use the Second Partials Test to determine the nature of the function $f(x, y)$ at the point $(x_0, y_0)$ if $f_{xx}(x_0, y_0) = 2$, $f_{yy}(x_0, y_0) = 8$, and $f_{xy}(x_0, y_0) = 4$. Assume $f_x(x_0, y_0) = f_y(x_0, y_0) = 0$.

(a) Relative maximum (b) Relative minimum

(c) Saddle point (d) Test is inconclusive

**11.** Use Lagrange multipliers to find the point on the plane $2x + 3y - z = 1$ which is closest to the origin.

(a) $\left(\frac{1}{7}, \frac{3}{14}, -\frac{1}{14}\right)$ (b) $\left(-\frac{1}{7}, -\frac{3}{14}, \frac{1}{14}\right)$ (c) $\left(\frac{2}{9}, \frac{1}{3}, \frac{1}{9}\right)$

(d) $(0, 0, -1)$ (e) None of these

**Test Form D** **Name** ____________ **Date** ____________

**Chapter 12** **Class** ____________ **Section** ____________

1. Show that $f(x, y) = \ln(x^2 + y^2)$ satisfies Laplaces' equation: $\dfrac{\partial^2 f}{\partial x^2} + \dfrac{\partial^2 f}{\partial y^2} = 0$.

2. Find $\dfrac{\partial w}{\partial x}$ for $w^2 + w\sin(xyz) = 0$.

3. Use the appropriate Chain Rule to find $\dfrac{\partial z}{\partial u}$ for $z = \cos x \sin y$, $x = u - v$, and $y = u^2 + v^2$. Write your answer as a function of $u$ and $v$.

4. Let $f(x, y) = \dfrac{x}{x^2 + y^2}$. Use a total differential to approximate the change in $f(x, y)$ as $(x, y)$ varies from the point $(1, 2)$ to the point $(0.98, 2.01)$.

5. Find the directional derivative of $f(x, y) = 3x^2y$ at the point $(1, 2)$ in the direction of $\mathbf{v} = 3\mathbf{i} + 4\mathbf{j}$.

6. For the function $f(x, y, z) = e^{xyz}$, find the maximum value of the directional derivative at the point $(1, 2, 3)$.

7. Find a set of symmetric equations for the normal line to the surface $z = \frac{1}{7}(4x^2 - 2y^2)$ at the point $(-2, -1)$.

8. Determine the relative extrema of $f(x, y) = x^2 + x - 3xy + y^3 - 5$.

9. Use Lagrange Multipliers to find three positive numbers whose sum is 33 and whose product is a maximum.

10. Find the least squares regression line for the points $(-2, -1)$, $(-1, 1)$, $(0, 1)$, $(1, 2)$, and $(2, 2)$.

**Test Form E** **Name** ______________ **Date** ______________

**Chapter 12** **Class** ______________ **Section** ______________

1. Find the domain of the function $f(x, y) = \ln(x^2 + y^2 - 4)$.

2. Find the limit: $\displaystyle\lim_{(x,y)\to(-2,1)} \frac{x^2 - 4y^2}{x + 2y}$.

3. Discuss the continuity of the function $f(x, y, z) = \dfrac{5}{x + 2y + 3z - 6}$.

4. Find $f_y(-2, -1, 3)$ for $f(x, y, z) = \dfrac{x - z}{y}$.

5. The height and radius of a right circular cylinder are approximately 15 centimeters and 8 centimeters, respectively. The maximum error in each measurement is $\pm 0.02$ centimeters.

   a. Find the approximate volume of the cylinder.

   b. Use the total differential to estimate the propagated error in the calculated volume of the cylinder.

   c. Use the total differential to estimate the relative error in the calculated volume of the cylinder.

6. The length, width, and height of a rectangular chamber are changing at the rate of 3 feet per minute, 2 feet per minute and $\frac{1}{2}$ foot per minute, respectively. Find the rate at which the volume is changing at the instant the length is 10 feet, the width is 6 feet, and the height is 4 feet.

7. Find the first partial derivative of $z$ with respect to $x$: $x^2 - (y + z)^2 = 8$.

8. Let $f(x, y) = \ln(x^2 + y^2 + 1) + e^{2xy}$.

   a. Find the gradient of $f$ at the point $(0, -2)$.

   b. Find the directional derivative of $f$ at the point $(0, -2)$ in the direction of the vector $\mathbf{v} = 5\mathbf{i} - 12\mathbf{j}$.

   c. Find the maximum value of the directional derivative at the point $(0, -2)$.

9. Consider the surface given by $x^2 + 4y^2 - 10z = 0$.

   a. Identify the given quadric surface.

   b. Find an equation for the tangent plane to the surface at the point $(2, -2, 2)$.

   c. Find symmetric equations for the normal line to the surface at the point $(2, -2, 2)$.

**10.** Use the Second Partials Test to determine whether the function $f$ at the point $(x_0, y_0)$ has a relative maximum, a relative minimum, a saddle point, or if the test is inconclusive given that $f_{xx}(x_0, y_0) = 5$, $f_{yy}(x_0, y_0) = -2$, and $f_{xy}(x_0, y_0) = 1$.

**11.** Use Lagrange multipliers to find all points on the circle $x^2 + y^2 = 18$ at which the product $xy$ is a maximum.

**Test Form A** **Name** ______________ **Date** __________

**Chapter 13** **Class** ______________ **Section** __________

**1.** Evaluate the integral: $\int_0^{\pi}\int_0^{\cos y} x \sin y \, dx \, dy$.

(a) $-\frac{1}{3}$ (b) $\frac{1}{2}$ (c) $\frac{1}{3}$

(d) 0 (e) None of these

**2.** Evaluate the integral: $\int\int_R (x^2 + 4y) \, dA$ where $R$ is the region bounded by the graphs of $y = 2x$ and $y = x^2$.

(a) $\frac{77}{30}$ (b) $-\frac{152}{15}$ (c) $\frac{16}{3}$

(d) $\frac{152}{15}$ (e) None of these

**3.** Evaluate the integral: $\int_0^2\int_{y/2}^1 e^{x^2} \, dx \, dy$ by reversing the order of integration.

(a) $e - 1$ (b) $e + 1$ (c) $e^2$

(d) $e^4 - 1$ (e) None of these

**4.** Use a double integral to find the volume of the solid in the first octant bounded above by the plane $x + y + z = 4$ and below by the rectangle on the $xy$-plane: $\{(x, y) : 0 \le x \le 1, 0 \le y \le 2\}$.

(a) 2 (b) 4 (c) 8

(d) 5 (e) None of these

**5.** Use polar coordinates to find the volume of the solid under the plane $z = 4x$ and above the circle $x^2 + y^2 = 16$ in the $xy$-plane.

(a) $\frac{1024}{3}$ (b) $\frac{512}{3}$ (c) 64

(d) $\frac{256}{3}$ (e) None of these

**6.** Find the limits of integration for calculating the volume of the solid $Q$ enclosed by the graphs of $y = x^2$, $z = 0$ and $y + z = 2$ if $V = \int\int_Q\int dz \, dy \, dx$.

(a) $\int_{-\sqrt{2}}^{\sqrt{2}}\int_0^2\int_0^2 dz \, dy \, dx$ (b) $\int_{-\sqrt{2}}^{\sqrt{2}}\int_{x^2}^2\int_0^{2-y} dz \, dy \, dx$

(c) $\int_0^{\sqrt{2}}\int_{x^2}^2\int_0^2 dz \, dy \, dx$ (d) $\int_{-\sqrt{2}}^{\sqrt{2}}\int_{x^2}^2\int_0^2 dz \, dy \, dx$

(e) None of these

**7.** Use spherical coordinates to find the mass of the conical solid bounded by the graphs of $z = \sqrt{x^2 + y^2}$ and $z = 4$ if the density is $k\sqrt{x^2 + y^2}$.

(a) $\dfrac{256\pi k}{3}$ (b) $\dfrac{512\pi k}{3}$ (c) $\dfrac{128\pi k}{3}$

(d) $16\pi k(\pi - 2)$ (e) None of these

**8.** Evaluate the integral $\displaystyle\int_{-5}^{5}\int_{0}^{\sqrt{25-x^2}}\int_{0}^{1/(x^2+y^2)} \sqrt{x^2 + y^2}\, dz\, dy\, dx$ using cylindrical coordinates.

(a) $5\pi$ (b) $\pi \ln 5$ (c) $10\pi$

(d) $\dfrac{5\pi}{3}$ (e) None of these

**9.** Find the surface area for that portion of the surface $z = xy$ that is inside the cylinder $x^2 + y^2 = 1$.

(a) $\dfrac{4\pi\sqrt{2}}{3}$ (b) $\dfrac{3\pi}{2}$ (c) $\dfrac{2\pi}{3}(2\sqrt{2} - 1)$

(d) $2\pi$ (e) None of these

**10.** Suppose the transformation $T$ is defined by $x = \dfrac{u}{v}$ and $y = v$. Find the Jacobian, $\dfrac{\partial(x, y)}{\partial(u, v)}$, of $T$.

(a) $-\dfrac{u}{v^2}$ (b) 1 (c) $\dfrac{v + 1}{v}$

(d) $\dfrac{1}{v}$ (e) None of these

**Test Form B** **Name** ________________ **Date** __________

**Chapter 13** **Class** ________________ **Section** __________

1. Evaluate the integral: $\int_0^{\sqrt{\pi}} \int_{\pi/6}^{y^2} 2y \cos x \, dx \, dy$.

(a) $\frac{4-\pi}{2}$ (b) $-\frac{\pi}{2}$ (c) $\frac{\pi-4}{2}$

(d) $-\pi$ (e) None of these

2. Evaluate the integral $\int_R \int \frac{x}{\sqrt{1+y^2}} \, dA$ where $R$ is the region in the first quadrant bounded by the graphs of $y = x^2$, $y = 4$, and $x = 0$.

(a) $\frac{1}{2}\sqrt{17} - 1$ (b) $4\sqrt{17} - \frac{10\sqrt{5}}{3} + 1$ (c) $68\sqrt{17}$

(d) $34\sqrt{17}$ (e) None of these

3. Evaluate the integral $\int_0^1 \int_{2x}^2 e^{y^2} \, dy \, dx$ by reversing the order of integration.

(a) 0 (b) $\frac{1}{4}(e^4 - 1)$ (c) $\frac{3e^4 + 1}{8}$

(d) $\frac{1}{4}e^4$ (e) None of these

4. Use a double integral to find the volume of the solid in the first octant bounded above by the plane $z = 5 - 2y$ and below by the rectangle in the $xy$-plane: $\{(x, y) : 0 \le x \le 3,\ 0 \le y \le 2\}$.

(a) 12 (b) 6 (c) 18

(d) 9 (e) None of these

5. Use polar coordinates to evaluate $\int_R \int \sqrt{x^2 + y^2} \, dA$ where $R$ is the region in the $xy$-plane enclosed by the graphs of $x^2 + y^2 = 9$.

(a) $\frac{512\sqrt{2} - 40}{15}$ (b) $6\pi$ (c) $9\pi$

(d) $18\pi$ (e) None of these

**6.** Find the limits of integration for calculating the volume of the solid $Q$ enclosed by the graph of $y^2 = x$, $z = 0$ and $x + z = 1$ if $V = \iiint_Q dz\, dy\, dx$.

(a) $\int_0^1 \int_0^1 \int_0^1 dz\, dy\, dx$

(b) $\int_0^1 \int_0^{\sqrt{x}} \int_0^{1-x} dz\, dy\, dx$

(c) $\int_0^1 \int_{-\sqrt{x}}^{\sqrt{x}} \int_0^{1-x} dz\, dy\, dx$

(d) $\int_0^{y^2} \int_{-\sqrt{x}}^{\sqrt{x}} \int_0^{1-x} dz\, dy\, dx$

(e) None of these

**7.** Use spherical coordinates to find the mass of the sphere $x^2 + y^2 + z^2 = 9$ if its density at a point is proportional to the distance from the point to the origin. $\left(\rho = k\sqrt{x^2 + y^2 + z^2}\right)$

(a) $\dfrac{81\pi k}{2}$

(b) $81\pi k$

(c) $9\pi^2 k$

(d) $18\pi^2 k$

(e) None of these

**8.** Evaluate the integral $\int_{-2}^{2} \int_{-\sqrt{4-x^2}}^{\sqrt{4-x^2}} \int_0^{(x^2+y^2)/2} (x^2 + y^2)\, dz\, dy\, dx$ using cylindrical coordinates.

(a) $\dfrac{32\pi}{5}$

(b) $\dfrac{32\pi}{3}$

(c) $\dfrac{16\pi}{3}$

(d) $16$

(e) None of these

**9.** Find the surface area of that portion of the paraboloid $z = 1 - x^2 - y^2$ that lies above the $xy$-plane.

(a) $\dfrac{\pi\left(5\sqrt{5} - 1\right)}{6}$

(b) $2\pi$

(c) $\dfrac{5\sqrt{5}\pi}{6}$

(d) $\dfrac{\pi}{2}$

(e) None of these

**10.** Suppose the transformation $T$ is defined by $x = \dfrac{1}{5}(3u + 2v)$ and $y = \dfrac{1}{5}(2u - v)$. Find the Jacobian, $\dfrac{\partial(x, y)}{\partial(u, v)}$, of $T$.

(a) $-\dfrac{1}{25}$

(b) $\dfrac{6}{5}$

(c) $-\dfrac{1}{v}$

(d) $-\dfrac{7}{25}$

(e) None of these

**Test Form C** **Name** ______________ **Date** ______________

**Chapter 13** **Class** ______________ **Section** ______________

**A graphing calculator/utility is recommended for this test.**

1. Evaluate the integral: $\int_0^{\pi} \int_0^{x} \sin y \, dy \, dx$.

(a) $-\pi$ (b) $\pi$ (c) $0$

(d) $\pi + 1$ (e) None of these

2. Use an iterated integral to calculate the area of the region bounded by the graphs of $y = \dfrac{\ln x}{x}$, $y = x - 1$, and $x = 5$.

(a) $8 - \frac{1}{2}(\ln 5)^2$ (b) $7 - \frac{1}{2}(\ln 5)^2$ (c) $\dfrac{15 - (\ln 5)^2}{2}$

(d) $8 - \ln 5$ (e) None of these

3. Evaluate the integral: $\int_0^{1} \int_{2x}^{2} e^{y^2} \, dy \, dx$.

(a) $\frac{1}{4}e^4$ (b) $\frac{1}{4}(e^4 - 1)$ (c) $0$

(d) $-1$ (e) None of these

4. Use a double integral to calculate the volume of the solid under the surface $f(x, y) = x^2y^2$ and above the closed region bounded by the lines $y = 1, y = 2, x = 0$, and $x = y$.

(a) $\frac{7}{2}$ (b) $\frac{11}{6}$ (c) $\frac{16}{9}$

(d) $\frac{56}{3}$ (e) None of these

5. Find the volume of the solid bounded by the $xy$-plane, the cylinder $x^2 + y^2 = 9$ and the paraboloid $z = 2(x^2 + y^2)$.

(a) $6561\pi$ (b) $\frac{81}{2}\pi$ (c) $81\pi$

(d) $27\pi$ (e) None of these

6. A lamina has the shape of a closed region bounded by the graphs of $x + y = 3$, $3x + y = 3$, and $y = 0$, and has density $\rho(x, y) = 2xy$. Determine the iterated integral for the moment of inertia about the $y$-axis.

(a) $\int_0^{3} \int_{(3-y)/3}^{3-y} 2xy^3 \, dx \, dy$ (b) $\int_0^{3} \int_{(3-y)/3}^{3-y} 2x^3y \, dx \, dy$ (c) $\int_0^{3} \int_0^{3} 2x^3y \, dy \, dx$

(d) $\int_0^{3} \int_0^{3-y} 2x^3y \, dx \, dy$ (e) None of these

7. Use the integration capabilities of a graphing utility to approximate the surface area of that portion of the surface $z = x + x^2/2$ that lies over the square region in the $xy$-plane having vertices $(0, 0)$, $(1, 0)$, $(0, 1)$ and $(1, 1)$. Round your answer to three decimal places.

(a) 0.318 (b) 1.578 (c) 1.810

(d) 1.148 (e) None of these

8. Use spherical coordinates to calculate the volume of the solid bounded above by the sphere $x^2 + y^2 + z^2 = 16$ and below by the half-cone $z = \sqrt{3x^2 + 3y^2}$.

(a) $\frac{16\pi}{3}(2 - \sqrt{3})$ (b) $16\pi(2 - \sqrt{3})$ (c) $64\pi(2 - \sqrt{3})$

(d) $\frac{64\pi}{3}(2 - \sqrt{3})$ (e) None of these

9. Use cylindrical coordinates to calculate the volume of the solid bounded above by the plane $z = y$ and below by the paraboloid $x^2 + y^2 = z$.

(a) 0 (b) $\frac{\pi}{16}$ (c) $\frac{2\pi}{96}$

(d) $\frac{\pi}{32}$ (e) None of these

10. Let $R$ be the region bounded by the graphs of $x + y = 1$, $x + y = 2$, $2x - 3y = 2$, and $2x - 3y = 5$. Use the change of variables $x = \frac{1}{5}(3u + v)$, $y = \frac{1}{5}(2u - v)$, to evaluate the integral $\int_R\int 10\,dA$.

(a) 6 (b) 30 (c) $\frac{6}{5}$

(d) $\frac{1}{5}$ (e) None of these

**Test Form D** **Name** ______________ **Date** __________

**Chapter 13** **Class** ______________ **Section** __________

1. Evaluate the integral: $\int_{-1}^{2}\int_{0}^{1-x} (4 - y)\,dy\,dx$.

2. Evaluate the integral $\int_R\int xy\,dA$ where $R$ is the region bounded by the graphs of $y = \sqrt{x}$, $y = \frac{1}{2}x$, $x = 2$, and $x = 4$.

3. Evaluate the integral $\int_0^1\int_{\sqrt{y}}^{1} \sin \pi x^3\,dx\,dy$ by reversing the order of integration.

4. Find the volume inside the paraboloid $z = x^2 + y^2$ below the plane $z = 4$.

5. Evaluate the integral $\int_0^1\int_0^{\sqrt{1-x^2}} e^{-(x^2+y^2)}\,dy\,dx$ using polar coordinates.

6. Find the volume of the solid in the first octant bounded by the graphs of $z = 1 - y^2$, $y = 2x$, and $x = 3$.

7. Let $S$ be the solid in the first octant bounded by the cylinder $x^2 + y^2 = 4$ and $z = 4$. Use cylindrical coordinates to calculate the mass if the density at a point is proportional to the distance from the $yz$-plane ($\rho = kx$).

8. Evaluate the integral $\iiint_Q \frac{1}{\sqrt{x^2 + y^2 + z^2}}\,dV$ using spherical coordinates if $Q$ is the sphere $x^2 + y^2 + z^2 = 25$.

9. Find the surface area of that portion of the sphere $x^2 + y^2 + z^2 = 4$ that is above the $xy$-plane and within the cylinder $x^2 + y^2 = 1$.

10. Show that the Jacobian for the change of variables from rectangular coordinates to polar coordinates is $r$.

**Test Form E** **Name** ______________ **Date** ______________

**Chapter 13** **Class** ______________ **Section** ______________

**A graphing calculator/utility is recommended for this test.**

**1.** Evaluate the integral: $\int_0^1 \int_{3y}^3 e^x \, dx \, dy$.

**2.** Let $R$ be the region bounded by the graphs of $y = x - \sin x$, $y = \pi$ and $x = 0$.

**a.** Sketch the region $R$.

**b.** Use an iterated integral to calculate the area of the region $R$.

**3.** Evaluate the integral: $\int_0^1 \int_y^1 \sin x^2 \, dx \, dy$.

**4.** Use the integration capabilities of a graphing utility to approximate the volume of the solid below the surface given by $f(x, y) = \dfrac{x}{\sqrt{1 - y^2}}$ and above the region in the $xy$-plane bounded the graphs of $x = 0$, $y = \dfrac{1}{2}$, and $y = x$. Round your answers to three decimal places.

**5.** Consider the solid bounded above by the plane $z = 4x$ and below by the circle $x^2 + y^2 = 16$ in the $xy$-plane.

**a.** Write the double integral in rectangular coordinates to calculate the volume of the solid.

**b.** Write the double integral in polar coordinates to calculate the volume of the solid.

**c.** Evaluate part **a** or part **b**.

**6.** A lamina has the shape of a closed region bounded by the graphs of $x^2 + y^2 = 4$ and $x + y = 2$, and has density $\rho(x, y) = xy$. Write the iterated integral for the moment of inertia about the $y$-axis.

**7.** Use the integration capabilities of a graphing utility to approximate the surface area of that portion of the surface $z = e^x$ that lies over the region in the $xy$-plane bounded by the graphs of $y = 0$, $y = x$, and $x = 1$. Round your answer to three decimal places.

**8.** Let $Q$ be the sphere $x^2 + y^2 + z^2 = a^2$.

**a.** Use cylindrical coordinates to set up the integral to calculate the volume of $Q$.

**b.** Use spherical coordinates to set up the integral to calculate the volume of $Q$.

**9.** Consider the solid inside the surface $x^2 + y^2 + z^2 = 9$ and outside the surface $x^2 + y^2 + z^2 = 1$.

**a.** Use spherical coordinates to write the integral to calculate the volume of the solid.

**b.** Calculate the integral from part **a.**

**10.** Let $R$ be the region bounded by the graphs of $x + y = 1, x + y = 2, 2x - 3y = 2$, and $3x - 3y = 5$. Use the change of variables, $x = \frac{1}{5}(3u + v), y = \frac{1}{5}(2u - v)$, to evaluate the integral for $\int\int_R (2x - 3y)dA$.

**Test Form A** **Name** ______________ **Date** ______________

**Chapter 14** **Class** ______________ **Section** ______________

1. Let $\mathbf{F}(x, y, z) = (x^3 \ln z)\mathbf{i} + (xe^{-y})\mathbf{j} - (y^2 + 2z)\mathbf{k}$. Calculate the divergence of $\mathbf{F}$ at the point $(2, \ln 2, 1)$.

   (a) $-3$ (b) $-2 \ln 2\mathbf{i} + 8\mathbf{j} + \frac{1}{2}\mathbf{k}$ (c) $\frac{1}{2}$

   (d) $2$ (e) None of these

2. Let $\mathbf{F}(x, y, z) = \cos x\mathbf{i} + \sin y\mathbf{j} + e^{xy}\mathbf{k}$. Calculate **curl F** at the point $(1, 1, 1)$.

   (a) $\cos 1 - \sin 1$ (b) $e\mathbf{i} - e\mathbf{j}$ (c) $-\frac{\pi}{2}$

   (d) $e\mathbf{i} + e\mathbf{j}$ (e) None of these

3. Let $\mathbf{F}(x, y) = e^y\mathbf{i} + (xe^y + y)\mathbf{j}$. Find the potential function of $\mathbf{F}$.

   (a) $2xe^y + \frac{y^2}{2} + C$ (b) $4xe^y + y^2 + C$ (c) $xe^y + \frac{y^2}{2} + C$

   (d) $xe^y + 1 + C$ (e) None of these

4. Determine which of the following vector fields is *not* conservative.

   (a) $(2xy + z^2)\mathbf{i} + x^2\mathbf{j} + (2xz + \pi \cos \pi z)\mathbf{k}$ (b) $e^y\mathbf{i} + (xe^y + y)\mathbf{j}$

   (c) $yz\mathbf{i} + xz\mathbf{j} + xy\mathbf{k}$ (d) $(4x^2 - 4y^2)\mathbf{i} + (8xy - \ln y)\mathbf{j}$

   (e) None of these

5. Let $C$ be the line segment from the point $(0, 0, 0)$ to the point $(1, -3, 2)$. Find $\int_C (x + y^2 - 2z)\,ds$.

   (a) $\frac{3}{2}$ (b) $\frac{3}{2}\sqrt{14}$ (c) $10$

   (d) $-\frac{25}{2}$ (e) None of these

6. Let $\mathbf{F}(x, y, z) = (2x - y)\mathbf{i} + 2z\mathbf{j} + (y - z)\mathbf{k}$. Find the work done by the force $\mathbf{F}$ on an object moving along the straight line from the point $(0, 0, 0)$ to the point $(1, 1, 1)$.

   (a) $1$ (b) $3$ (c) $\frac{3}{2}$

   (d) $3\sqrt{3}$ (e) None of these

7. Evaluate $\int_C (3x^2y - 2y^3 + 5)\,dx + (x^3 - 6xy^2 + 2y)\,dy$ where $C$ is the ellipse $\frac{(x-2)^2}{9} + \frac{(y+1)^2}{16} = 1$.

   (a) $218$ (b) $103$ (c) $-190$

   (d) $0$ (e) None of these

**8.** Which of the following integrals calculates the area of the region bounded by $y = 4x^2$ and $y = 16x$?

(a) $\frac{1}{2}\int_0^4 4x^2\,dx$

(b) $\int_0^{64} (4x^2 - 16x)\,dy$

(c) $\int_0^4 (4x^2 - 16x)\,dx$

(d) $\int_0^4 \left(\frac{1}{2}\sqrt{y} - \frac{y}{16}\right) dy$

(e) None of these

**9.** Let $Q$ be the solid bounded by the cylinder $x^2 + y^2 = 1$ and the planes $z = 0$ and $z = 1$. Use the Divergence Theorem to calculate $\int_S\int \mathbf{F} \cdot \mathbf{N}\,dS$ where $S$ is the surface of $Q$ and $\mathbf{F}(x, y, z) = x\mathbf{i} + y\mathbf{j} + z\mathbf{k}$.

(a) $6\pi$

(b) 1

(c) $3\pi$

(d) 0

(e) None of these

**10.** Let $C$ be the triangle from $(6, 0, 0)$ to $(0, 3, 0)$ to $(0, 0, 6)$ to $(6, 0, 0)$ which lies in the plane $x + 2y + z = 6$. Use Stokes's Theorem to evaluate $\int_C \mathbf{F} \cdot d\mathbf{r}$ where $\mathbf{F}(x, y, z) = z\mathbf{i} - x^2\mathbf{j} + y\mathbf{k}$.

(a) 19

(b) $-9$

(c) 11

(d) $-54$

(e) None of these

**Test Form B** **Name** ______________ **Date** ______________

**Chapter 14** **Class** ______________ **Section** ______________

1. Let $\mathbf{F}(x, y, z) = \cos x\mathbf{i} + \sin y\mathbf{j} + z\mathbf{k}$. Calculate the divergence of $\mathbf{F}$ at the point $\left(\frac{\pi}{2}, \pi, 1\right)$.

   (a) $-\mathbf{i} - \mathbf{j} + \mathbf{k}$ (b) 0 (c) 1

   (d) $-1$ (e) None of these

2. Let $\mathbf{F}(x, y, z) = (x^3 \ln z)\mathbf{i} + xe^{-y}\mathbf{j} - (y^2 + 2z)\mathbf{k}$. Calculate **curl F** at (1, 1, 1).

   (a) $-2\mathbf{i} + \mathbf{j} + \frac{1}{e}\mathbf{k}$ (b) $-3$ (c) $\frac{-1 - 2e}{e}$

   (d) $-2\mathbf{i} - \mathbf{j} + \frac{1}{e}\mathbf{k}$ (e) None of these

3. Let $\mathbf{F}(x, y, z) = (2xy + z^2)\mathbf{i} + x^2\mathbf{j} + (2xz + \pi \cos \pi z)\mathbf{k}$. Find the potential function of $\mathbf{F}$.

   (a) $2x^2y + 2xz^2 + \sin \pi z + C$ (b) $x^2y + xz^2 + \sin \pi z + C$

   (c) $2x^2y + 2xz^2 - \pi \sin \pi z + C$ (d) $x^2y + xz^2 - \sin \pi z + C$

   (e) None of these

4. Determine which of the following fields is conservative.

   (a) $(4x^2 - 4y^2)\mathbf{i} + (8xy - \ln y)\mathbf{j}$ (b) $(8xy - \ln y)\mathbf{i} + (4y^2 - 4x^2)\mathbf{j}$

   (c) $e^y\mathbf{i} + (xe^y + y)\mathbf{j}$ (d) $(2xy^3 + x + z)\mathbf{i} + (3x^2y^2 - y)\mathbf{j} + (y + \sin z)\mathbf{k}$

   (e) None of these

5. Let $C$ be the line segment from the point (0, 0, 0) to the point (1, 3, −2). Find $\int_C (x + y^2 - 2z)\,ds$.

   (a) $\frac{11}{2}\sqrt{14}$ (b) $\frac{3}{2}\sqrt{14}$ (c) $\frac{3}{2}$

   (d) $\frac{11}{2}$ (e) None of these

6. A particle moves upward along the circular helix given by $\mathbf{r}(t) = (\cos t)\mathbf{i} + (\sin t)\mathbf{j} + t\mathbf{k}$ for $0 \le t \le 2\pi$ under a force given by $\mathbf{F}(x, y\ z) = (-zy)\mathbf{i} + (zx)\mathbf{j} + (xy)\mathbf{k}$. Find the work done on the particle by the force.

   (a) $2\pi^2 + 1$ (b) $2\pi^2$ (c) 0

   (d) $\frac{1}{2}$ (e) None of these

7. Evaluate $\int_C xy\,dx - 5x^2y\,dy$ where the curve $C$ is represented by $\mathbf{r}(t) = t\mathbf{i} + t^2\mathbf{j}$, for $0 \le t \le 1$.

(a) $-\frac{17}{12}$ (b) $-2$ (c) $-\frac{3}{4}$

(d) $-4$ (e) None of these

8. Let $C$ be the triangle path from $(0, 0)$ to $(1, 1)$ to $(0, 1)$ and back to $(0, 0)$. Then $\int_C 2y\,dx - 3x\,dy =$

(a) $\int_0^1 \int_0^y [2 - (-3)]dx\,dy.$ (b) $\int_0^1 \int_0^1 [2 - (-3)]dx\,dy.$

(c) $\int_0^1 \int_0^y [(-3) - 2]dx\,dy.$ (d) $\int_0^1 \int_0^1 [(-3) - 2]dx\,dy.$

(e) None of these

9. Let $S$ be the surface bounded by the $xy$-plane and the top hemisphere of $x^2 + y^2 + z^2 = 4$. Let $\mathbf{F}(x, y, z) = x^3\mathbf{i} + y^3\mathbf{j} + z^3\mathbf{k}$. Which of the following integrals uses the Divergence Theorem to calculate $\int_S\int \mathbf{F} \cdot \mathbf{N}\,dS$?

(a) $\int_S\int (x^3\mathbf{i} + y^3\mathbf{j} + z^3\mathbf{k}) \cdot (\mathbf{i} + \mathbf{j} + \mathbf{k})ds$

(b) $\int_S\int (3x^2 + 3y^2 + 3z^2)ds$

(c) $\iiint_Q (x^3 + y^3 + z^3)dV$, where $Q$ is the solid with surface $S$.

(d) $\iiint_Q (3x^2 + 3y^2 + 3z^2)dV$, where $Q$ is the solid with surface $S$.

(e) None of these

10. Let $C$ be the triangle from $(0, 0, 0)$ to $(2, 0, 0)$ to $(0, 2, 1)$ to $(0, 0, 0)$ which lies in the plane $z = \frac{y}{2}$.

If $\mathbf{F}(x, y, z) = -3y^2\mathbf{i} + 4z\mathbf{j} + 6x\mathbf{k}$, calculate $\int_C \mathbf{F} \cdot d\mathbf{r}$ using Stokes's Theorem.

(a) $\frac{2}{3}$ (b) 14 (c) 2

(d) 0 (e) None of these

**Test Form C** **Name** ______________ **Date** ______________

**Chapter 14** **Class** ______________ **Section** ______________

**1.** Let $\mathbf{F}(x, y, z) = z \sin x\mathbf{i} - \cos xy\mathbf{j} + xyz\mathbf{k}$. Calculate **curl F** at the point $\left(\frac{\pi}{2}, 1, 1\right)$.

(a) $\frac{\pi}{2}\mathbf{i} + \mathbf{j} + \mathbf{k}$ (b) $\frac{\pi}{2}\mathbf{i} + \mathbf{k}$ (c) $\frac{\pi}{2}\mathbf{i} - \mathbf{j} + \mathbf{k}$

(d) $\mathbf{j} + \mathbf{k}$ (e) None of these

**2.** Find div $(\mathbf{F} \times \mathbf{G})$ if $\mathbf{F} = x\mathbf{i} + xy\mathbf{j} + \mathbf{k}$ and $\mathbf{G} = (x + y)\mathbf{i} - 2yz\mathbf{k}$.

(a) $1 + 2xz - 2y^2z$ (b) $-2y^2z - 2xz - 1$ (c) $-4xyz + x + y + x^2y$

(d) $-2xz + 2y^2z$ (e) None of these

**3.** Let $C$ be the triangle in the plane from $(0, 0)$ to $(1, 1)$ to $(0, 1)$, back to $(0, 0)$. Evaluate the integral $\int_C (x + 2y)\, ds$.

(a) $\frac{7}{2}$ (b) 4 (c) $\frac{1}{2}(3\sqrt{2} + 5)$

(d) $\frac{1}{2}(3\sqrt{2} + 7)$ (e) None of these

**4.** Let $C$ be a piecewise smooth path from the point $(3, 2)$ to the point $(6, 5)$. Let $f(x, y) = x^2y$ be a potential function for the vector field $\mathbf{F}(x, y)$. Then, $\int_C \mathbf{F} \cdot d\mathbf{r} =$ _______.

(a) $(6^2)(5) - (3^2)(2)$ (b) $(6 - 3)^2(5 - 2)$ (c) $(6^2)(3) - (5^2)(2)$

(d) Varies with the path covered (e) None of these

**5.** Use the Fundamental Theorem of Line Integrals to evaluate $\int_C 2 \cos 2x \cos 2y\mathbf{i} - 2 \sin 2x \sin 2y\mathbf{j}$, where $C$ is a smooth curve from $\left(0, \frac{\pi}{8}\right)$ to $\left(\frac{\pi}{4}, \pi\right)$.

(a) 2 (b) $-2$ (c) 0

(d) 1 (e) None of these

6. Use Green's Theorem to calculate $\int_C \mathbf{F} \cdot d\mathbf{r}$ where $\mathbf{F}(x, y) = (-16y + \sin x^2)\mathbf{i} + (4e^y + 3x^2)\mathbf{j}$ and $C = C_1 + C_2 + C_3$ as shown.

(a) $2\sqrt{3} + 8\pi$ (b) $4\pi$

(c) $8\pi$ (d) $2\sqrt{3} + 4\pi$

(e) None of these

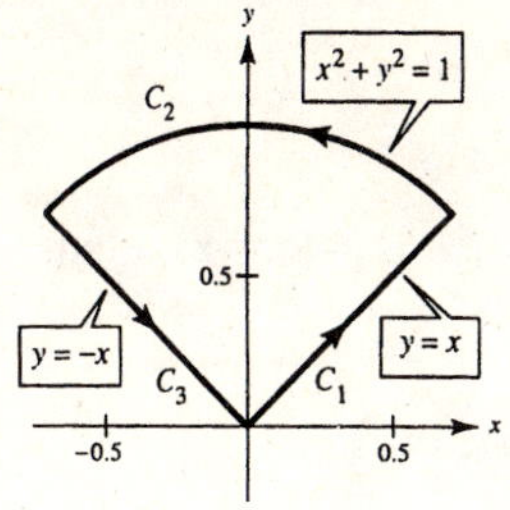


7. Find the surface area of that part of the cylinder given by $\mathbf{r}(u, v) = 3\cos u\,\mathbf{i} + 3\sin u\,\mathbf{j} + v\mathbf{k}$ over the region where $0 \le u \le 2\pi$ and $0 \le v \le 2$.

(a) $6\pi$ (b) 3 (c) $12\pi$

(d) $12\pi - 2$ (e) None of these

8. Evaluate the surface integral $\int_S\int z\, dS$ where $S$ is the hemisphere $z = \sqrt{16 - x^2 - y^2}$.

(a) $8\pi$ (b) $64\pi$ (c) $32\pi$

(d) $4\pi$ (e) None of these

9. Use the Divergence Theorem to evaluate $\int_S\int \mathbf{F} \cdot \mathbf{N}\, dS$ where $\mathbf{F}(x, y, z) = x^2\mathbf{i} + y\mathbf{j} + z\mathbf{k}$ and $S$ is the surface of the cube $\{(x, y, z) : 0 \le x \le 2, 0 \le y \le 2, 0 \le z \le 2\}$.

(a) 24 (b) 8 (c) 32

(d) 18 (e) None of these

10. Use Stokes's Theorem to evaluate $\int_C \mathbf{F} \cdot d\mathbf{r}$ where $\mathbf{F}(x, y, z) = (3z - 2y)\mathbf{i} + (4x - 3y)\mathbf{j} + (z + 2y)\mathbf{k}$ and $C$ is the unit circle in the plane $z = 3$.

(a) $6\pi$ (b) $\pi$ (c) $3\pi$

(d) $12\pi$ (e) None of these

**Test Form D** **Name** ______________ **Date** ______________

**Chapter 14** **Class** ______________ **Section** ______________

1. Let $\mathbf{F}(x, y, z) = \cos x\mathbf{i} + \sin y\mathbf{j} + e^{xy}\mathbf{k}$. Calculate the divergence of $\mathbf{F}$ at the point $\left(\frac{\pi}{6}, \frac{\pi}{6}, 32\right)$.

2. Compute **curl F** if $\mathbf{F}(x, y, z) = xy\mathbf{i} + (y^2 - 2z)\mathbf{j} + \sin yz\mathbf{k}$.

3. Let $\mathbf{F}(x, y) = \left(4x^3y^3 + \frac{1}{x}\right)\mathbf{i} + \left(3x^4y^2 - \frac{1}{y}\right)\mathbf{j}$. Find the potential function of $\mathbf{F}$.

4. Let $C$ be the portion of the parabola $y = x^2$ with the initial point $(0, 0)$ and the terminal point $(2, 4)$ and let $\mathbf{F}(x, y) = y^3\mathbf{i} + 3xy^2\mathbf{j}$. Find $\int_C \mathbf{F} \cdot d\mathbf{r}$.

5. Evaluate $\int_C (1 + xy)\, ds$ where $C$ is the closed curve, $C_1 + C_2$, shown.

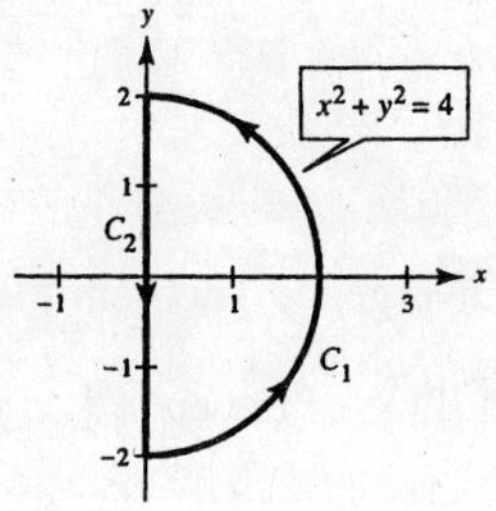


6. Evaluate $\int_C e^x \sin y\, dx + e^x \cos y\, dy$ where $C$ is the square with vertices $(0, 0)$, $(1, 1)$, $(0, 2)$, and $(-1, 1)$ oriented counterclockwise.

7. Let $C$ be the closed curve shown, oriented clockwise. Evaluate $\int_C 2y^3\, dx + (x^4 + 6y^2x)\, dy$ using Green's Theorem.

8. A particle moves along a curve given by $\mathbf{r}(t) = t\mathbf{i} + t^2\mathbf{j} + t^3\mathbf{k}$ from the point (0, 0, 0) to the point (1, 1, 1) under a force given by $\mathbf{F}(x, y, z) = 2xz\mathbf{i} - yz\mathbf{j} + yz^2\mathbf{k}$. Calculate the work done on the particle by the force.

9. Let $Q$ be the cube in the first octant as shown and let $\mathbf{F}(x, y, z) = x\mathbf{i} + y^2\mathbf{j} + z\mathbf{k}$. Use the Divergence Theorem to calculate $\int_S\int \mathbf{F} \cdot \mathbf{N}\, dS$.

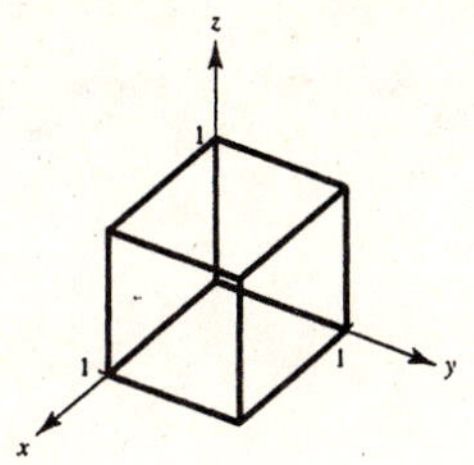


10. Let $S$ be that part of the paraboloid $z = 9 - x^2 - y^2$ where $z \geq 0$ and let $C$ be the trace in the $xy$ plane oriented counterclockwise. Use Stokes's Theorem to calculate $\int_C \mathbf{F} \cdot d\mathbf{r}$ where $\mathbf{F}(x, y, z) = 2z\mathbf{i} + 4x\mathbf{j} + 3y\mathbf{k}$.

**Test Form E** **Name** ______________ **Date** ______________

**Chapter 14** **Class** ______________ **Section** ______________

1. Find the curl of the vector field $\mathbf{F}(x, y, z) = xz\mathbf{i} + xy^2z\mathbf{j} - e^{2y}\mathbf{k}$.

2. Determine whether the vector field $\mathbf{F}(x, y, z) = (2xz + 1)\mathbf{i} + (2yz + 2y)\mathbf{j} + (x^2 + y^2 + 3z^2)\mathbf{k}$ is conservative. If it is, find a potential function for the vector field.

3. Evaluate the integral $\int_C (3x - 2y + z)\, ds$ where $C$ is the curve, $C_1 + C_2$, as shown.

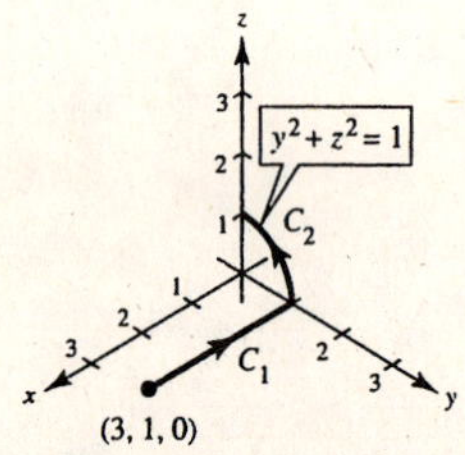


4. Let $C$ be the twisted cubic curve parameterized by $\mathbf{r}(t) = t\mathbf{i} + t^2\mathbf{j} + t^3\mathbf{k}, 0 \le t \le 1$.

   Evaluate the integral $\int_C xy\, dx + 3xz\, dy - 5x^2yz\, dz$.

5. Evaluate $\int_C \left(3x^2y + \frac{1}{x}\right) dx + \left(2x^3y - \frac{1}{y}\right) dy$ along the unit circle oriented counterclockwise.

6. Use Green's Theorem to calculate $\int_C \mathbf{F} \cdot d\mathbf{r}$ where $\mathbf{F}(x, y) = (-16y + \sin x^2)\mathbf{i} + (4e^y + 3x^2)\mathbf{j}$ and $C = C_1 + C_2 + C_3$ as shown.

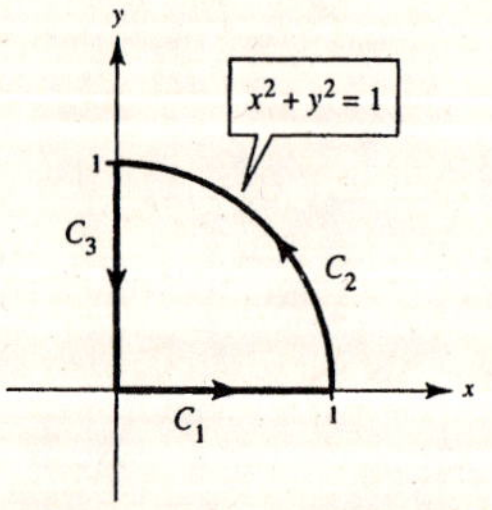


7. Find the surface area of that part of the cone given by $\mathbf{r}(u, v) = 4u \cos v\mathbf{i} + 4u \sin v\mathbf{j} + u\mathbf{k}$ over the region where $0 \le u \le 3$ and $0 \le v \le 2\pi$.

8. Evaluate the surface integral $\int_S\int xy\, ds$ where $S$ is the first octant portion of the plane $z = 6 - 2x - y$.

9. Use the Divergence Theorem to evaluate $\int_S\int \mathbf{F} \cdot \mathbf{N}\, dS$ where $\mathbf{F}(x, y, z) = x\mathbf{i} + 3y\mathbf{j} - z\mathbf{k}$ and $S$ is the spherical surface $x^2 + y^2 + z^2 = 4$.

10. Use Stokes's Theorem to evaluate $\int_C \mathbf{F} \cdot d\mathbf{r}$ where $\mathbf{F}(x, y, z) = xy\mathbf{i} + 2yz\mathbf{j} + xz\mathbf{k}$ and $C$ is the boundary of the plane $z = 1 - y$, $0 \le x \le 2$, as shown.

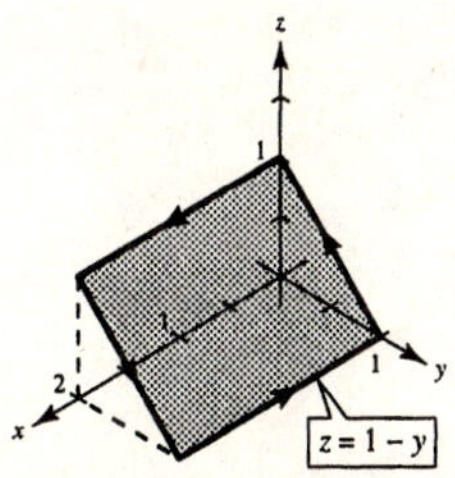

# Answers to CHAPTER P Tests

## Test Form A

| | | | |
|---|---|---|---|
| 1. d | 2. c | 3. c | 4. a |
| 5. d | 6. a | 7. b | 8. b |
| 9. d | 10. c | 11. b | 12. a |
| 13. d | 14. c | 15. d | 16. d |
| 17. b | 18. b | 19. b | 20. c |

## Test Form B

| | | | |
|---|---|---|---|
| 1. a | 2. b | 3. d | 4. c |
| 5. a | 6. c | 7. b | 8. c |
| 9. b | 10. e | 11. c | 12. c |
| 13. b | 14. a | 15. a | 16. a |
| 17. c | 18. a | 19. d | 20. a |

## Test Form C

| | | | |
|---|---|---|---|
| 1. a | 2. b | 3. b | 4. c |
| 5. d | 6. b | 7. b | 8. d |
| 9. c | 10. d | 11. c | 12. b |
| 13. a | 14. d | 15. b | 16. d |
| 17. a | 18. b | 19. c | 20. d |

## Test Form D

**1.** $f(-x) = \dfrac{-x}{(-x)^2+1} = -\dfrac{x}{x^2+1} = -f(x)$

**2.** $\left(0, -\frac{1}{3}\right), \left(\frac{1}{2}, 0\right)$ **3.** (0, 0), (2, 4)

**4.** $3x - 2y + 11 = 0$ **5.** $2x + \Delta x + 3$ **6.** $\dfrac{1}{x} - 5$

**7.** All reals except $1 \pm \sqrt{3}$

**8.**

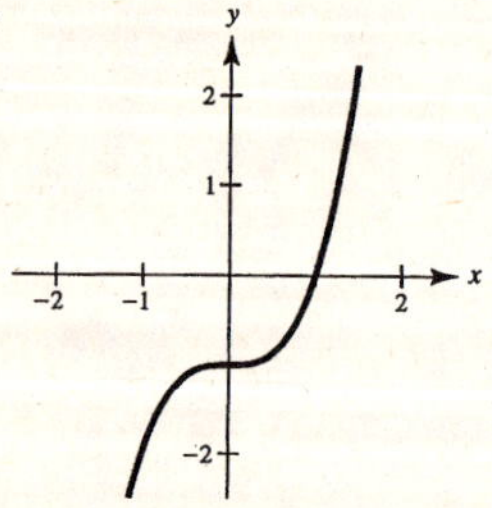

**9.**

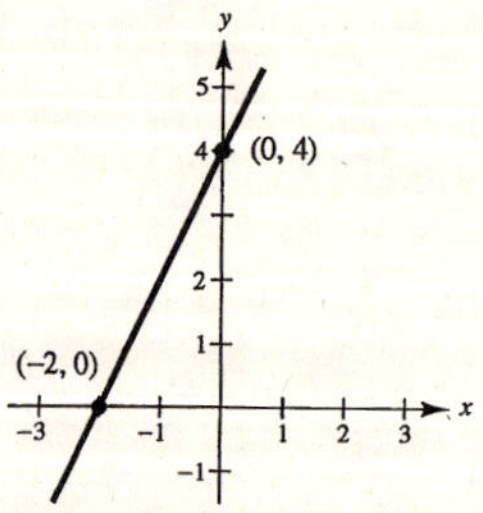


**10.** a and b **11.** Neither

**12.**

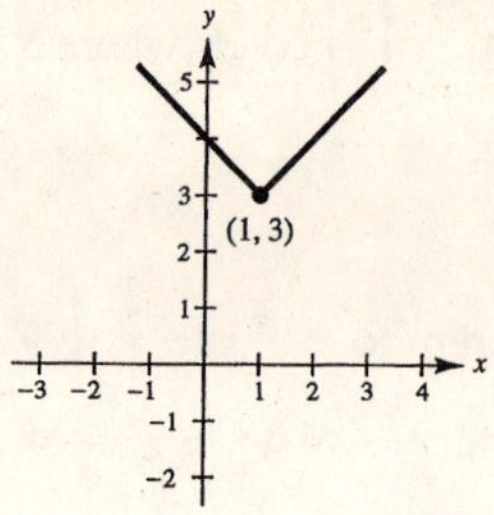


**13.** Slope: $-\frac{5}{4}$, $y$-intercept (0, 3)

**14. a.** $-4$ **b.** $-1$ **c.** $-3$

**15. a.** $S = 24 + 1.50x$ **b.** \$45.00 **c.** 51

**16.** Yes; $\dfrac{1-bx}{ax}$ **17.** $\dfrac{x}{\sqrt{1-x^2}}$

**18.** $f^{-1}(x) = \sqrt[3]{x-5}$ **19.** $\dfrac{1}{5-e^3}$ **20.** 6

## Test Form E

**1. a.** $y = 3 \pm \sqrt{1-(x+2)^2}$

**b.**

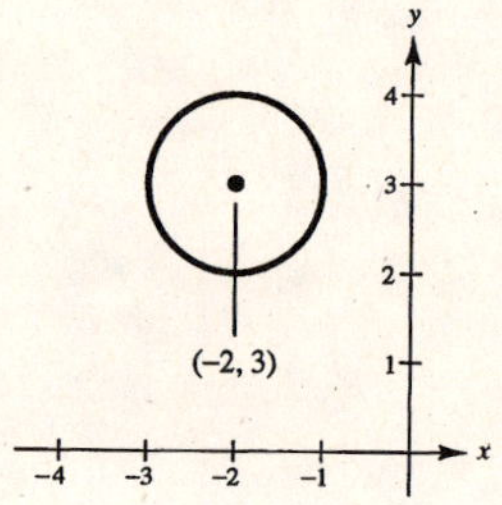


**2.** One such equation is $y = x^3 - 25x$.

**3. a.**

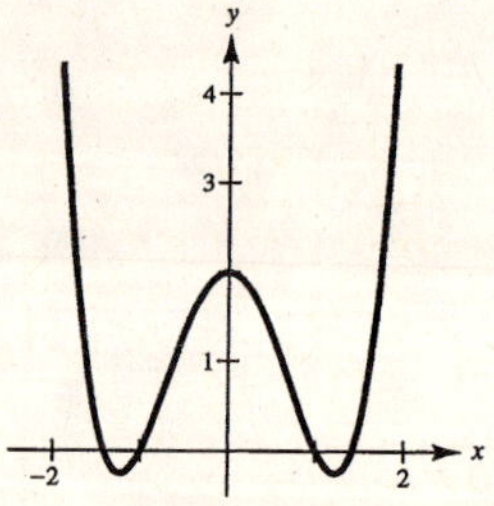

**b.** $\left(-\sqrt{2}, 0\right), (-1, 0), (0, 2), (1, 0), \left(\sqrt{2}, 0\right)$

**c.** Symmetric with respect to the $y$-axis.

**4.** Symmetric to the origin

**5.** $(-2, 2)$ and $(5, 9)$

**6.**

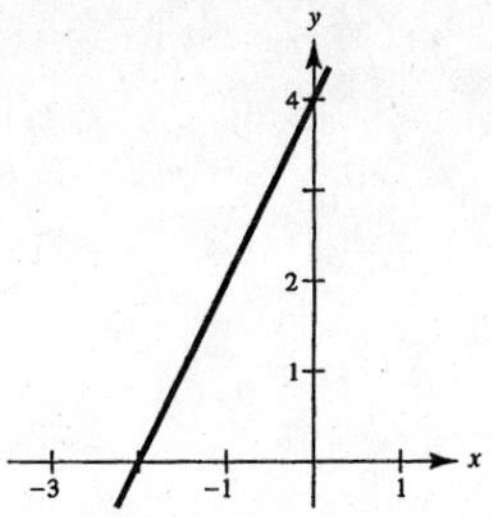


**7. a.** 3

**b.** 2

**c.** 0

**d.** −1

**8. a.**

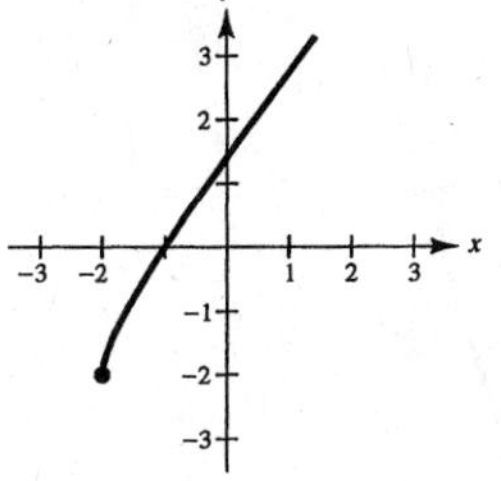


**b.** Domain: $[-2, \infty)$; Range: $[-2, \infty)$

**c.** $(-1, 0), (0, 1.4)$

**d.** $(-1, 0)$ and $\left(0, \sqrt{2}\right)$

**9.** $-2.4,\ 0.41,\ 2$ **10. a, d,** and **e** **11.** 3

**12.** $2x + \Delta x + 3$

**13.** Even. $f(-x) = -(-x)^4 + 2(-x)^2 - 1$

$$= -x^4 + 2x^2 - 1$$

$$= f(x)$$

**14.**

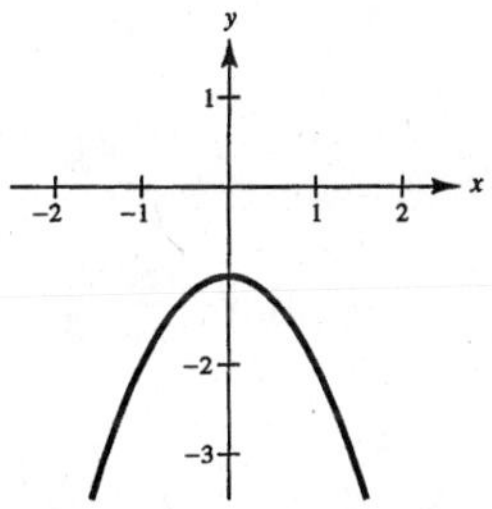


**15.** $\dfrac{x + 3}{3}$

**16. a.** $S = -12{,}500t + 224{,}000$

**c.** \$160,000

**d.** 2011

**17. a.** 5.61

**b.** 21.2825

**18.** No **19.** $\dfrac{10}{\sqrt{109}}$ **20.** $\dfrac{1}{4}(\ln 3 - 1)$

# Answers to CHAPTER 1 Tests

## Test Form A

| | | | |
|---|---|---|---|
| **1.** b | **2.** d | **3.** a | **4.** a |
| **5.** a | **6.** c | **7.** b | **8.** b |
| **9.** d | **10.** a | **11.** c | **12.** a |
| **13.** b | **14.** a | **15.** d | **16.** a |
| **17.** d | **18.** b | **19.** b | **20.** c |

## Test Form B

| | | | |
|---|---|---|---|
| **1.** c | **2.** c | **3.** d | **4.** b |
| **5.** b | **6.** b | **7.** b | **8.** d |
| **9.** e | **10.** c | **11.** b | **12.** b |
| **13.** a | **14.** c | **15.** a | **16.** c |
| **17.** b | **18.** a | **19.** c | **20.** b |

## Test Form C

| | | | |
|---|---|---|---|
| **1.** c | **2.** d | **3.** b | **4.** b |
| **5.** b | **6.** a | **7.** a | **8.** b |
| **9.** d | **10.** c | **11.** b | **12.** c |
| **13.** a | **14.** a | **15.** c | **16.** d |
| **17.** a | **18.** c | **19.** a | **20.** a |

## Test Form D

**1.** $-3$ **2.** 0.001 **3.** 5 **4.** 0 **5.** $\sqrt{e}$

**6.** $-\frac{7}{6}$ **7.** $\frac{1}{12}$ **8.** $\frac{1}{2\sqrt{x}}$ **9.** Does not exist

**10.** $-\sqrt{3}$ **11.** $\frac{1}{3}$ **12.** $\sqrt{3}$ **13.** $-\infty$

**14.** $+\infty$ **15.** $\infty$

**16.** $x = 2$, removable; $x = -2$, nonremovable

**17. a.** $\frac{5}{x^4 - 1}$ **b.** $-1, 1$ **18.** 27

**19.** $x = \frac{-1 - \sqrt{5}}{2}$, $x = \frac{-1 + \sqrt{5}}{2}$ **20.** 1

## Test Form E

**1.** The limit does not exist.

**2.** False. $f(x) = \begin{cases} x^2, & x \neq 3 \\ 0, & x = 3 \end{cases}$

**3.** $-3$ **4.** 0 **5.** $-\pi$ **6.** 5

**7. a.**

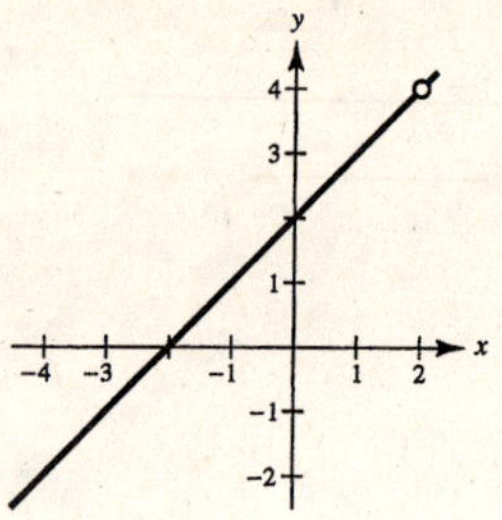

**b.** 4 **c.** 4

**8. a.** 4 **b.** $g(x) = x^2$

**c.**

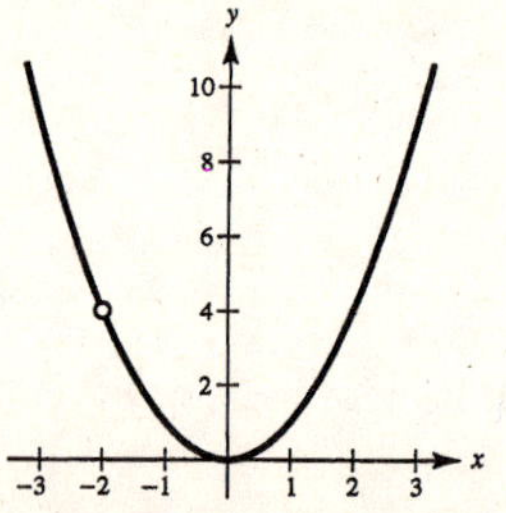

**9.** $\frac{1}{2\sqrt{x + 2}}$ **10.** $2x - 2$ **11.** $-1$ **12.** 1

**13.** $((2n - 1)\pi, (2n + 1)\pi)$ where $n$ is an integer.

**14.** $f(x)$ is continuous on the interval $(-\infty, \infty)$. $f(-2) = 2 > 0$ and $f(-1) = -4 < 0$. Therefore, there exists a real number $c$, $-2 < c < -1$, such that $f(c) = 0$.

**15.** $-2$ and 2

**16. a.** 1 **b.** $-3$ **c.** The limit does not exist.

**17.** 3; $\lim_{x \to 0} \frac{\sin 3x}{x} = \lim_{x \to 0} \frac{3 \sin 3x}{3x} = 3\left(\lim_{x \to 0} \frac{\sin 3x}{3x}\right)$
$= 3(1) = 3$

**18.** $x = -3$ **19.** $+\infty$ **20.** $-\infty$

# Answers to CHAPTER 2 Tests

## Test Form A

| | | | |
|---|---|---|---|
| **1.** c | **2.** d | **3.** b | **4.** a |
| **5.** a | **6.** b | **7.** b | **8.** d |
| **9.** e | **10.** c | **11.** a | **12.** d |
| **13.** b | **14.** b | **15.** d | **16.** b |
| **17.** d | **18.** e | | |

## Test Form B

| | | | |
|---|---|---|---|
| **1.** b | **2.** d | **3.** a | **4.** c |
| **5.** d | **6.** a | **7.** c | **8.** b |
| **9.** b | **10.** d | **11.** b | **12.** c |
| **13.** c | **14.** c | **15.** b | **16.** c |
| **17.** a | **18.** a | | |

## Test Form C

| | | | |
|---|---|---|---|
| **1.** d | **2.** b | **3.** d | **4.** c |
| **5.** a | **6.** c | **7.** c | **8.** d |
| **9.** c | **10.** b | **11.** c | **12.** a |
| **13.** a | **14.** c | **15.** c | **16.** b |
| **17.** c | **18.** d | | |

## Test Form D

**1.** $-\frac{1}{x^2}$ **2.** $\frac{6x^2+2}{(1-3x^2)^2}$ **3.** $\frac{x^2(7x+3)}{\sqrt{2x+1}}$

**4.** $\frac{-3\cot^2\sqrt{x}\left(\csc^2\sqrt{x}\right)}{2\sqrt{x}}$ **5.** 42 $\text{ft/sec}^2$

**6.** $\frac{y}{(x+y)^2+x}$ **7.** $\frac{1-ye^x}{e^x-2y}$ **8.** $2\theta \sec\theta^2 \tan\theta^2$

**9.** $2\sin 2x$ **10.** $2x - 4y + \pi - 2 = 0$

**11.** $\frac{1}{3}, 1$ **12.** $4x - \frac{1}{x^2}$

**13. a.** $s = -16t^2 - 26t + 220$ **b.** $-58$ ft/sec

**14.** $\frac{1}{\pi}$ in./sec **15.** $\frac{1}{2x} + \frac{1}{5-x}$ **16.** $x^x(1 + \ln x)$

**17.** $-\frac{e^{1/x}}{x^2}$ **18.** $e^x x^{e-1}(x+1)$

## Test Form E

**1. a.** no derivative **b.** negative **c.** zero
**d.** positive **e.** zero

**2. a.** $-\frac{4}{x^2}$ **b.** $-1$ **c.** $y = -x - 4$

**d.**

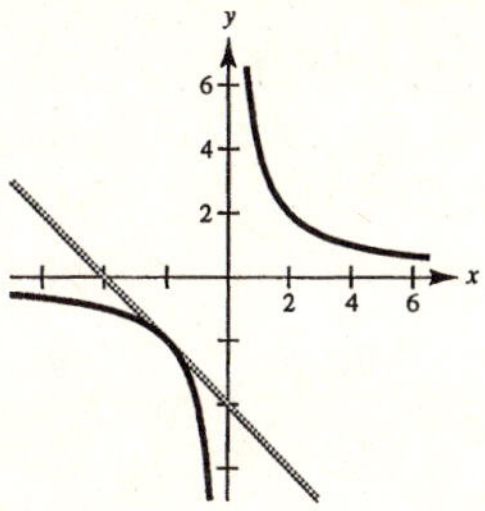


**3.** $\left(-\frac{3}{2}, -\frac{2}{3}\right)$ and $\left(\frac{3}{2}, \frac{2}{3}\right)$

**4. a.** $-64$ ft/sec **b.** $-96$ ft/sec

**c.** $\frac{5\sqrt{30}}{4}$ sec $\approx 6.85$ sec

**d.** $-40\sqrt{30}$ ft/sec $\approx -219.09$ ft/sec

**5.** $4\pi \text{ ft}^2$ **6.** $\frac{-8e^{2x}}{(4+e^{2x})^5}$ **7.** $\frac{6+2\sqrt{3}\pi}{3}$

**8. a.** $5x^4 - 5$

**c.** $(-1, 4)$ and $(1, -4)$

**d.** 0

**9.**

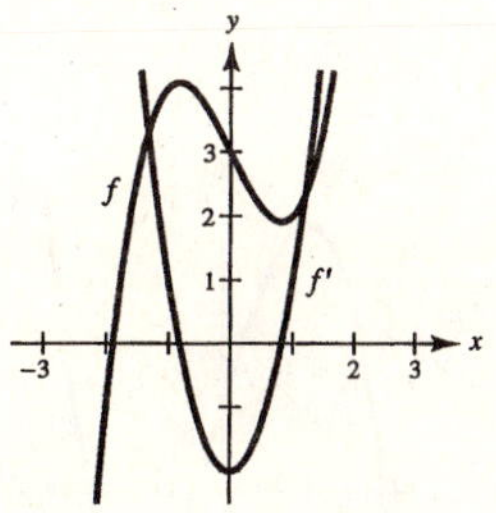


The derivative of $f$ is zero if its tangent line is horizontal.

**10.** $\frac{10}{(2x^2+5)^{3/2}}$ **11.** $\left(\sqrt{2}, -2\sqrt{2}\right)$ and $\left(-\sqrt{2}, 2\sqrt{2}\right)$

**12.** $\frac{96}{845}$ rad/sec **13.** $\frac{2x+6}{x^2+6x+9}$

**14.** $y\left[\frac{3}{x} + \frac{1}{2x+3} - \frac{2}{x-2}\right] = \frac{x^3\sqrt{2x+3}}{(x-2)^2}\left[\frac{3}{x} + \frac{1}{2x+3} - \frac{2}{x-2}\right]$ **15.** $\frac{-2t}{\sqrt{1-t^4}}$

# Answers to CHAPTER 3 Tests

## Test Form A

**1.** c **2.** d **3.** b **4.** c

**5.** a **6.** d **7.** c **8.** b

**9.** d **10.** a **11.** a **12.** d

**13.** b **14.** d

## Test Form B

**1.** d **2.** d **3.** a **4.** b

**5.** c **6.** b **7.** a **8.** c

**9.** b **10.** c **11.** b **12.** c

**13.** e **14.** b

## Test Form C

**1.** b **2.** c **3.** d **4.** c

**5.** c **6.** c **7.** a **8.** b

**9.** e **10.** c **11.** d **12.** c

**13.** b **14.** a **15.** b

## Test Form D

**1.** Increasing $(-\infty, 0)$; Decreasing $(0, \infty)$

**2.** $-\frac{1}{3}, -\frac{1}{2}$ **3.** $\left(2, \frac{1}{54}\right)$, relative maximum

**4.** $(0, 0)$ **5.** $(-\infty, 0)$ and $(2, \infty)$

**6.** $x = 0$, relative maximum; $x = \frac{2}{3}$, relative minimum

**7.** 2

**8.**

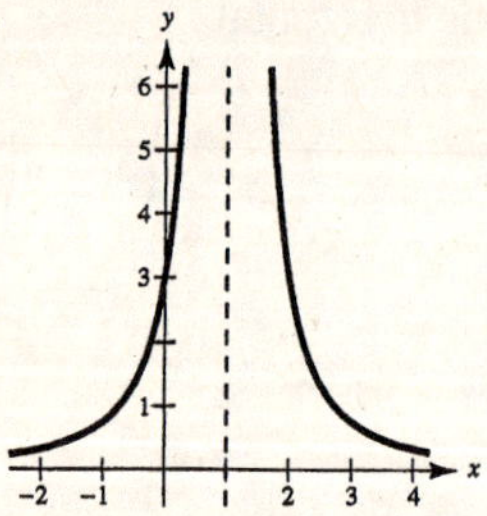

**9.**

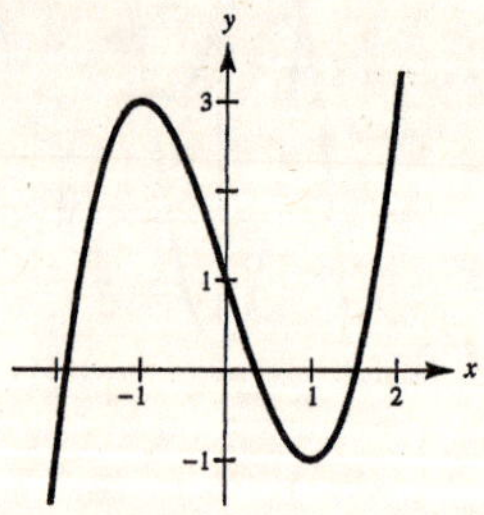

**10.** $\left(-1, \frac{1}{\sqrt{2\pi}} e^{-1/2}\right), \left(1, \frac{1}{\sqrt{2\pi}} e^{-1/2}\right)$

**11.** 320,000 sq ft **12.** $y = -1$

**13.** $\pm 0.001$ in.

## Test Form E

**1.** The domain of $y_1$ is $(-\infty, 0) \cup (0, \infty)$, whereas the domain of $y_2$ is just $(0, \infty)$. The functions $y_3$ and $y_4$ have the same domain.

**2. a.**

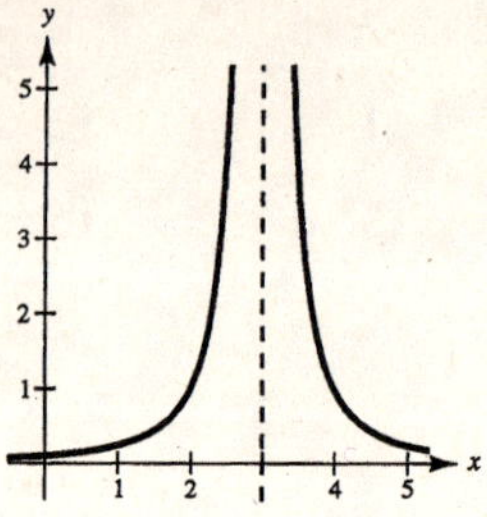

**b.** $f(2) = f(4) = 1$

**c.** $f$ is not continuous on $[2, 4]$.

**3.** $\frac{1}{4}$

**4. a.** Relative minimum at $x = \pm 1.4$
Relative maximum at $x = -0.3$

**b.** Relative minimum at $x = \pm\sqrt{2}$
Relative maximum at $x = -\frac{1}{3}$

**5.** Increasing $(-\infty, -3)$ and $(1, \infty)$
Decreasing $(-3, 1)$

**6.** Relative maximum

**7.** Concave downward: $(-\infty, 2)$
Concave upward: $(2, \infty)$

**8.**

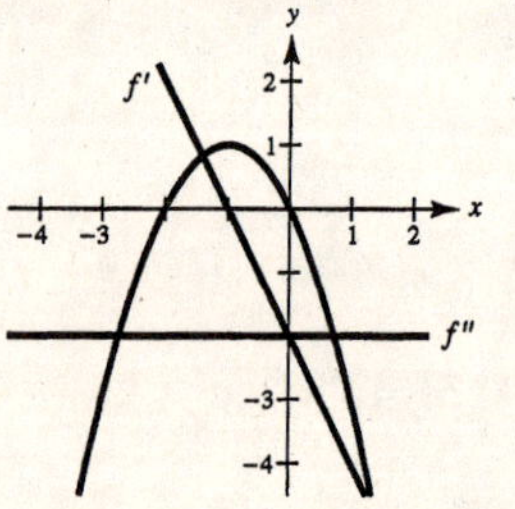


**9. a.** 2 **b.** $-2$ **c.** The limit does not exist.

**10. a.** $y = 1$

**b.**

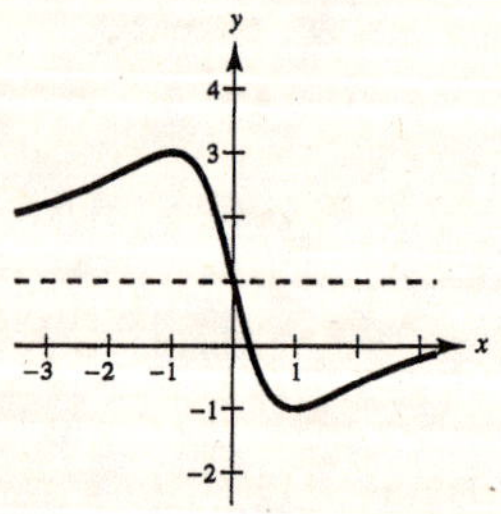

**c.** $(0, 1)$

**11.** $f(x) = \dfrac{x^2 - x + 5}{x}$ (The answer is not unique.)

**12. a.** $V = x(7 - 2x)(3 - 2x)$

**b.** 0.65 in. by 0.65 in.; $V = 6.3$ cubic inches

**13. a.** 0.5 **b.** 0.4459

**14. a.** $T(x) = 12x - 16$

**c.** $f(2.1) \approx 9.5,\ T(2.1) \approx 9.4$

**d.** $f(2.1) = 9.261,\ T(2.1) = 9.2$

**15. a.** 9

**b.** 33

**c.** 300

# Answers to CHAPTER 4 Tests

## Test Form A

| | | | |
|---|---|---|---|
| **1.** a | **2.** b | **3.** c | **4.** a |
| **5.** b | **6.** d | **7.** c | **8.** b |
| **9.** d | **10.** b | **11.** a | **12.** b |
| **13.** d | **14.** a | **15.** c | **16.** a |
| **17.** d | **18.** c | **19.** b | **20.** b |

## Test Form B

| | | | |
|---|---|---|---|
| **1.** b | **2.** c | **3.** b | **4.** d |
| **5.** a | **6.** a | **7.** b | **8.** c |
| **9.** c | **10.** d | **11.** b | **12.** a |
| **13.** b | **14.** c | **15.** a | **16.** d |
| **17.** a | **18.** b | **19.** c | **20.** b |

## Test Form C

| | | | |
|---|---|---|---|
| **1.** c | **2.** d | **3.** a | **4.** d |
| **5.** d | **6.** b | **7.** d | **8.** d |
| **9.** e | **10.** b | **11.** b | **12.** d |
| **13.** b | **14.** d | **15.** b | **16.** c |

## Test Form D

**1.** $\frac{2}{5}x^{5/2} + C$ **2.** $-3 \csc x + C$

**3.** $\dfrac{x^2}{2} - x + C$ **4.** $\dfrac{1}{2} \sin \theta + C$

**5.** $y = x^2 - x + 3$ **6.** $-186.6$ ft/sec **7.** 4

**8.** $\displaystyle\int_0^4 (4x - x^2)\,dx$ **9.** $(2x^2 + 5)^2$ **10.** $-4$

**11.** $\dfrac{2\sqrt{2}}{\pi}$ **12.** $2\sqrt{\tan x} + C$

**13.** $\frac{2}{3}\sqrt{x - 1}(x + 2) + C$ **14.** 1

**15.** $\frac{1}{2} \ln|\sec 2x + \tan 2x| + C$ **16.** $-95e^{-t/5} + C$

**17.** $2\sqrt{e^x + 1} + C$

**18.** $\dfrac{5}{2} \ln(x^2 + 9) + \dfrac{16}{3} \arctan \dfrac{x}{3} + C$

**19.** $\arcsin(x - 2) + C$ **20.** $2 \tanh 2x$

## Test Form E

**1.** $\dfrac{2a}{5}x^{5/2} + \dfrac{2b}{7}x^{7/2} + C$ **2.** $\dfrac{1 - 2x}{2(x - 1)^2} + C$

**3.** $y = x^2 - x + 3$

**4. a.** $P = 10t^4 - 30t^{7/2} + 16{,}000$

**b.** 4.8 **c.** 53,120

**5.**

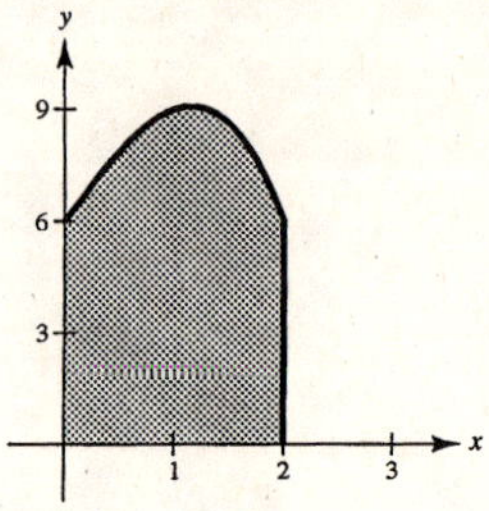


**6.** No, because $f(x)$ is not continuous at $x = \frac{3}{2}$.

**7.** $\sqrt{2} - 1 \approx 0.414$

**8.** $F'(x) = \dfrac{1}{1 + x^4}$, $F'(2) = \dfrac{1}{17}$ **9.** $2 \sin\sqrt{x} + C$

**10.** $\sqrt{3} - 1$ **11.** $\frac{1}{3}[2^{3/2} - 1]$

**12.** $\ln|1 + \sin \theta| + C$ **13.** $-\frac{1}{3} e^{3/x} + C$

**14.** $x - \arctan x + C$ **15.** $\dfrac{2\pi}{3}$

**16.** $\operatorname{arcsec} \dfrac{x - 1}{5} + C$ **17.** $\dfrac{-2}{\sqrt{1 - t^4}}$

# Answers to CHAPTER 5 Tests

## Test Form A

| | | | |
|---|---|---|---|
| **1.** c | **2.** c | **3.** d | **4.** a |
| **5.** d | **6.** b | **7.** c | **8.** c |
| **9.** a | **10.** d | | |

## Test Form B

| | | | |
|---|---|---|---|
| **1.** b | **2.** a | **3.** b | **4.** b |
| **5.** e | **6.** c | **7.** c | **8.** d |
| **9.** b | **10.** a | | |

## Test Form C

| | | | |
|---|---|---|---|
| **1.** a | **2.** d | **3.** a | **4.** b |
| **5.** b | **6.** c | **7.** a | **8.** b |
| **9.** b | **10.** c | | |

## Test Form D

**1.** $x^2 + \sec^2 y = C$ **2.** $x^2 = y^2 \ln(Cy^2)$ **3.** 4.64

**4.** $Q(t) = 10e^{-0.0231t}$

**5.** $y' = 2x + 2 + Ce^x$

Then, $y' - y + x^2 =$

$(2x + 2 + Ce^x) - (x^2 + 2x + 2 + Ce^x) + x^2 = 0.$

**6.** $y = Cx^2$; parabolas
$x^2 + 2y^2 = C$; orthogonal ellipses

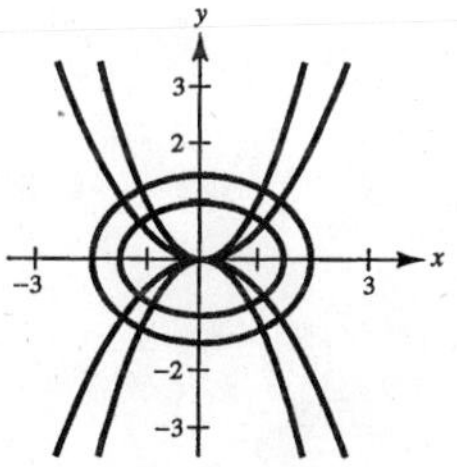


**7.** $e^y + \sin y = \frac{1}{2}(x^2 + 1)$ **8.** 0.0302961

**9.** $4x^2y = x^4 + C$ **10.** $y = 1 + e^{\cos x}$

## Test Form E

**1.** $y = Cx$ **2.** $y^2 = e^{x^2} + C$

**3.** $y = 500 - 493e^{-x}$

**4. a.** $\frac{dy}{dx} = \frac{k}{\sqrt{y}}$ **b.** $y = \left(\frac{3}{2}kx\right)^{2/3} + C$

**5.** $4y^3y' = -4xe^{-x^2} - 4x\,Ce^{-2x^2}$

$\Rightarrow y' + xy^3 = xe^{-x^2}y^{-3}$

**6.** $y = Cx^4$

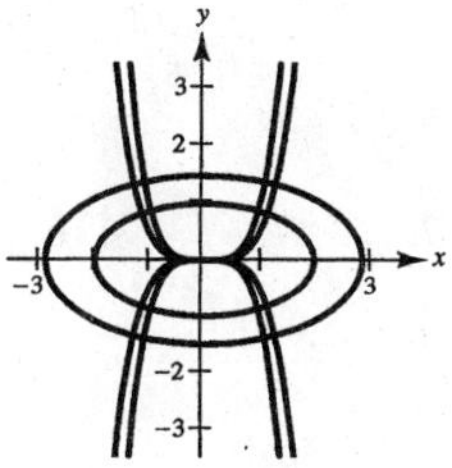


**7.** $y = x\tan(C - \ln|x|)$ **8.** $-1.143172$

**9.** $y = 1 + C\cos x$ **10.** $y = \frac{5x - 4}{5x}$

# Answers to CHAPTER 6 Tests

## Test Form A

| | | | |
|---|---|---|---|
| **1.** c | **2.** b | **3.** d | **4.** c |
| **5.** a | **6.** b | **7.** d | **8.** a |

## Test Form B

| | | | |
|---|---|---|---|
| **1.** a | **2.** c | **3.** b | **4.** b |
| **5.** c | **6.** a | **7.** a | **8.** d |

## Test Form C

| | | | |
|---|---|---|---|
| **1.** c | **2.** a | **3.** c | **4.** b |
| **5.** b | **6.** c | **7.** c | **8.** d |

## Test Form D

**1.** $\frac{15}{8} - 2\ln 2$ **2.** $\frac{\pi}{2}(e^2 - 1)$

**3.** $2\pi \int_{1/2}^{2} \left(y - \frac{1}{2}\right)\left(\frac{5 - 2y}{2} - \frac{1}{y}\right) dy$

**4.** $\frac{256}{5}\pi$ **5.** $\int_0^{\pi} \sqrt{1 + 4\cos^2 2x}\, dx$

**6.** $2\pi \int_0^4 x\sqrt{1 + \frac{1}{4x}}\, dx$ **7.** $\frac{1280}{3}$ in-lb

**8.** $\bar{x} = \frac{4}{5}, \bar{y} = \frac{2}{7}$

## Test Form E

**1. a.**

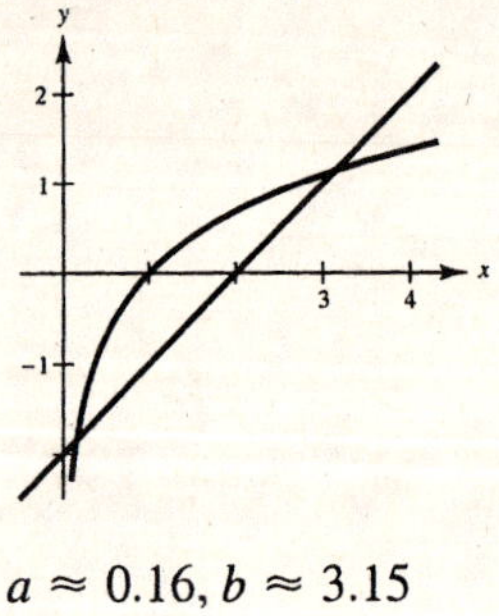


**b.** $a \approx 0.16, b \approx 3.15$

**c.** 1.95

**2.** 0.4216

**3. a.**

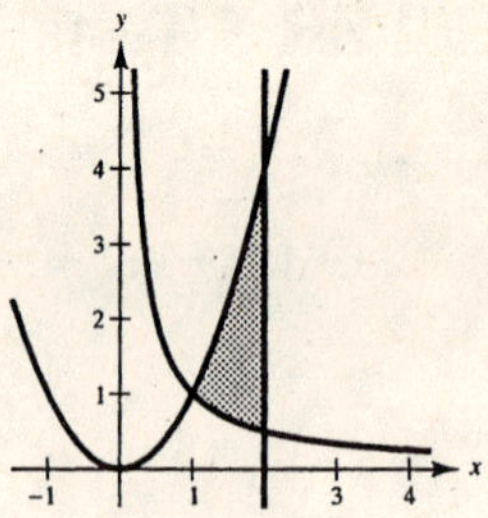


**b.** $(1, 1), \left(2, \frac{1}{2}\right)$, and $(2, 4)$

**c.** $2\pi\left[\frac{23}{12} - 2\ln 2\right]$

**4.** 104.22 ft **5.** $1.014 \times 10^{11}$ ft-lb **6.** $640\pi$ lb

**7. a.** $\frac{28}{15}$ **b.** $\left(0, \frac{7}{5}\right)$ **c.** $\frac{56\pi}{15}$ **d.** $\frac{56\pi}{15}$

**8.** (1.324, 0.611)

# Answers to CHAPTER 7 Tests

### Test Form A

| | | | |
|---|---|---|---|
| **1.** c | **2.** b | **3.** b | **4.** c |
| **5.** a | **6.** d | **7.** b | **8.** c |
| **9.** d | **10.** b | **11.** c | **12.** b |

### Test Form B

| | | | |
|---|---|---|---|
| **1.** b | **2.** a | **3.** c | **4.** a |
| **5.** d | **6.** c | **7.** a | **8.** a |
| **9.** b | **10.** c | **11.** b | **12.** a |

### Test Form C

| | | | |
|---|---|---|---|
| **1.** d | **2.** a | **3.** d | **4.** c |
| **5.** b | **6.** a | **7.** b | **8.** c |
| **9.** d | **10.** a | **11.** b | **12.** b |

### Test Form D

**1.** $\ln|1 - \cos x| + C$ **2.** $\sec x + \tan x - x + C$

**3.** $\frac{1}{4}x^2[2\ln x - 1] + C$

**4.** $2x + \frac{2}{3}\arctan\frac{x+1}{3} + C$

**5.** $-t + \frac{2}{3}\cos^3 t - \frac{1}{5}\cos^5 t + C$

**6.** $2\left[\frac{1}{5}\sec^5\frac{x}{2} - \frac{1}{3}\sec^3\frac{x}{2}\right] + C$

**7.** $\ln|x + \sqrt{x^2 - 4}| + C$ **8.** $\ln(2 + \sqrt{3}) - \frac{\sqrt{3}}{2}$

**9.** $5\ln|x| + \frac{1}{6}\arctan\frac{2x}{3} + C$ **10.** $\infty$

**11.** 1 **12.** The integral diverges.

### Test Form E

**1.** $2 - \ln 4$ **2.** $-\frac{4x+1}{16}e^{-4x} + C$

**3.** $\pi\left[3(\ln 3)^2 - 6(\ln 3) + 4\right]$

**4.** $\frac{2\sqrt{\sin x}}{5}(5 - \sin^2 x) + C$

**5.** $x = \frac{5}{2}\sec\theta + \frac{1}{2}$ **6.** $2\left(\sqrt{3} - \frac{\pi}{3}\right)$

**7.** $\ln\left|\frac{x+2}{x-1}\right| - \frac{3}{x+2} + C$ **8.** $-\frac{1}{2\pi}$

**9. a.** 1 **b.** 1

**10.** $f(x) = 7x^2,\ g(x) = x^2 + 1$

(There are many correct answers.)

**11. a.** 2 **b.** $\infty$ **c.** $\infty$

**12. a.**

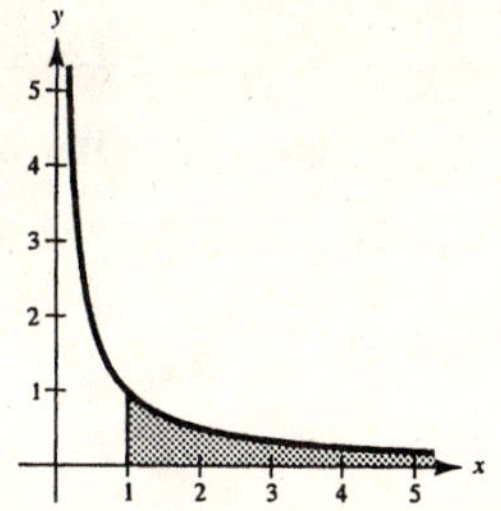


**b.** $\infty$ **c.** $\pi$

# Answers to CHAPTER 8 Tests

**Test Form A**

| | | | |
|---|---|---|---|
| **1.** b | **2.** b | **3.** d | **4.** c |
| **5.** d | **6.** b | **7.** d | **8.** d |
| **9.** c | **10.** a | **11.** c | **12.** b |
| **13.** c | **14.** d | **15.** b | **16.** a |
| **17.** a | **18.** a | **19.** c | **20.** d |

**Test Form B**

| | | | |
|---|---|---|---|
| **1.** d | **2.** d | **3.** a | **4.** a |
| **5.** b | **6.** a | **7.** b | **8.** c |
| **9.** c | **10.** b | **11.** c | **12.** d |
| **13.** b | **14.** a | **15.** c | **16.** b |
| **17.** b | **18.** b | **19.** d | **20.** e |

**Test Form C**

| | | | |
|---|---|---|---|
| **1.** b | **2.** a | **3.** b | **4.** a |
| **5.** b | **6.** d | **7.** d | **8.** c |
| **9.** d | **10.** b | **11.** d | **12.** b |
| **13.** c | **14.** b | **15.** d | **16.** c |

**Test Form D**

**1.** $\frac{243}{7}$ **2.** Diverges **3.** Converges to $e$

**4.** 2 **5.** $\frac{5}{12}$ **6.** Diverges, $p$-Series

**7.** Diverges, Limit Comparison Test

**8.** Converges, Geometric Series Test

**9.** Converges, Alternating Series Test

**10.** Converges, Telescopic Series Test

**11.** Diverges, Limit Comparison Test

**12.** Diverges, Ratio Test **13.** Converges, Root Test

**14.** Diverges, $n$th-Term Test for Divergence

**15.** 1000 **16.** $\frac{4}{3}$ **17.** $\left[\frac{7}{2}, \frac{9}{2}\right)$

**18.** $1 - \frac{x}{2!} + \frac{x^2}{4!} - \frac{x^3}{6!} + \frac{x^4}{8!} - \cdots$

**19.** $-\frac{1}{4}(x-1)^4$ **20.** 0.615

**Test Form E**

**1.** Converges to 0 **2. a.** No **b.** Yes **c.** Yes

**3.** Diverges by the $n$th-Term Test for Divergence

**4. a.** $0.21 + 0.0021 + 0.000021 + 0.00000021 + \cdots$

**b.** $a = 0.21,\ r = 0.01$

**c.** $\sum_{n=0}^{\infty} (0.21)(0.01)^n$

**5.** $\sum_{n=0}^{\infty} \left(\frac{x}{3}\right)^n$

**6. a.** $\int_1^{\infty} x^{-3}\,dx = \frac{1}{2}$

Therefore the series converges also.

**b.** 1.178 **c.** 0.031 **d.** 1.209

**7.** Diverges using $\sum_{n=1}^{\infty} \frac{1}{n}$

**8. a.** $\sum_{n=2}^{\infty} \frac{1}{n}$ diverges

**b.** Diverges because $\frac{1}{\ln n} > \frac{1}{n}, n \ge 2$.

**9.** $\lim_{x\to\infty} \frac{1}{n(\ln n)} = 0$. For $f(x) = \frac{1}{x \ln x}$,

$f'(x) = -\frac{1 + \ln x}{(x \ln x)^2} < 0, x > 0$.

Since the corresponding function is decreasing, the series is also.

**10. a.** $\sum_{n=1}^{\infty} \left|\frac{(-1)^n}{2^n}\right| = \sum_{n=1}^{\infty} \frac{1}{2^n}$, a geometric series with

$r = \frac{1}{2} < 1$. Therefore, the series converges and the original series converges absolutely.

**b.** $-0.328$

**c.** 9

**11.** Diverges by Ratio Test

**12. a.** $P_3(x) = x - \frac{x^3}{3}$ **b.** $-0.291$

**13.** $\int f(x)\,dx = C + \sum_{n=0}^{\infty} \frac{(-1)^n x^{n+1}}{3^n (n+1)}, \quad -3 < x \le 3$

**14.** $1 - 2x + \frac{2^2x^2}{2!} - \frac{2^3x^3}{3!} + \cdots + \frac{(-2x)^n}{n!} + \cdots$

**15.** $1 + \sum_{n=1}^{\infty} \frac{(-1)^n 2^{2n-1} x^{2n}}{(2n)!}$

# Answers to CHAPTER 9 Tests

## Test Form A

| | | | |
|---|---|---|---|
| **1.** a | **2.** c | **3.** d | **4.** c |
| **5.** b | **6.** c | **7.** d | **8.** a |
| **9.** c | **10.** a | **11.** d | **12.** b |
| **13.** a | **14.** c | **15.** b | **16.** d |
| **17.** c | **18.** c | **19.** b | **20.** d |

## Test Form B

| | | | |
|---|---|---|---|
| **1.** c | **2.** d | **3.** c | **4.** b |
| **5.** d | **6.** a | **7.** c | **8.** d |
| **9.** a | **10.** b | **11.** c | **12.** a |
| **13.** c | **14.** b | **15.** a | **16.** c |
| **17.** b | **18.** b | **19.** a | **20.** b |

## Test Form C

| | | | |
|---|---|---|---|
| **1.** a | **2.** c | **3.** d | **4.** b |
| **5.** d | **6.** a | **7.** c | **8.** b |
| **9.** c | **10.** d | **11.** c | **12.** c |
| **13.** b | **14.** a | **15.** c | |

## Test Form D

**1.** $(x-4)^2 = \frac{9}{2}y$ or $y^2 = -\frac{4}{3}(x-4)$

**2.** $\frac{(y-1)^2}{4} - \frac{(x+2)^2}{16} = 1$

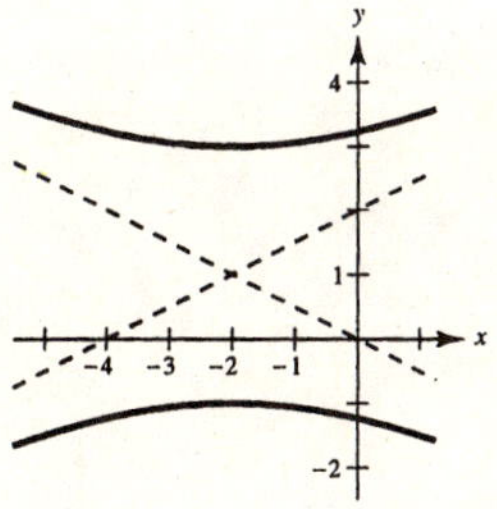

**3.** $2y = x + 4$

**4.**

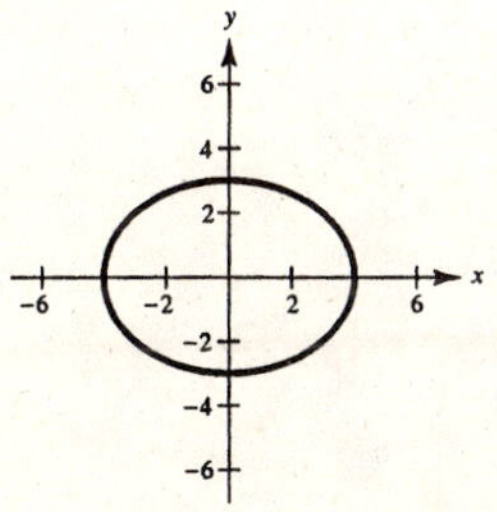

**5.** $y^2 - x - 4y + 3 = 0$

**6.** $x^3 - y^2 - 4x + 2y + 2 = 0$

**7.** $y = -\frac{1}{2}e^t + 2$  **8.** $-\frac{1}{2}\cot\theta$

**9.** $y = 16x - 24$  **10.** $\frac{-2(t+1)}{9t^5}$  **11.** 12

**12.**

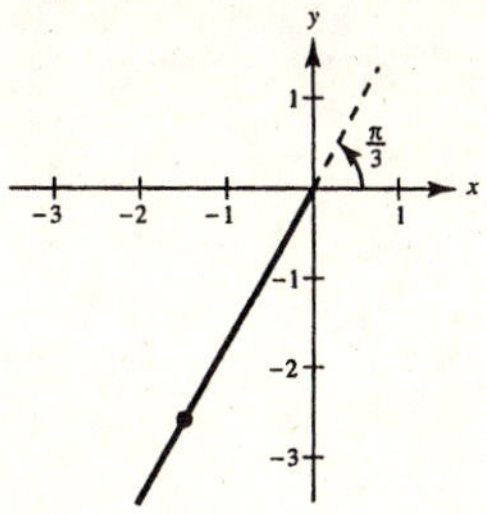

**13.** $(\sqrt{3}, 1)$  **14.** $r = \frac{2}{2\sin\theta - 3\cos\theta}$

**15.** $x^2 + y^2 - 3x = 0$  **16.** $\frac{\pi}{2}, \frac{3\pi}{2}$

**17.**

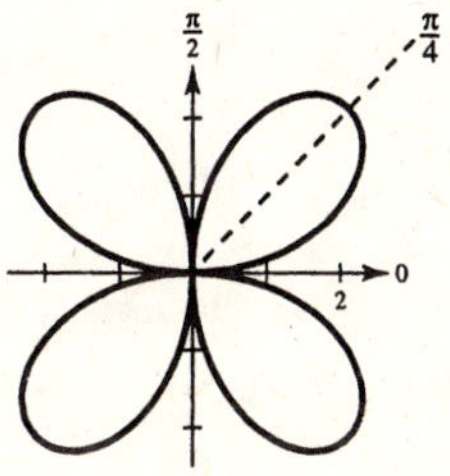

**18.** $\frac{\pi}{3}$  **19.** $\frac{\pi}{6}, \frac{5\pi}{6}$  **20.** $\frac{3\pi}{4}$

## Test Form E

**1. a.** $y = 2x + 1, y = -2x + 9$

**b.**

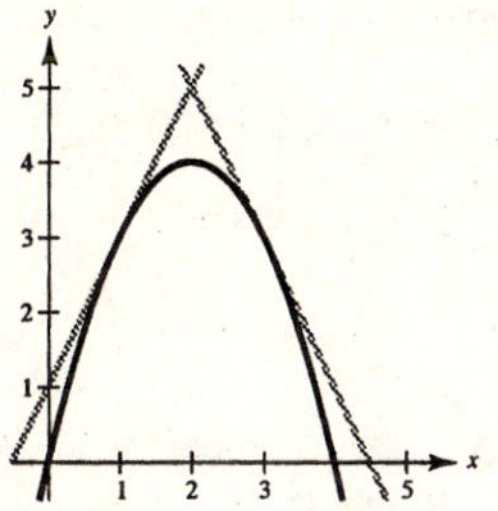

**c.** $\frac{2}{3}$

**2.** $49x^2 + 144y^2 = 2{,}822{,}400$

**3. a.** $y = 3x^2 - 1$

**b.**

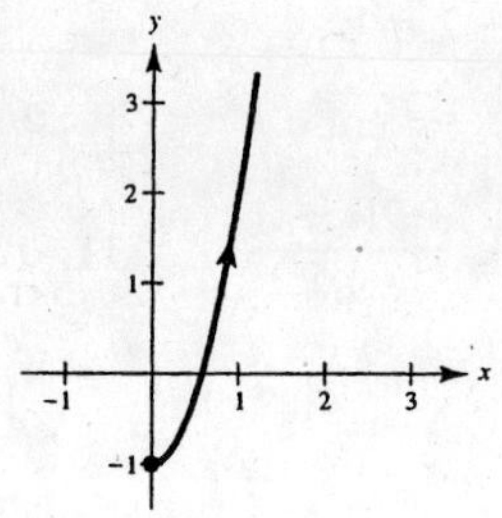

**c.** $x \geq 0$

**4. a.** $(y - 1)^2 + \left(\frac{x}{2}\right)^2 = 1$

**b.**

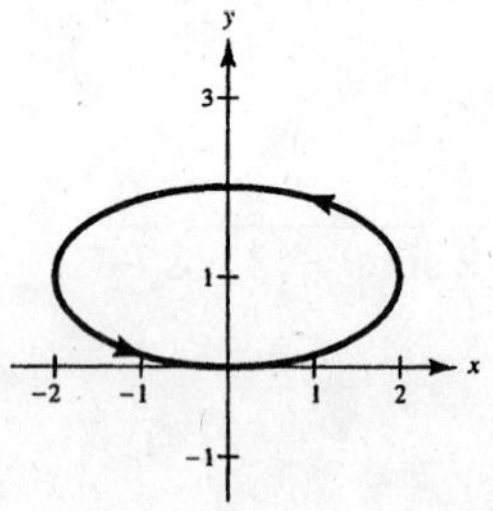

**5. a.**

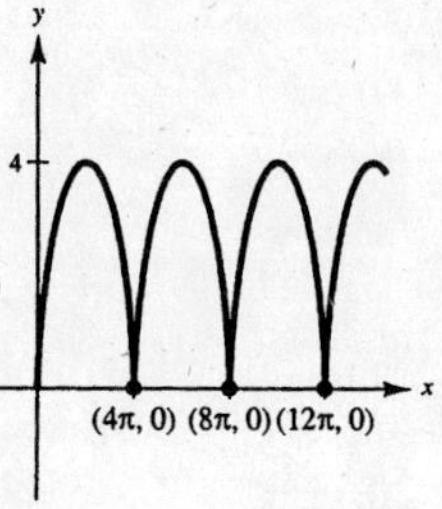


**b.** $(4\pi n, 0), n = 0, 1, 2, \cdots$

**6. a.**

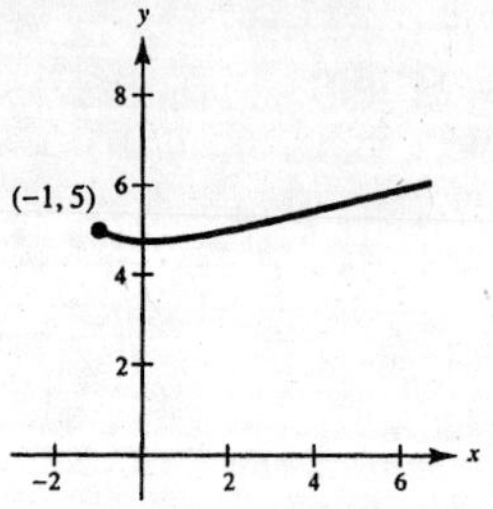


**b.** $\int_1^5 \sqrt{(3t^3 + 2)^2 + (2t - 1)^2}\, dt$

**c.** 133.566

**7.** Relative minimum, $\left(\frac{1}{8}, \frac{19}{4}\right)$

**8.** 2; concave upward

**9. a.**

$$
\begin{aligned}
r &= \cos\theta + \sin\theta \\
r^2 &= r\cos\theta + r\sin\theta \\
x^2 + y^2 &= x + y \\
\left(x^2 - x + \tfrac{1}{4}\right) + \left(y^2 - y + \tfrac{1}{4}\right) &= \tfrac{1}{4} + \tfrac{1}{4} \\
\left(x - \tfrac{1}{2}\right)^2 + \left(y - \tfrac{1}{2}\right)^2 &= \tfrac{1}{2}
\end{aligned}
$$

**b.** center: $\left(\frac{1}{2}, \frac{1}{2}\right)$; radius $= \sqrt{\frac{1}{2}}$

**10.** $4y = 2x + 1$

**11. a.**

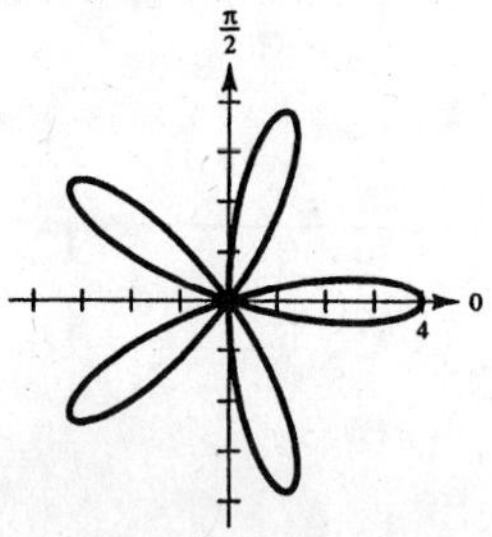

**b.** $\theta = \dfrac{2\pi}{5} = 72°$

**12.**

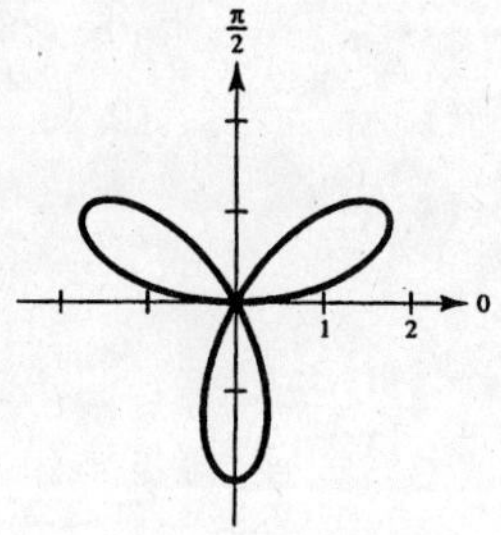

Area: $\pi$

**13.** 0 and $\dfrac{3\pi}{2}$

# Answers to CHAPTER 10 Tests

## Test Form A

| | | | |
|---|---|---|---|
| **1.** d | **2.** b | **3.** a | **4.** b |
| **5.** a | **6.** b | **7.** b | **8.** c |
| **9.** d | **10.** c | **11.** a | **12.** b |
| **13.** d | **14.** c | **15.** b | **16.** e |
| **17.** a | **18.** d | **19.** c | **20.** c |

## Test Form B

| | | | |
|---|---|---|---|
| **1.** c | **2.** d | **3.** b | **4.** a |
| **5.** c | **6.** b | **7.** d | **8.** b |
| **9.** b | **10.** a | **11.** c | **12.** d |
| **13.** c | **14.** b | **15.** a | **16.** c |
| **17.** c | **18.** b | **19.** b | **20.** d |

## Test Form C

| | | | |
|---|---|---|---|
| **1.** b | **2.** c | **3.** a | **4.** a |
| **5.** b | **6.** d | **7.** d | **8.** b |
| **9.** c | **10.** c | **11.** d | **12.** c |
| **13.** c | **14.** a | **15.** d | **16.** b |
| **17.** d | **18.** b | | |

## Test Form D

**1.** $\left\langle -\frac{3}{\sqrt{59}}, -\frac{1}{\sqrt{59}}, \frac{7}{\sqrt{59}} \right\rangle$ **2.** $-\mathbf{j}$ **3.** $-1$

**4.** $3\mathbf{i} - \mathbf{j} - 5\mathbf{k}$ **5.** 14 **6.** $-\frac{1}{3}\mathbf{i} - \frac{1}{6}\mathbf{j} - \frac{1}{6}\mathbf{k}$

**7.** 1.74 **8.** $c(\mathbf{i} + 2\mathbf{j} + 3\mathbf{k}), c \in R, c \neq 0$

**9.** $(-4)(1) + (5)(-2) + (7)(2) = 0$

**10.** $x = 4 + t, y = -1 + 2t, z = 3 + 5t$

**11.** $x - 2z - 6 = 0$

**12.**

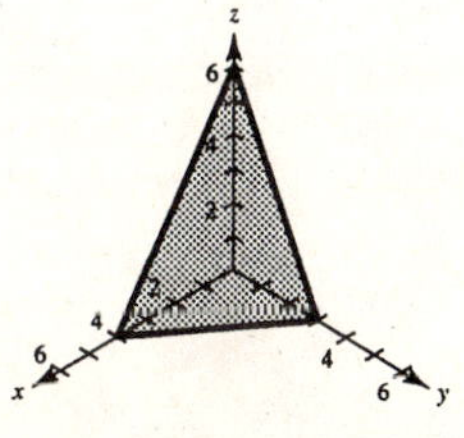

**13.** $x = -\frac{2}{5},\ y = \frac{13}{5}$

**14.** $\left(x - \frac{1}{2}\right)^2 + (y + 2)^2 + (z - 1)^2 = 3$

Center: $\left(\frac{1}{2}, -2, 1\right)$, Radius: $\sqrt{3}$

**15.** Elliptic paraboloid

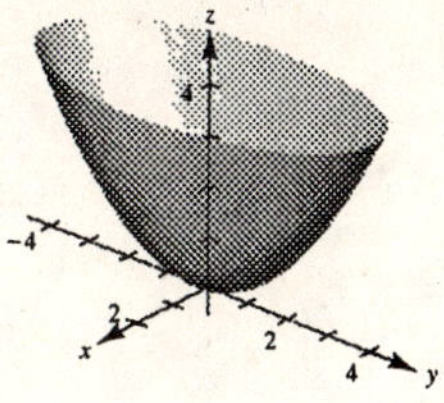

**16.**

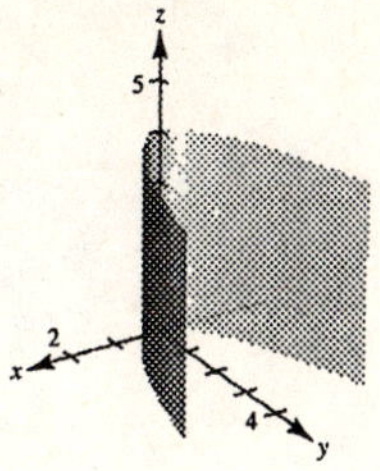

**17.** $4x^2 - 9y^2 - 9z^2 - 4x + 1 = 0$

**18.** $\left(3\sqrt{2}, \frac{7\pi}{4}, 7\right)$

**19.** $y = 9$, plane **20.** $2\sqrt{285}$

## Test Form E

**1.** $\frac{1}{\sqrt{2}}(-\mathbf{i} + \mathbf{j})$ **2.** $\left\langle \frac{3}{\sqrt{5}}, \frac{6}{\sqrt{5}} \right\rangle$

**3.** Direction $\approx 17.4°$; magnitude $\approx 44.4$

**4.** $\left(x - \frac{1}{2}\right)^2 + (y - 6)^2 + (z + 2)^2 = \frac{33}{4}$

**5.** **a.** No **b.** Yes, $c = 2$ **c.** Yes, $c = -\frac{3}{4}$ **d.** No

**6.** **a.** $\langle -6, 1, -4 \rangle$

**b.** $-6\mathbf{i} + \mathbf{j} - 4\mathbf{k}$

**c.** $\sqrt{53}$

**d.** $\frac{1}{\sqrt{53}}\langle -6, 1, -4 \rangle$

**e.** $\frac{1}{\sqrt{53}}\langle 6, -1, 4 \rangle$

**7.** $42\sqrt{2}$

**8.** **a.** $-\frac{15}{14}\mathbf{i} + \frac{5}{7}\mathbf{j} - \frac{5}{14}\mathbf{k}$ **b.** $\frac{1}{14}\mathbf{i} + \frac{9}{7}\mathbf{j} + \frac{33}{14}\mathbf{k}$

**9.** $\cos\alpha = \frac{1}{\sqrt{21}}, \cos\beta = \frac{2}{\sqrt{21}}, \cos\gamma = -\frac{4}{\sqrt{21}}$

**10.** $-3$

**11.** **a.** $6\mathbf{i} + 7\mathbf{j} - 4\mathbf{k}$

**b.**
$(\mathbf{u} \times \mathbf{v}) \cdot \mathbf{v} = (6\mathbf{i} + 7\mathbf{j} - 4\mathbf{k}) \cdot (3\mathbf{i} - 2\mathbf{j} + \mathbf{k}) = 0$

**12.** $\frac{5}{2}\sqrt{6}$

**13.** $x = 2 - 5t, y = 4 + 11t, z = -1 + 19t$

**14.** $\left(-\frac{17}{7}, \frac{33}{14}, \frac{11}{14}\right)$

15.

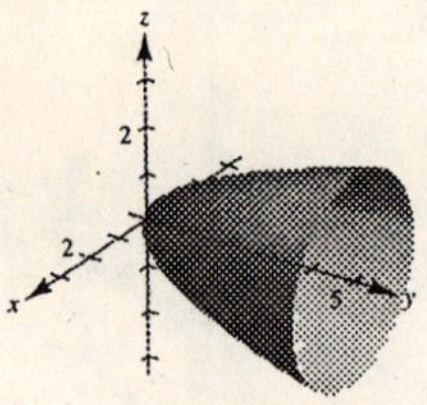

16. **a.** Sphere **b.** Cone

**c.** Hyperboloid of two sheets

17. **a.** $r^2 = 1 + z^2$

**b.** $\rho^2 = -\sec 2\phi$

18.

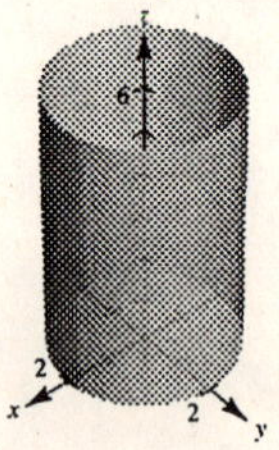

# Answers to CHAPTER 11 Tests

### Test Form A

**1.** d **2.** a **3.** c **4.** c

**5.** a **6.** b **7.** a **8.** b

**9.** c **10.** d

### Test Form B

**1.** c **2.** b **3.** d **4.** b

**5.** d **6.** a **7.** c **8.** c

**9.** a **10.** d

### Test Form C

**1.** b **2.** d **3.** a **4.** a

**5.** d **6.** d **7.** b **8.** b

**9.** d **10.** b

### Test Form D

**1.** $(0, \infty)$ **2.** $\mathbf{i} - 4\mathbf{j} - \mathbf{k}$ **3.** $-\pi\mathbf{i} - \mathbf{j} + \mathbf{k}$

**4.** $5\sqrt{41}$ **5.** $2\pi n,\ n = 0, \pm1, \pm2, \cdots$

**6.** $\mathbf{r}(t) = 3t\mathbf{i} + (5 - 2t)\mathbf{j} + (2 + t - 16t^2)\mathbf{k}$

**7.** $\dfrac{11}{3}$ **8.** $K = \dfrac{1}{2}$ **9.** $-\dfrac{1}{\sqrt{3}}(\mathbf{i} + \mathbf{j} + \mathbf{k})$

**10.** $\dfrac{6\sqrt{10}t}{\sqrt{10 + 9t^4}}$

### Test Form E

**1.**

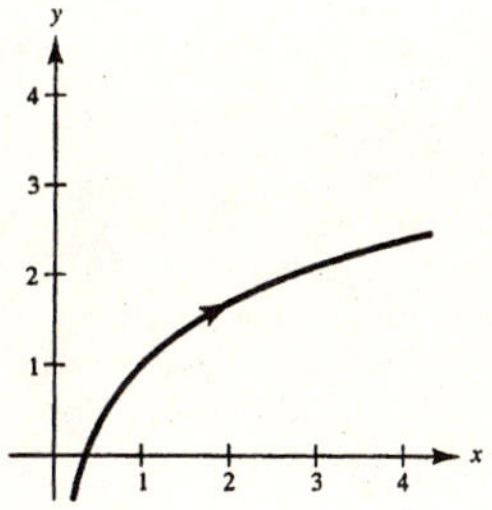


**2.** $-\mathbf{i} + 2\mathbf{k}$ **3.** $\dfrac{3}{(9t^2 + 5)^{3/2}}\langle 5, -3, 6t\rangle$

**4.** $(2n\pi,\ 2(n + 1)\pi),\ n = 0, 1, 2, 3, \cdots$

**5.** $\mathbf{v}(t) = -\dfrac{\sqrt{2}}{2}\mathbf{i} + \dfrac{\sqrt{2}}{2}\mathbf{j} - 8\pi\mathbf{k}$

$\|\mathbf{v}(t)\| = \sqrt{1 + 64\pi^2}$

$\mathbf{a}(t) = -\dfrac{\sqrt{2}}{2}\mathbf{i} - \dfrac{\sqrt{2}}{2}\mathbf{j} - 32\mathbf{k}$

**6. a.** $\dfrac{75\sqrt{3}}{2}t\mathbf{i} + \left(\dfrac{75}{2}t - 16t^2\right)\mathbf{j}$

**b.**

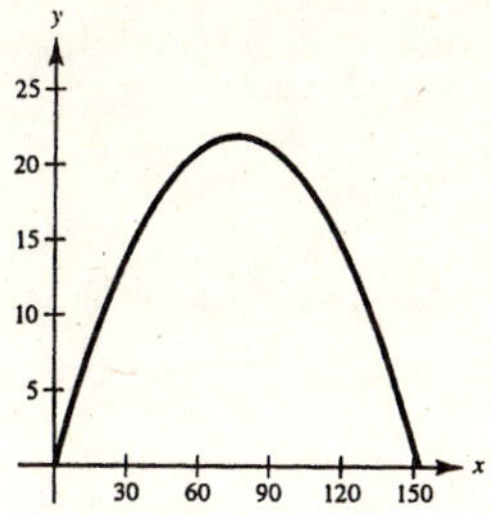


**c.** 21.9 ft, 152 ft

**7. a.** $\dfrac{1}{\sqrt{154}}(\mathbf{i} + 12\mathbf{j} + 3\mathbf{k})$

**b.** $x = 2 + s,\ y = 8 + 12s,\ z = 6 + 3s$

**8. a.** $\mathbf{T}(t) = \dfrac{1}{\sqrt{1 + 4t^2}}(\mathbf{i} + 2t\mathbf{j})$

$\mathbf{T}(1) = \dfrac{1}{\sqrt{5}}(\mathbf{i} + 2\mathbf{j})$

$\mathbf{N}(1) = \dfrac{1}{\sqrt{5}}(-2\mathbf{i} + \mathbf{j})$

**b.**

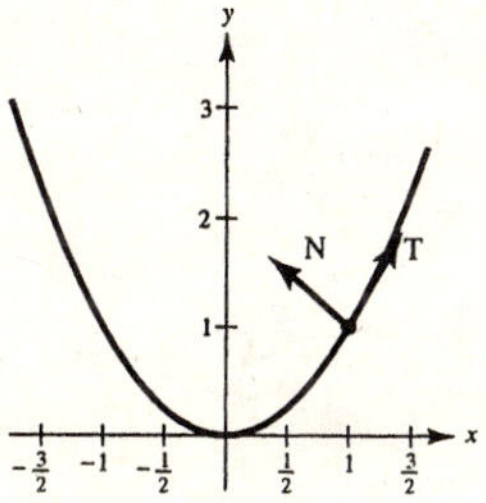


**9.** $\dfrac{\sqrt{2}}{3}$ **10.** 7.763

# Answers to CHAPTER 12 Tests

## Test Form A

1. b 2. d 3. a 4. a
5. c 6. b 7. c 8. d
9. d 10. b

## Test Form B

1. c 2. c 3. b 4. d
5. a 6. d 7. b 8. b
9. d 10. a

## Test Form C

1. c 2. b 3. a 4. c
5. c 6. a 7. c 8. d
9. c 10. d 11. a

## Test Form D

1. $\dfrac{\partial f}{\partial x} = \dfrac{2x}{x^2 + y^2}; \dfrac{\partial^2 f}{\partial x^2} = \dfrac{2y^2 - 2x^2}{(x^2 + y^2)^2};$

$\dfrac{\partial f}{\partial y} = \dfrac{2y}{x^2 + y^2}; \dfrac{\partial^2 f}{\partial y^2} = \dfrac{2x^2 - 2y^2}{(x^2 + y^2)^2};$

$\dfrac{\partial^2 f}{\partial x^2} + \dfrac{\partial^2 f}{\partial y^2} = \dfrac{2y^2 - 2x^2}{(x^2 + y^2)^2} + \dfrac{2x^2 - 2y^2}{(x^2 + y^2)^2} = 0$

2. $-\dfrac{yzw\cos(xyz)}{2w + \sin(xyz)}$

3. $-\sin(u - v)\sin(u^2 + v^2) + 2u\cos(u - v)\cos(u^2 + v^2)$

4. $-0.004$ 5. $\frac{48}{5}$ 6. $7e^6$

7. $\dfrac{x + 2}{16} = \dfrac{y + 1}{-4} = \dfrac{z - 2}{7}$

8. $f(1, 1)$ is a relative minimum

9. 11, 11, 11 10. $y = \frac{7}{10}x + 1$

## Test Form E

1. $\{(x, y) : x^2 + y^2 > 4\}$ 2. $-4$

3. $f$ is continuous at each point in space except at the points on the plane given by $x + 2y + 3z = 6$.

4. 5

5. a. $960\pi \approx 3015.9 \text{ cm}^3$

b. $\pm 6.08\pi \approx \pm 19.1 \text{ cm}^3$

c. 0.6%

6. $182 \text{ ft}^3/\text{min}$ 7. $\dfrac{x}{y + z}$

8. a. $-4\mathbf{i} - \frac{4}{5}\mathbf{j}$

b. $-\dfrac{4}{5}$

c. $\dfrac{4}{5}\sqrt{26}$

9. a. Elliptic paraboloid

b. $2x - 8y - 5z - 10 = 0$

c. $\dfrac{x - 2}{2} = \dfrac{y + 2}{-8} = \dfrac{z - 2}{-5}$

10. Saddle point 11. $(3, 3)$ and $(-3, -3)$

# Answers to CHAPTER 13 Tests

## Test Form A

1. c 2. d 3. a 4. d

5. b 6. b 7. c 8. a

9. c 10. d

## Test Form B

1. a 2. e 3. b 4. c

5. d 6. c 7. b 8. b

9. a 10. d

## Test Form C

1. b 2. a 3. b 4. a

5. c 6. b 7. c 8. d

9. d 10. a

## Test Form D

1. $\frac{9}{2}$ 2. $\frac{11}{6}$ 3. $\frac{2}{3\pi}$ 4. $8\pi$

5. $\frac{\pi(e-1)}{4e}$ 6. $\frac{15}{8}$ 7. $\frac{32}{3}k$ 8. $50\pi$

9. $4\pi(2-\sqrt{3})$

10. $\frac{\partial(x, y)}{\partial(r, \theta)} = \begin{vmatrix} \frac{\partial x}{\partial r} & \frac{\partial x}{\partial \theta} \\ \frac{\partial y}{\partial r} & \frac{\partial y}{\partial \theta} \end{vmatrix} = \begin{vmatrix} \cos\theta & -r\sin\theta \\ \sin\theta & r\cos\theta \end{vmatrix} = r$

## Test Form E

1. $\frac{1}{3}(2e^3+1)$

2. a.

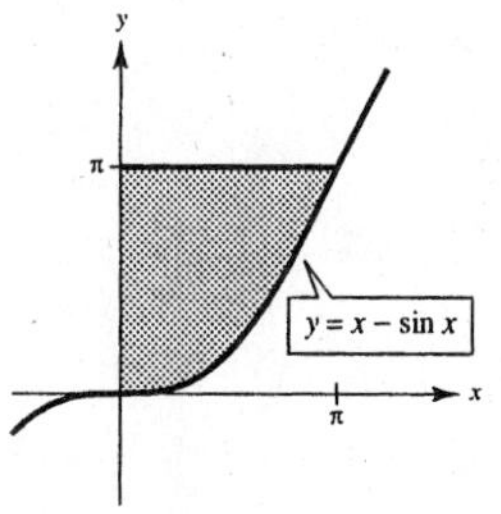


b. $2 + \frac{\pi^2}{2}$

3. $\frac{1}{2}(1-\cos 1)$ 4. 0.0226

5. a. $2\int_0^4 \int_0^{\sqrt{16-x^2}} 4x\,dy\,dx$

b. $8\int_0^{\pi/2} \int_0^4 r^2 \cos\theta\,dr\,d\theta$

c. $\frac{512}{3}$

6. $\int_0^2 \int_{2-x}^{\sqrt{4-x^2}} x^3y\,dy\,dx$ or $\int_0^2 \int_{2-y}^{\sqrt{4-y^2}} x^3y\,dx\,dy$

7. 1.123

8. a. $8\int_0^{\pi/2} \int_0^a \int_0^{\sqrt{a^2-r^2}} r\,dz\,dr\,d\theta$

b. $8\int_0^{\pi/2} \int_0^{\pi/2} \int_0^a \rho^2 \sin\phi\,d\rho\,d\phi\,d\theta$

9. a. $\int_0^{2\pi} \int_0^{\pi} \int_1^3 \rho^2 \sin\phi\,d\rho\,d\phi\,d\theta =$

$8\int_0^{\pi/2} \int_0^{\pi/2} \int_1^3 \rho^2 \sin\phi\,d\rho\,d\phi\,d\theta$

b. $\frac{104\pi}{3}$

10. $\frac{21}{10}$

# Answers to CHAPTER 14 Tests

## Test Form A

**1.** a **2.** b **3.** c **4.** d

**5.** b **6.** c **7.** d **8.** a

**9.** c **10.** b

## Test Form B

**1.** d **2.** a **3.** b **4.** c

**5.** a **6.** b **7.** a **8.** c

**9.** d **10.** b

## Test Form C

**1.** b **2.** a **3.** d **4.** a

**5.** d **6.** b **7.** c **8.** b

**9.** c **10.** a

## Test Form D

**1.** $\frac{\sqrt{3}-1}{2}$ **2.** $(z\cos yz + 2)\mathbf{i} - x\mathbf{k}$

**3.** $x^4y^3 + \ln|x| - \ln|y| + C$ **4.** 128

**5.** $2\pi + 4$ **6.** 0 **7.** $\frac{4}{5}$ **8.** $\frac{149}{385}$

**9.** 3 **10.** $36\pi$

## Test Form E

**1.** $-(2e^{2y} + xy^2)\mathbf{i} + x\mathbf{j} + y^2z\mathbf{k}$

**2.** Conservative;
$f(x, y, z) = x^2z + y^2z + z^3 + y^2 + x + C$

**3.** $\frac{13}{2}$ **4.** $-\frac{1}{4}$ **5.** 0 **6.** $2 + 4\pi$

**7.** $36\pi\sqrt{17}$ **8.** $\frac{27\sqrt{6}}{2}$ **9.** $32\pi$ **10.** $-3$

**Test Form A** **Name** ____________________ **Date** ____________

**FINAL EXAM** **Class** ____________________ **Section** ____________

**Chapters P–5**

**1.** Find the limit: $\lim_{\Delta x \to 0} \frac{\sqrt{(x + \Delta x) - 9} - \sqrt{x - 9}}{\Delta x}$.

(a) $\frac{1}{\sqrt{x - 9}}$ (b) $\sqrt{x - 9}$ (c) $\frac{1}{2\sqrt{x - 9}}$

(d) 0 (e) None of these

**2.** Find the limit: $\lim_{x \to 4^+} \frac{x^2 - x}{(x - 4)^2}$.

(a) 0 (b) $+\infty$ (c) $-\infty$

(d) 4 (e) None of these

**3.** Find an equation of the tangent line to the graph of $f(x) = \frac{1}{x - 1}$ at the point (2, 1).

(a) $x + y + 3 = 0$ (b) $x - y = 1$ (c) $y - 1 = -\frac{(x - 2)}{(x - 1)^2}$

(d) $x + y = 6$ (e) None of these

**4.** Find $f'(x)$: $f(x) = \frac{x^2 - 4x}{\sqrt{x}}$.

(a) $\frac{3x - 4}{2\sqrt{x}}$ (b) $\frac{2x - 4}{\sqrt{x}}$ (c) $\frac{2x - 4}{1/(2\sqrt{x})}$

(d) $x^{3/2} - 4x^{1/2}$ (e) None of these

**5.** The position function for a particular object is $s = -12t^2 + 51t + 38$. Which statement is true?

(a) The velocity at time $t = 1$ is 27. (b) The velocity is a constant.

(c) The initial position is 51. (d) The initial velocity is $-24$.

(e) None of these

**6.** Find the derivative: $s(t) = \sec\sqrt{t}$.

(a) $\tan^2\sqrt{t}$ (b) $\frac{\sec\sqrt{t}\tan\sqrt{t}}{2\sqrt{t}}$ (c) $\sec\frac{1}{2\sqrt{t}}\tan\frac{1}{2\sqrt{t}}$

(d) $\sec\sqrt{t}\tan\sqrt{t}$ (e) None of these

**7.** Find $\frac{dy}{dx}$ for $5x^2 - 2xy + 7y^2 = 0$.

(a) $\frac{5x + 7y}{x}$ (b) $\frac{y - 5x}{7y}$ (c) $10x - 2y + 14y$

(d) $\frac{y - 5x}{7y - x}$ (e) None of these

**8.** A machine is rolling a metal cylinder under pressure. The radius of the cylinder is decreasing at a constant rate of 0.05 inches per second and the volume $V$ is $128\pi$ cubic inches. At what rate is the length $h$ changing when the radius $r$ is 2.5 inches? [Hint: $V = \pi r^2h$]

(a) 20.48 in/sec (b) $-0.8192$ in/sec (c) $-16.38$ in/sec

(d) 0.8192 in/sec (e) None of these

**9.** Determine whether the Mean Value Theorem applies to $f(x) = -\frac{1}{x}$ on the interval $\left[-3, -\frac{1}{2}\right]$.

If the Mean Value Theorem applies, find all values of $c$ in the interval such that $f'(c) = \frac{f(-1/2) - f(-3)}{-1/2 - (-3)}$. If the Mean Value Theorem does not apply, state why.

(a) Mean Value Theorem applies; $c = -\sqrt{\frac{3}{2}}$.

(b) Mean Value Theorem applies; $c = \pm\sqrt{\frac{3}{2}}$.

(c) The Mean Value Theorem does not apply because $f$ is not continuous at $x = 0$.

(d) The Mean Value Theorem does not apply because $f\left(-\frac{1}{2}\right) \neq f(-3)$.

(e) None of these

**10.** Given that $f(x) = -x^2 + 12x - 34$ has a relative maximum at $x = 6$, choose the correct statement.

(a) $f'$ is positive on the interval $(6, \infty)$. (b) $f'$ is positive on the interval $(-\infty, \infty)$.

(c) $f'$ is negative on the interval $(6, \infty)$. (d) $f'$ is negative on the interval $(-\infty, 6)$.

(e) None of these

**11.** Give the sign of the second derivative of $f$ at the indicated point.

(a) Positive

(b) Negative

(c) Zero

(d) The sign cannot be determined.

(e) None of these

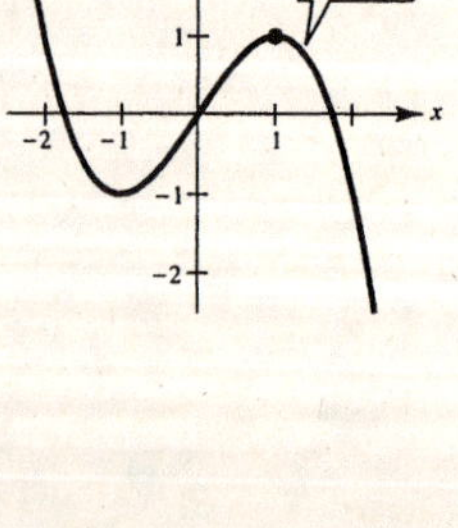

**12.** Find the horizontal asymptotes (if any) for $f(x) = \dfrac{ax^3}{b + cx + dx^2}$.

(a) $y = \dfrac{a}{b}$ (b) $y = 0$ (c) $y = \dfrac{a}{d}$

(d) There are no horizontal asymptotes. (e) None of these

**13.** Evaluate the integral: $\displaystyle\int \frac{4 + 5x^{3/2}}{\sqrt{x}}\,dx$.

(a) $2x^{-3/2} + \dfrac{5}{2}x^2 + C$ (b) $8\sqrt{x} + \dfrac{5}{2}x^2 + C$ (c) $-2x^{-3/2} + 5 + C$

(d) $2\sqrt{x} + \dfrac{5}{2}x^2 + C$ (e) None of these

**14.** Determine which of the following is *not* equal to $\displaystyle\int_2^7 ax\,f(x)\,dx$.

(a) $\displaystyle a\int_2^7 x\,f(x)\,dx$ (b) $\displaystyle x\int_2^7 a\,f(x)\,dx$ (c) $\displaystyle -\int_7^2 ax\,f(x)\,dx$

(d) $\displaystyle\int_2^4 ax\,f(x)\,dx + \int_4^7 ax\,f(x)\,dx$ (e) None of these

**15.** Evaluate the integral: $\displaystyle\int \frac{1}{\sqrt{2x + 1}}\,dx$.

(a) $\sqrt{x^2 + x} + C$ (b) $\dfrac{1}{\sqrt{x^2 + x}} + C$ (c) $\sqrt{2x + 1} + C$

(d) $\dfrac{1}{2}\sqrt{2x + 1} + C$ (e) None of these

**16.** Solve the differential equation: $\dfrac{ds}{dt} = \dfrac{\sec t \tan t}{\sec t + 5}$.

(a) $s = \ln|\sec t + 5| + C$ (b) $s = \dfrac{1}{5}\ln|\sec t| + C$ (c) $s = 2\sec^3 t - \sec t + C$

(d) $s = \frac{1}{5}\tan t + C$ (e) None of these

**17.** Evaluate the indefinite integral: $\displaystyle\int \frac{1}{x^2 e^{2/x}}\,dx$.

(a) $\dfrac{1}{2}xe^{-2/x} + C$ (b) $\dfrac{1}{2}xe^{2/x} + C$ (c) $\dfrac{1}{2}e^{2/x} + C$

(d) $\dfrac{1}{2}e^{-2/x} + C$ (e) None of these

**18.** Find the derivative: $g(x) = \operatorname{arcsec} \dfrac{x}{2}$.

(a) $\dfrac{1}{\sqrt{4 - x^2}}$ (b) $\dfrac{1}{\sqrt{x^2 - 4}}$ (c) $\dfrac{2}{x\sqrt{x^2 - 4}}$

(d) $\dfrac{4}{x\sqrt{x^2 - 4}}$ (e) None of these

**19.** Evaluate the integral: $\displaystyle\int \frac{(\arctan x)^3}{1 + x^2}\,dx$.

(a) $\dfrac{1}{4}(\arctan x)^4 + C$ (b) $\dfrac{1}{4}(\arctan x)^5 + C$ (c) $\dfrac{3(\arctan x)^2}{(1 + x^2)^2} + C$

(d) $\dfrac{3(\arctan x)^2}{2x(1 + x^2)} + C$ (e) None of these

**20.** Evaluate the integral: $\displaystyle\int \frac{3}{\sqrt{3 - 2x - x^2}}\,dx$.

(a) $6\sqrt{3 - 2x - x^2} + C$ (b) $3 \arcsin \dfrac{x + 1}{2} + C$

(c) $\dfrac{3}{2} \arcsin \dfrac{x + 1}{2} + C$ (d) $3 \ln\sqrt{3 - 2x - x^2} + C$

(e) None of these

**Test Form B** **Name** ________________ **Date** ________

**FINAL EXAM** **Class** ________________ **Section** ________

**Chapters P–5**

1. Find the limit: $\lim_{x \to 1} \dfrac{1 - \sqrt{2x^2 - 1}}{x - 1}$

2. Find the limit: $\lim_{x \to 1^+} \dfrac{x^2 - x - 2}{x - 1}$.

3. Find an equation of the tangent line to the graph of $f(x) = x^2 - 2x - 3$ at the point $(-2, 5)$.

4. Find $f'(x)$: $f(x) = \dfrac{x^2 - 3x}{x^2}$.

5. The position function for an object is given by $s(t) = 6t^2 + 240t$, where $s$ is measured in feet and $t$ is measured in seconds. Find the velocity of the object when $t = 2$ seconds.

6. Find $f'(x)$ if $f(x) = \sin x(\sin x + \cos x)$.

7. Find $\dfrac{dy}{dx}$ for the equation $x^3 - 2x^2y + 3xy^2 = 38$.

8. Sand is falling off a conveyor onto a conical pile at the rate of 15 cubic feet per minute. The diameter of the base of the cone is approximately twice the altitude. At what rate is the height of the pile changing when it is 10 feet high?

9. Determine whether the Mean Value Theorem applies to $f(x) = 3x - x^2$ on the interval $[2, 3]$.

   If the Mean Value Theorem can be applied, find all values of $c$ in the interval such that $f'(c) = \dfrac{f(3) - f(2)}{3 - 2}$. If the Mean Value Theorem does not apply, state why.

10. Let $f(x) = \dfrac{x}{1 - x}$. Use calculus to show that $f$ is increasing wherever it is defined.

11. Sketch the graph of a function $f$ such that $f'$ is a decreasing function.

12. Find the horizontal asymptotes (if any) for $f(x) = \dfrac{ax^3}{bx^3 + cx + d}$.

**13.** A right circular cylinder is to be designed to hold 22 cubic inches of a soft drink. The cost for the material for the top and bottom of the can is twice the cost for the material of the sides. Let $r$ represent the radius and $h$ the height of the cylinder.

**a.** Write the equation for the surface area, $SA$, in $r$ and $h$.

**b.** Write the cost function, $C$.

**c.** Write the cost function as a function of one variable, $r$.

**d.** Find the radius that minimizes cost.

**14.** Evaluate the integral: $\displaystyle\int \frac{7 + 3x^{3/2}}{\sqrt{x}}\,dx.$

**15.** Sketch the region whose areas is indicated by the integral: $\displaystyle\int_0^3 \sqrt{9 - x^2}\,dx.$

**16.** Find the indefinite integral: $\displaystyle\int \left(x - \frac{1}{x}\right)^2 dx.$

**17.** Solve the differential equation: $\dfrac{dy}{dx} = \dfrac{3x}{1 - x^2}.$

**18.** Differentiate: $f(x) = \ln \dfrac{e^x}{e^x + 1}.$

**19.** Find the derivative: $y = \arctan e^x.$

**20.** Evaluate the integral: $\displaystyle\int \frac{2 - x}{\sqrt{4 - x^2}}\,dx.$

**Test Form A** **Name** ____________ **Date** ____________

**FINAL EXAM** **Class** ____________ **Section** ____________

**Chapters 6–9**

**1.** Find the area of the region bounded by the graphs of $f(x) = x^3 + 4x^2 - 12x$ and $g(x) = -x^2 + 2x$.

(a) $\frac{3901}{12}$ (b) $\frac{32}{3}$ (c) $\frac{3773}{6}$

(d) $\frac{1215}{4}$ (e) None of these

**2.** Identify the definite integral that computes the volume of the solid generated by revolving the region bounded by the graph of $y = x^3$ and the line $y = x$, between $x = 0$ and $x = 1$, about the $y$-axis.

(a) $\pi\int_0^1 (x^2 - x^4)\,dx$ (b) $\pi\int_0^1 (y^{1/3} - y)^2\,dy$ (c) $\pi\int_0^1 (x^4 - x^2)\,dx$

(d) $\pi\int_0^1 (y^{2/3} - y^2)\,dy$ (e) None of these

**3.** Evaluate the integral: $\displaystyle\int \frac{x}{(x+6)^{2/3}}\,dx.$

(a) $3(x+6)^{1/3} + C$ (b) $\frac{3}{4}(x+6)^{2/3} - 18(x+6)^{1/6} + C$

(c) $\frac{3}{4}(x+6)^{4/3} - 18(x+6)^{1/3} + C$ (d) $\frac{3}{4}(x+6)^{1/3}(x+18) + C$

(e) None of these

**4.** Evaluate the integral: $\displaystyle\int x\cos 2x\,dx.$

(a) $\frac{1}{4}\sin 2x + \frac{1}{2}x\cos 2x + C$ (b) $\frac{1}{4}\cos 2x - \frac{1}{2}x\sin 2x + C$

(c) $\frac{1}{4}\cos 2x + \frac{1}{2}x\sin 2x + C$ (d) $\frac{1}{4}\sin 2x - \frac{1}{2}x\cos 2x + C$

(e) None of these

**5.** Evaluate the integral: $\displaystyle\int \sec^5 x\tan^3 x\,dx.$

(a) $\frac{1}{7}\sec^7 x - \frac{1}{5}\sec^5 x + C$ (b) $\frac{1}{24}\sec^6 x\tan^4 x + C$

(c) $\frac{1}{4}\sec^4 x\tan x + C$ (d) $\frac{1}{8}\sec^8 x - \frac{1}{6}\sec^6 x + C$

(e) None of these

**6.** Evaluate the integral: $\displaystyle\int \frac{dx}{(9 - x^2)^{3/2}}.$

(a) $\frac{-2}{\sqrt{9-x^2}} + C$ (b) $\frac{2}{3\sqrt{9-x^2}} + C$ (c) $\frac{x}{\sqrt{9-x^2}} + C$

(d) $\frac{x}{9\sqrt{9-x^2}} + C$ (e) None of these

**7.** Evaluate the integral: $\int \frac{\cos x}{\sin x(1 - \sin x)}\, dx$.

(a) $\frac{\sin^2 x}{2} - \frac{\sin^3 x}{3} + C$ (b) $\ln|\sin x - \sin^2 x| + C$ (c) $\ln\left|\frac{\sin x}{1 - \sin x}\right| + C$

(d) $\frac{\csc^2 x}{2} - \frac{\csc^3 x}{3} + C$ (e) None of these

**8.** Evaluate the limit using L'Hôspital's Rule, if applicable: $\lim_{x\to 0} \frac{8e^{x/8} - (8 + x)}{x^2}$.

(a) $\frac{1}{16}$ (b) $\infty$ (c) $\frac{1}{2}$

(d) 0 (e) None of these

**9.** Evaluate the improper integral: $\int_{-5}^{-3} \frac{dx}{(x + 4)^2}$.

(a) $-2$ (b) The integral diverges. (c) 2

(d) 0 (e) None of these

**10.** Find the sum of the infinite geometric series: $\sum_{n=3}^{\infty} 2\left(\frac{1}{3}\right)^n$.

(a) 3 (b) $\frac{1}{3}$ (c) $\frac{1}{9}$

(d) $\frac{4}{3}$ (e) None of these

**11.** Determine which of the following tests could be used to show that the harmonic series diverge.

(a) Geometric Series Test (b) $n$th-Term Test for Divergence

(c) Telescopic Series Test (d) Integral Test

(e) None of these

**12.** Investigate the series $\sum_{n=1}^{\infty} \frac{\sqrt{n}}{4n^3 - 6n^2 + 5}$ for convergence or divergence.

(a) Converges by $n$th-Term Test for Divergence

(b) Diverges by $n$th-Term Test for Divergence

(c) Converges by Root Test

(d) Converges by Limit Comparison Test

(e) None of these

**13.** Find the interval of convergence of the power series: $\sum_{n=1}^{\infty} \frac{(-1)^n}{n}(x - 3)^{n-1}$.

(a) $(-1, 1)$ (b) $(-1, 1]$ (c) $(2, 4)$

(d) $[2, 4]$ (e) None of these

**14.** Find the Maclaurin series for $f(x) = e^{-2x}$.

(a) $1 - 2x + \frac{4x^2}{2!} - \frac{8x^3}{3!} + \frac{16x^4}{4!} - \cdots$

(b) $1 + 2x + \frac{4x^3}{2!} + \frac{8x^3}{3!} + \frac{16x^4}{4!} + \cdots$

(c) $-2\left(1 + x + \frac{x^2}{2!} + \frac{x^3}{3!} + \frac{x^4}{4!} + \cdots\right)$

(d) $-2 + x + \frac{x^2}{2!} + \frac{x^3}{3!} + \frac{x^4}{4!} + \cdots$

(e) None of these

**15.** Find the corresponding rectangular equation by eliminating the parameter: $x = 1 + \sec\theta$, $y = 2 + \tan\theta$.

(a) $x^2 + y^2 - 2x - 4y + 4 = 0$

(b) $x^2 - y^2 - 2x + 4y - 2 = 0$

(c) $x^2 - y^2 - 2x + 4y - 4 = 0$

(d) $y = 2 + \frac{x+1}{x-1}$

(e) None of these

**16.** Find an equation for the tangent line at the point where $t = 1$ for the curve given by the parametric equations $x = 3t - 1$ and $y = t^2$.

(a) $3x - 2y = 1$

(b) $2x - 3y = 1$

(c) $2x + 3y = 1$

(d) $\frac{2}{3}$

(e) $y - 1 = \frac{2}{3}t(x - 2)$

**17.** Find the slope of the tangent line for the curve $r = 2\cos 3\theta$ at the point where $\theta = \frac{\pi}{6}$.

(a) $-6$

(b) $\frac{1}{\sqrt{3}}$

(c) $-\sqrt{3}$

(d) 1

(e) None of these

**18.** Find the arclength of the curve given by $r = \sin\theta$, $0 \le \theta \le \pi$.

(a) $\pi$

(b) $\frac{\pi}{2}$

(c) $2\pi$

(d) $\pi^2$

(e) 0

**19.** Calculate the area inside the curve given by $r = 5\sin\theta$ and outside the curve given by $r = 2 + \sin\theta$.

(a) $\frac{8\pi}{3}$

(b) 21

(c) $\frac{8\pi}{3} + \sqrt{3}$

(d) $\frac{8\pi}{3} - \sqrt{3}$

(e) $\frac{25\pi}{4}$

**20.** Identify the graph: $r = \frac{4}{2 + \cos\theta}$.

(a) Ellipse

(b) Parabola

(c) Hyperbola

(d) Circle

(e) None of these

**Test Form B** **Name** ________________ **Date** ____________

**FINAL EXAM** **Class** ________________ **Section** ____________

**Chapters 6–9**

1. Find the area of the region bounded by the graphs of $f(x) = \sin x$ and $g(x) = \cos x$, $\pi/4 \le x \le 5\pi/4$.

2. Find the volume of the solid formed by revolving the region bounded by the graphs of $y = \sqrt{x-2}$, $y = 0$, and $x = 6$, about the $y$-axis.

3. Evaluate the integral: $\displaystyle\int \frac{dx}{x\sqrt{\ln x}}$.

4. Evaluate the integral: $\displaystyle\int x^2 e^{3x}\,dx$.

5. Evaluate the integral: $\displaystyle\int \csc^4 3\theta \tan^4 3\theta\,d\theta$.

6. Evaluate the integral: $\displaystyle\int_{1/2}^{1} \frac{\sqrt{1-x^2}}{x}\,dx$.

7. Evaluate the integral: $\displaystyle\int \frac{3}{x(x+3)}\,dx$.

8. Find the limit: $\displaystyle\lim_{x\to 1} \frac{\ln x}{x-1}$.

9. Evaluate the improper integral: $\displaystyle\int_0^{\infty} \frac{1}{4+x^2}\,dx$.

10. Determine the convergence or divergence of the following series, and state the test used: $\displaystyle\sum_{n=1}^{\infty} \frac{1}{1+e^{-n}}$.

11. Use the Integral Test to determine convergence or divergence: $\displaystyle\sum_{n=1}^{\infty} ne^{-n}$.

12. Determine the convergence or divergence of the following series, and state the test used: $\displaystyle\sum_{n=1}^{\infty} \frac{n}{\sqrt{n^3+2n}}$.

13. Find the interval of convergence of the power series: $\displaystyle\sum_{n=1}^{\infty} \frac{(x+4)^n}{n \cdot 2^n}$.

**14.** Write the first four terms and the general term for the Taylor series expansion of $f(x) = \sin x$ about $x = \pi$.

**15.** Eliminate the parameter and find a corresponding rectangular equation: $x = 1 + \cos\theta$, $y = 2 - \sin\theta$.

**16.** Find an equation for the tangent line at the point where $t = 1$ for the curve given by the parametric equations $x = 2t$ and $y = t^2 + 5$.

**17.** Find the slope of the tangent line for the curve $r = \sin 2\theta$ at the point where $\theta = \dfrac{\pi}{4}$.

**18.** Write an integral for the arclength of the curve given by $r = \theta$, $0 \le \theta \le \pi$.

**19.** Find the area enclosed by the curve $r = 2 + \sin\theta$.

**20.** Identify the conic, then sketch the graph of the polar equation $r = \dfrac{4}{1 + \sin\theta}$.

**Test Form A**

**FINAL EXAM**
**Chapters 10–14**

**Name** ____________________ **Date** ____________

**Class** ____________________ **Section** ____________

**1.** The vector **v** has magnitude 8 and direction $\theta = 120°$. Find its component form.

(a) $\left\langle \frac{8\sqrt{3}}{3}, -8\sqrt{3} \right\rangle$
(b) $\left\langle 8\sqrt{3}, \frac{8\sqrt{3}}{3} \right\rangle$
(c) $\langle 4\sqrt{3}, -4 \rangle$
(d) $\langle -4, 4\sqrt{3} \rangle$
(e) None of these

**2.** Which of the following pairs of vectors are orthogonal?

(a) $\mathbf{v} = 3\mathbf{i} - 2\mathbf{j}, \mathbf{w} = -\mathbf{i} + 2\mathbf{j}$
(b) $\mathbf{v} = -2\mathbf{i}, \mathbf{w} = 5\mathbf{j}$
(c) $\mathbf{v} = -\mathbf{i} + 2\mathbf{j}, \mathbf{w} = -\frac{1}{2}\mathbf{j}$
(d) $\mathbf{v} = 2\mathbf{i} - 3\mathbf{j}, \mathbf{w} = -2\mathbf{i} + 3\mathbf{j}$
(e) None of these

**3.** Calculate the angle between $\mathbf{u} = 2\mathbf{i} - \mathbf{j} + \mathbf{k}$ and $\mathbf{v} = -3\mathbf{i} + 2\mathbf{j} + 2\mathbf{k}$.

(a) 126.4°
(b) 98.3°
(c) 2.2°
(d) 92°
(e) None of these

**4.** Find a vector orthogonal to the two given lines.

$x = -1 + 3t$ $\quad$ $x = 4 + 5t$

$y = 3 - 2t$ $\quad$ $y = 2 - t$

$z = 1 + t$ $\quad$ $z = -1 - 2t$

(a) $5\mathbf{i} - 3\mathbf{j} + 14\mathbf{k}$
(b) $-5\mathbf{i} - 11\mathbf{j} - 7\mathbf{k}$
(c) $4\mathbf{i} + 11\mathbf{j} + 7\mathbf{k}$
(d) $-5\mathbf{i} + 3\mathbf{j} + 10\mathbf{k}$
(e) None of these

**5.** Find an equation for the surface of revolution generated by revolving the curve given by $x = z^2$ in the $xz$-plane about the $x$-axis.

(a) $y = x^2 + z^2$
(b) $z = x^2 + y^2$
(c) $x = y^2 + z^2$
(d) $x = y + z^2$
(e) None of these

**6.** Represent the parabola $x = y^2 - 1$ by a vector-valued function.

(a) $t\mathbf{i} + (t^2 - 1)\mathbf{j}$
(b) $t\mathbf{i} + (t^2 + 1)\mathbf{j}$
(c) $t^2\mathbf{i} + (t - 1)\mathbf{j}$
(d) $(t^2 + 1)\mathbf{i} + t\mathbf{j}$
(e) None of these

**7.** Evaluate the integral: $\int_0^1 \left(e^{-t}\mathbf{i} + \frac{1}{t+1}\mathbf{j}\right) dt$.

(a) $-\frac{1}{e}\mathbf{i} + \ln 2\mathbf{j}$
(b) $\left(1 - \frac{1}{e}\right)\mathbf{i} + \ln 2\mathbf{j}$
(c) $(1 - e)\mathbf{i} + \ln 1\mathbf{j}$
(d) $\frac{1}{e}\mathbf{i} + \ln 2\mathbf{j}$
(e) None of these

**8.** An object starts from rest at the point (0, 1, 1) and moves with an acceleration of $\mathbf{i} + \mathbf{j}$. Find the position at $t = 4$.

(a) $8\mathbf{i} + 9\mathbf{j} + \mathbf{k}$ (b) $8\mathbf{i} + 8\mathbf{j}$ (c) $4\mathbf{i} + 5\mathbf{j} + \mathbf{k}$

(d) $9\mathbf{i} + 8\mathbf{j} + \mathbf{k}$ (e) None of these

**9.** Find the unit tangent vector to the curve given by $\mathbf{r}(t) = 4\cos t\mathbf{i} - 3\sin t\mathbf{j} + \mathbf{k}$ when $t = \dfrac{3\pi}{2}$.

(a) $\mathbf{i}$ (b) $\mathbf{j}$ (c) $-\mathbf{i}$

(d) $-\mathbf{j}$ (e) None of these

**10.** Find the length of the curve $\mathbf{r}(t) = e^t\mathbf{i} + e^{-t}\mathbf{j} + \sqrt{2}t\mathbf{k}$ over the interval [0, 2].

(a) $e^{2t} - e^{-2t}$ (b) $e^{2t} - 1$ (c) $e^{2t} + e^{-2t}$

(d) $e^{2t} - e^{-2t} + 1$ (e) None of these

**11.** Find $f_{xy}(x, y)$ for $f(x, y) = x^2y + 2y^2x^2 + 4x$.

(a) $2x + 4yx$ (b) $4x^2$ (c) $4xy$

(d) $2x + 8yx + 4$ (e) None of these

**12.** The radius of a right circular cylinder is decreasing at the rate of 4 inches per minute and the height is increasing at the rate of 8 inches per minute. What is the rate of change of the volume when $r = 4$ inches and $h = 8$ inches?

(a) $128\pi$ in$^3$/min (b) $384\pi$ in$^3$/min (c) $-256\pi$ in$^3$/min

(d) $-128\pi$ in$^3$/min (e) None of these

**13.** Find the directional derivative of $f(x, y) = x^2y$ at the point $(1, -3)$ in the direction of $-2\mathbf{i} + \mathbf{j}$.

(a) 13 (b) $\dfrac{13}{\sqrt{5}}$ (c) $-\dfrac{11}{\sqrt{5}}$

(d) $-11$ (e) None of these

**14.** Find an equation of the tangent plane to the surface $2x^2 + 3y^2 + 4z^2 = 18$ at the point $(-1, 2, 1)$.

(a) $-x + 3y + 2z = 7$ (b) $-x + 3y + 2z = 9$ (c) $-4x + 12y + 8z = -36$

(d) $4x + 12y + 8z = 24$ (e) None of these

**15.** Find the minimum distance from the point (0, 0, 0) to the plane $x + 2y + z = 6$.

(a) 2 (b) $\sqrt{5}$ (c) $\sqrt{6}$

(d) $\sqrt{7}$ (e) None of these

**16.** Evaluate the integral: $\int_0^{\ln 2} \int_{e^y}^{2} x\, dx\, dy.$

(a) $4 \ln 2 - \frac{3}{2}$ (b) $2 \ln 2 - 1$ (c) $2 \ln 2 - \frac{1}{4}e^2$

(d) $2 \ln 2 - \frac{3}{4}$ (e) None of these

**17.** Evaluate the double integral $\int_{-1}^{1} \int_0^{\sqrt{1-x^2}} e^{x^2+y^2}\, dy\, dx$ by changing to polar coordinates.

(a) $\pi(e - 1)$ (b) $\frac{\pi}{2}(e - 1)$ (c) $\frac{\pi e}{2}$

(d) $\pi e$ (e) None of these

**18.** Determine the triple integral that calculates the volume of the solid bounded by the graph of $z = 0$, $z = x$, and $y^2 = 2 - x$.

(a) $\int_0^2 \int_{-\sqrt{2}}^{\sqrt{2}} \int_0^x dz\, dy\, dx$ (b) $\int_0^2 \int_{-\sqrt{2}}^{\sqrt{2}} \int_x^0 dz\, dy\, dx$ (c) $\int_0^2 \int_{-\sqrt{2-x}}^{\sqrt{2-x}} \int_0^x dz\, dx\, dy$

(d) $\int_{-\sqrt{2}}^{\sqrt{2}} \int_0^{2-y^2} \int_0^x dz\, dx\, dy$ (e) None of these

**19.** Evaluate $\int_C y\, ds$ where $C$ is the path given in the figure at the right.

(a) $\frac{1}{2}$ (b) 1

(c) $\frac{3}{2}$ (d) 2

(e) None of these

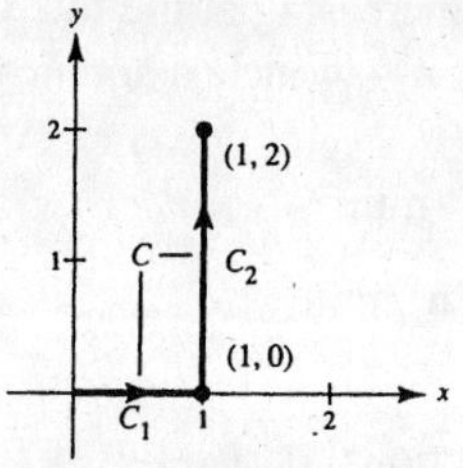


**20.** Use Green's Theorem to evaluate the line integral $\int_C (x - y^3)\, dx + x^3\, dy$, where $C$ is the circle of radius 2, $x^2 + y^2 = 4$.

(a) $6\pi$ (b) $12\pi$ (c) $18\pi$

(d) $24\pi$ (e) None of these

**Test Form B** **Name** ____________________ **Date** ____________

# FINAL EXAM
# Chapters 10–14

**Class** ____________________ **Section** ____________

1. If the point $(3, -5)$ is the initial point of $\mathbf{v} = 2\mathbf{i} - 4\mathbf{j}$, find the terminal point.

2. Determine if the vectors $\mathbf{v} = 3\mathbf{i} - 7\mathbf{j}$ and $\mathbf{w} = -2\mathbf{i} + \frac{14}{3}\mathbf{j}$ are orthogonal, parallel, or neither.

3. Find the direction cosines for $\mathbf{u} = \mathbf{i} - 2\mathbf{j} + 2\mathbf{k}$.

4. Find two unit vectors orthogonal to both $\mathbf{u} = \mathbf{i} + 2\mathbf{j}$ and $\mathbf{v} = 3\mathbf{i} - \mathbf{k}$.

5. Write parametric equations for the line perpendicular to the $xz$-plane and passing through the point $(1, 2, 3)$.

6. Consider the plane curve given by the equation $y^2 = x^3 + 1$. Represent the curve as a vector-valued function.

7. Evaluate the integral: $\int_0^{\pi} (\sin t\mathbf{i} + t^2\mathbf{j} + \mathbf{k})\,dt$.

8. Find $\mathbf{r}(t)$ given that $\mathbf{a}(t) = \mathbf{i} + e^t\mathbf{j}$, $\mathbf{r}'(0) = 2\mathbf{j}$ and $\mathbf{r}(0) = 2\mathbf{i}$.

9. Let $\mathbf{r}(t) = 4\cos t\mathbf{i} + 4\sin t\mathbf{j} + 2t\mathbf{k}$. Find the unit tangent vector.

10. Find the length of the space curve represented by the vector-valued function:

    $\mathbf{r}(t) = t\mathbf{i} + t\mathbf{j} + \frac{2}{3}t^{3/2}\mathbf{k},\ 2 \le t \le 7.$

11. Find the first partial derivative with respect to $y$: $F(x, y, z) = \arctan \dfrac{y}{xz}$.

12. Use the appropriate Chain Rule to find $\dfrac{dw}{dt}$ when $t = -1$ for $w = x^2 + y^2 - z^2$, $x = 1 - t^2$, $y = 2t + 3$, $z = t^2 + t$.

13. The surface of a mountain is described by the function $f(x, y) = 1500 - x^2 - 3y^2$. If a climber is at point $(10, 20, 200)$, what direction should she move in order to descend at the greatest rate?

14. Find symmetric equations for the normal line at the point $(4, 4, 1)$ on the surface $y = x(2z - 1)$.

15. A company manufactures two products. The total revenue from $x_1$ units of product 1 and $x_2$ units of product 2 is $R = -2x_1^2 - 3x_2^2 + 3x_1x_2 + 1000x_2 + 1600x_1$. Find $x_1$ and $x_2$ so as to maximize the revenue.

16. Evaluate the improper integral $\int_1^{\infty} \int_x^{2x} ye^{-x}\, dy\, dx$.

17. Use a double integral in polar coordinates to find the volume of the solid bounded by the graphs of the paraboloid $z = 1 - x^2 - y^2$, the cylinder $x^2 + y^2 = y$, and the $xy$-plane.

18. Set up a triple integral to calculate the volume of the solid bounded by $z = 9 - x^2 - y^2$ and $z = 0$. Use rectangular coordinates with the order $dz\, dy\, dx$.

19. Let $C$ be the curve $C_1 + C_2$ as shown. Evaluate the integral $\int_C (1 + xy)\, ds$.

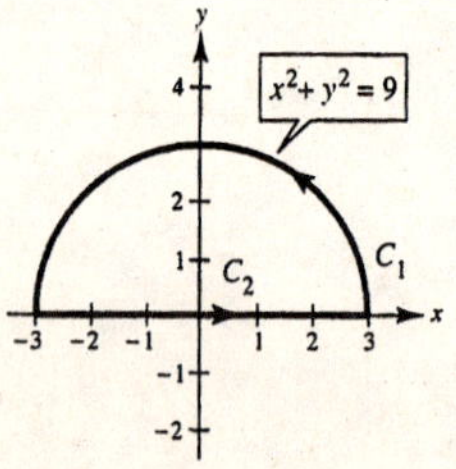


20. Use Green's Theorem to compute $\int_C 2y^3\, dx - x^2\, dy$ where $C$ is the square with vertices $(1, 1), (1, -1), (-1, -1)$ and $(-1, 1)$.

# FINAL EXAM Answers

## Chapters P–5 Tests

### Test Form A

| | | | |
|---|---|---|---|
| **1.** c | **2.** b | **3.** e | **4.** a |
| **5.** a | **6.** b | **7.** d | **8.** d |
| **9.** a | **10.** c | **11.** b | **12.** d |
| **13.** b | **14.** b | **15.** c | **16.** a |
| **17.** d | **18.** c | **19.** a | **20.** b |

### Test Form B

**1.** $-2$ **2.** $-\infty$ **3.** $6x + y + 7 = 0$

**4.** $\dfrac{3}{x^2}$ **5.** 264 ft/sec **6.** $\sin 2x + \cos 2x$

**7.** $\dfrac{3x^2 - 4xy + 3y^2}{2x^2 - 6xy}$ **8.** 0.048 ft/min

**9.** The Mean Value Theorem applies: $c = \frac{5}{2}$.

**10.** $f'(x) = \dfrac{1}{(1 - x)^2} > 0,\ x \neq 1.$

Therefore, $f$ is increasing for all $x \neq 1$.

**11.**

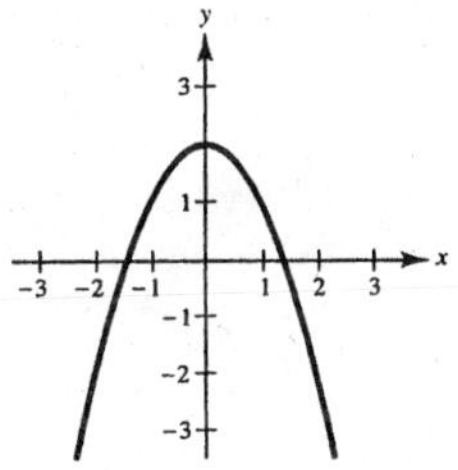

(The answer is not unique.)

**12.** $y = \dfrac{a}{b}$

**13. a.** $SA = 2\pi rh + 2\pi r^2$

**b.** $C = 2\pi krh + 4\pi kr^2$, $k$ constant

**c.** $C = \dfrac{44k}{4} + 4\pi kr^2$

**d.** $r = \sqrt[3]{\dfrac{11}{2\pi}} \approx 1.2$ in.

**14.** $\sqrt{x}\,[28 + 3x^{3/2}] + C$

**15.**

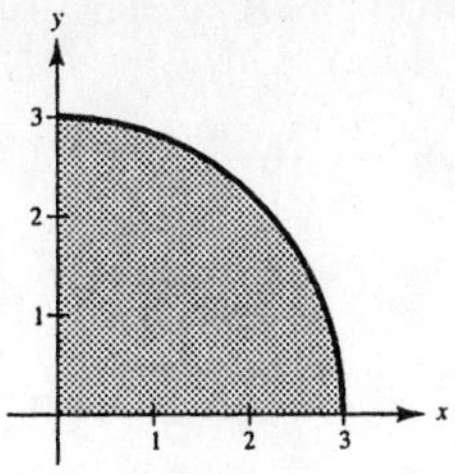

**16.** $\dfrac{x^3}{3} - 2x - \dfrac{1}{x} + C$

**17.** $y = -\dfrac{3}{2}\ln|1 - x^2| + C$ **18.** $\dfrac{1}{e^x + 1}$

**19.** $\dfrac{e^x}{1 + e^{2x}}$ **20.** $2\arcsin\dfrac{x}{2} + \sqrt{4 - x^2} + C$

## Chapters 6–9 Tests

### Test Form A

| | | | |
|---|---|---|---|
| **1.** a | **2.** d | **3.** c | **4.** c |
| **5.** a | **6.** d | **7.** c | **8.** a |
| **9.** b | **10.** c | **11.** d | **12.** d |
| **13.** e | **14.** a | **15.** c | **16.** b |
| **17.** b | **18.** a | **19.** c | **20.** a |

### Test Form B

**1.** $2\sqrt{2}$ **2.** $\dfrac{704}{15}\pi$ **3.** $2\sqrt{\ln x} + C$

**4.** $\dfrac{e^{3x}}{27}(9x^2 - 6x + 2) + C$

**5.** $\dfrac{1}{3}\tan 3\theta + \dfrac{1}{9}\tan^3 3\theta + C$

**6.** $\ln\left(2 + \sqrt{3}\right) - \dfrac{\sqrt{3}}{2}$ **7.** $\ln\left|\dfrac{x}{x + 3}\right| + C$

**8.** 1 **9.** $\dfrac{\pi}{4}$

**10.** Diverges, $n$th-Term Test for Divergence

**11.** Converges

**12.** Converges, Limit Comparison Test

**13.** $[-6, -2)$

**14.** $-(x - \pi) + \dfrac{(x - \pi)^3}{3!} - \dfrac{(x - \pi)^5}{5!} + \dfrac{(x - \pi)^7}{7!} - \cdots + \dfrac{(-1)^n(x - \pi)^{2n-1}}{(2n - 1)!} + \cdots$

**15.** $x^2 + y^2 - 2x - 4y + 4 = 0$ **16.** $y = x + 4$

**17.** $-1$ **18.** $\int_0^{\pi} \sqrt{\theta^2 + 1}\, d\theta$ **19.** $\frac{9\pi}{2}$

**20.** Parabola

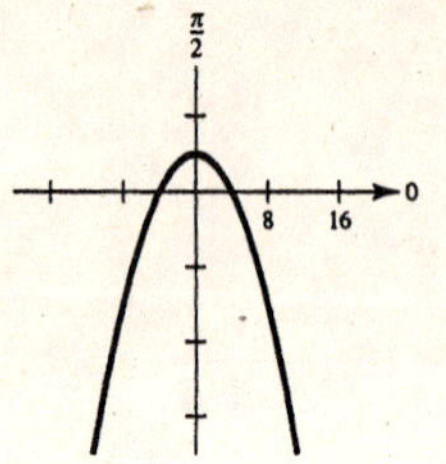


**15.** $x_1 = 840, x_2 = 587$ **16.** $\frac{15}{2e}$ **17.** $\frac{5\pi}{32}$

**18.** $\int_0^3 \int_0^{\sqrt{9-x^2}} \int_0^{\sqrt{9-x^2-y^2}} dx\, dy\, dx$

**19.** $3\pi + 6$ **20.** 8

# Chapters 10–14 Tests

## Test Form A

| | | | |
|---|---|---|---|
| **1.** d | **2.** b | **3.** a | **4.** b |
| **5.** c | **6.** e | **7.** b | **8.** a |
| **9.** a | **10.** e | **11.** e | **12.** d |
| **13.** b | **14.** b | **15.** c | **16.** d |
| **17.** b | **18.** d | **19.** a | **20.** d |

## Test Form B

**1.** $(5, -9)$ **2.** Parallel

**3.** $\cos\alpha = \frac{1}{3}, \cos\beta = -\frac{2}{3}, \cos\gamma = \frac{2}{3}$

**4.** $\pm\frac{1}{\sqrt{41}}(2\mathbf{i} - \mathbf{j} + 6\mathbf{k})$

**5.** $x = 1, y = 2 + t, z = 3$

**6.** $\mathbf{r}(t) = (t^2 - 1)^{1/3}\mathbf{i} + t\mathbf{j}$ **7.** $2\mathbf{i} + \frac{\pi^3}{3}\mathbf{j} + \pi\mathbf{k}$

**8.** $\mathbf{r}(t) = \left(\frac{t^2}{2} + 2\right)\mathbf{i} + (e^t + t - 1)\mathbf{j}$

**9.** $\mathbf{T}(t) = \frac{1}{\sqrt{5}}(-2\sin t\,\mathbf{i} + 2\cos t\,\mathbf{j} + \mathbf{k})$

**10.** $\frac{38}{3}$ **11.** $\frac{xz}{x^2z^2 + y^2}$ **12.** 4

**13.** $\nabla f = -20\mathbf{i} - 120\mathbf{j}$

**14.** $\frac{x-4}{1} = \frac{y-4}{-1} = \frac{z-1}{8}$

# GATEWAY TEST 1A

## Algebra

Name ______________________ Date __________

Class ______________________ Section __________

**1.** Factor and simplify. Express the answer as a fraction without negative exponents.

$$x(x - 1)^{-1/2} + 2(x - 1)^{1/2}$$

**2.** Express as a simple fraction.

$$\frac{\dfrac{1}{y - k} - \dfrac{1}{y}}{k}$$

**3.** Multiply.

$$\left(x^{3/2} + \frac{2}{\sqrt{3}}\right)^2$$

**4.** Solve for $x$.

$$x^2 - x = 5$$

**5.** Find the smallest value of $x$ that satisfies the equation.

$$|x + 5| = 3$$

**6.** Write the general form of the equation of the line passing through the point $(3, -1)$ with slope $\frac{5}{2}$.

**7.** Solve for $y'$.

$$xy' + y = 1 + y'$$

**8.** Write the equation of the circle in standard form and give the center and radius.

$$2x^2 + 2y^2 + 4x - 12y + 11 = 0$$

**9.** Solve for $x$.

$$2(x - 5)^{-1} + \frac{1}{x} = 0$$

**10.** Find the domain of $f$.

$$f(x) = \sqrt{2x + 3}$$

# Gateway Test 1B
# Algebra

Name ____________________ Date ______________

Class ____________________ Section ______________

**1.** Factor and simplify. Express the answer as a fraction without negative exponents.

$$3x(2x + 5)^{-1/2} + 3(2x + 5)^{1/2}$$

**2.** Express as a simple fraction.

$$\frac{\dfrac{3}{2(x + h)} - \dfrac{3}{2x}}{h}$$

**3.** Multiply.

$$\left(x^{5/2} + \frac{3}{\sqrt{2}}\right)^2$$

**4.** Solve for $x$.

$$x^2 - 4 = x$$

**5.** Find the smallest value of $x$ that satisfies the equation.

$$|2 - x| = 5$$

**6.** Write the general form of the equation of the line passing through the point $(-2, 5)$ with slope $-\frac{3}{4}$.

**7.** Solve for $p$.

$$hp - 1 = q + kp + 6p$$

**8.** Write the equation of the circle in standard form and give the center and radius.

$$5x^2 + 5y^2 - 20x + 10y + 21 = 0$$

**9.** Solve for $x$.

$$3(x + 2)^{-1} - \frac{4}{x} = 0$$

**10.** Find the domain of $f$.

$$f(x) = \sqrt{5 - 3x}$$

## GATEWAY TEST 1C

## Algebra

**Name** ____________________ **Date** ____________

**Class** ____________________ **Section** ____________

1. Factor and simplify. Express the answer as a fraction without negative exponents.

   $-4x(3x-2)^{-4/3} + 4(3x-2)^{-1/3}$

2. Express as a simple fraction.

   $$\frac{\dfrac{2}{x+h} - \dfrac{2}{x}}{h}$$

3. Multiply.

   $$\left(x^{4/3} - \frac{\sqrt{3}}{2}\right)^2$$

4. Solve for $x$.

   $2x^2 - x = 5$

5. Find the smallest value of $x$ that satisfies the equation.

   $|x-3| = 4$

6. Write the general form of the equation of the line passing through the point $(-1, 4)$ with slope $\frac{2}{3}$.

7. Solve for $t$.

   $$\frac{d}{t+r} = \frac{5}{t}$$

8. Write the equation of the circle in standard form and give the center and radius.

   $4x^2 + 4y^2 + 24x + 16y + 43 = 0$

9. Solve for $x$.

   $$4x^{-1} + \frac{3}{x+1} = 0$$

10. Find the domain of $f$.

    $f(x) = \sqrt{4x-3}$

# Gateway Test 1D
# Algebra

Name ____________________ Date ____________

Class ____________________ Section ____________

**1.** Factor and simplify. Express the answer as a fraction without negative exponents.

$$-\frac{5}{2}x(x+3)^{-3/2} + 5(x+3)^{-1/2}$$

**2.** Express as a simple fraction.

$$\frac{\dfrac{4}{x+h} - \dfrac{-4}{x}}{h}$$

**3.** Multiply.

$$\left(x^{5/4} - \frac{2}{\sqrt{5}}\right)^2$$

**4.** Solve for $x$.

$$x^2 - 2x = 5$$

**5.** Find the smallest value of $x$ that satisfies the equation.

$$|4 - x| = 2$$

**6.** Write the general form of the equation of the line passing through the point $(7, -3)$ with slope $-\frac{5}{3}$.

**7.** Solve for $y'$.

$$2x + 2xy' + 2y + 3y^2y' = 0$$

**8.** Write the equation of the circle in standard form and give the center and radius.

$$4x^2 + 4y^2 - 4x - 8y - 1 = 0$$

**9.** Solve for $x$.

$$5x^{-1} - \frac{2}{x-3} = 0$$

**10.** Find the domain of $f$.

$$f(x) = \sqrt{2 + 7x}$$

# GATEWAY TEST 2A

Name ______________________ Date ______________

## The Exponential and Logarithmic Functions

Class ______________________ Section ______________

**1.** Solve for $x$.

$$\ln(e^{7x}) = 15$$

**2.** Solve for $x$.

$$\frac{e^{x+5}}{e^5} = 3$$

**3.** Solve for $x$.

$$(e^3)^{2x} = e^3 e^{2x}$$

**4.** Solve for $x$.

$$e^{[2\ln x - \ln(x^2+x-3)]} = 1$$

**5.** Solve for $x$.

$$3^{2x} - 2 \cdot 3^{(x+5)} + 3^{10} = 0$$

**6.** Sketch the graph of the function.

$$f(x) = e^x$$

**7.** Find the $x$-intercept for the graph of the function.

$$f(x) = \ln x + 2$$

**8.** Use the properties of logarithms to expand the expression.

$$\ln\frac{(4x^5 - x - 1)\sqrt{x-7}}{(x^2+1)^3}$$

**9.** Solve for $x$.

$$\ln x - \ln(x+1) = 1$$

**10.** Find the domain of the function.

$$f(x) = \ln(3x + 2)$$

# GATEWAY TEST 2B

# The Exponential and Logarithmic Functions

Name ______________________ Date ______________

Class ______________________ Section ______________

**1.** Solve for $x$.

$$\ln(e^{-x}) = 3$$

**2.** Solve for $x$.

$$\frac{e^{2x+3}}{e^3} = 5$$

**3.** Solve for $x$.

$$(e^5)^{3x} = e^5 e^{3x}$$

**4.** Solve for $x$.

$$e^{[2 \ln x - \ln(x^2 + 2x - 5)]} = 1$$

**5.** Solve for $x$.

$$4^{2x} - 2 \cdot 4^{(x+4)} + 4^8 = 0$$

**6.** Sketch the graph of the function.

$$f(x) = \ln x$$

**7.** Find the $x$-intercept for the graph of the function.

$$f(x) = \ln x - 3$$

**8.** Use the properties of logarithms to expand the expression.

$$\ln\left[\frac{(3x^6 + 2)\sqrt{x + 8}}{(x - 1)^4}\right]$$

**9.** Solve for $x$.

$$\ln x - \ln(x - 1) = 1$$

**10.** Find the domain of the function.

$$f(x) = \ln(5 - 3x)$$

# GATEWAY TEST 2C

## The Exponential and Logarithmic Functions

**Name** ____________________ **Date** ____________

**Class** ____________________ **Section** ____________

**1.** Solve for $x$.

$$e^{\ln(2x+1)} = 5x$$

**2.** Solve for $x$.

$$\frac{e^{2x+2}}{e^2} = 4$$

**3.** Solve for $x$.

$$(e^2)^{3x} = e^2e^{3x}$$

**4.** Solve for $x$.

$$e^{[3\ln x - \ln(x^3 - 2x + 4)]} = 1$$

**5.** Solve for $x$.

$$5^{2x} - 2 \cdot 5^{(x+3)} + 5^6 = 0$$

**6.** Sketch the graph of the function.

$$f(x) = \frac{1}{x}$$

**7.** Find the $x$-intercept for the graph of the function.

$$f(x) = 2\ln x - 1$$

**8.** Use the properties of logarithms to expand the expression.

$$\ln\left[\frac{2x - 1}{\sqrt{3x + 1}\,(x^3 - 7)^9}\right]$$

**9.** Solve for $x$.

$$\ln x - \ln(x + 2) = 1$$

**10.** Find the domain of the function.

$$f(x) = \ln(2x + 1)$$

# GATEWAY TEST 2D

# The Exponential and Logarithmic Functions

**Name** ______________________ **Date** ____________

**Class** ______________________ **Section** ____________

**1.** Solve for $x$.

$$e^{\ln 3x} = 11$$

**2.** Solve for $x$.

$$\frac{e^{x+4}}{e^4} = 1$$

**3.** Solve for $x$.

$$(e^4)^{3x} = e^4 e^{3x}$$

**4.** Solve for $x$.

$$e^{[2 \ln x - \ln(x^2 - 3x + 1)]} = 1$$

**5.** Solve for $x$.

$$e^{2x} - 2 \cdot e^{(x+2)} + e^4 = 0$$

**6.** Sketch the graph of the function.

$$f(x) = \ln|x|$$

**7.** Find the $x$-intercept for the graph of the function.

$$f(x) = 2 \ln x - 4$$

**8.** Use the properties of logarithms to expand the expression.

$$\ln\left[\frac{3x + 2}{\sqrt[3]{2x + 1}\,(x^4 + x + 1)^2}\right]$$

**9.** Solve for $x$.

$$\ln x - \ln(2x + 1) = 1$$

**10.** Find the domain of the function.

$$f(x) = \ln(7 - 4x)$$

## GATEWAY TEST 3A

## Trigonometry

**Name** ______________________ **Date** ____________

**Class** ______________________ **Section** ____________

**1.** If $\csc \theta = \frac{13}{5}$ and $\theta$ is in the second quadrant, find $\sec \theta$.

**2.** Find all $\theta$ in the interval $[0, 2\pi)$ that satisfy the equation.

$\sin 2\theta = 0$

**3.** Write the expression $\sqrt{x^2 + 4}$ in terms of $\theta$ when $x = 2 \tan \theta$.

**4.** Simplify $\dfrac{\cot \theta}{\csc \theta}$.

**5.** Find $\sin 2A$ if $\sin A = \dfrac{1}{4}$ and $0 \le A \le \dfrac{\pi}{2}$.

**6.** Find all $\theta$ in the interval $[0, 2\pi)$ that satisfy the equation.

$2 \cos \theta \tan \theta + \tan \theta = 0$

**7.** If $\cos 2\theta = \frac{1}{3}$ and $0 \le 2\theta \le \pi$, find $\cos \theta$.

**8.** Rewrite the given equation using the substitutions $x = r \cos \theta$ and $y = r \sin \theta$. Simplify your answer.

$x^2 + y^2 + 3x = 0$

**9.** Write the given expression in algebraic form.

$\tan\left(\arccos \dfrac{x}{3}\right)$

**10.** Compute $\arcsin\left(-\frac{1}{2}\right)$.

# GATEWAY TEST 3B

## Trigonometry

**Name** ____________________ **Date** ______________

**Class** ____________________ **Section** ______________

1. If $\cot\theta = \frac{4}{3}$ and $\theta$ is in the third quadrant, find $\cos\theta$.

2. Find all $\theta$ in the interval $[0, 2\pi)$ that satisfy the equation.

   $\sin 2\theta = \frac{1}{2}$

3. Write the expression $\sqrt{9 - x^2}$ in terms of $\theta$ when $x = 3\sin\theta$.

4. Simplify $\dfrac{\sec\theta}{\tan\theta}$.

5. Find $\cos 2A$ if $\sin A = \dfrac{1}{4}$ and $0 \le A \le \dfrac{\pi}{2}$.

6. Find all $\theta$ in the interval $[0, 2\pi)$ that satisfy the equation.

   $\sin\theta - \sin\theta\cot\theta = 0$

7. If $\cos 2\theta = \frac{1}{3}$ and $0 \le 2\theta \le \pi$, find $\sin\theta$.

8. Rewrite the given equation using the substitutions $x = r\cos\theta$ and $y = r\sin\theta$. Simplify your answer.

   $x^2 + y^2 - 2y = 0$

9. Write the given expression in algebraic form.

   $\cot\left(\arcsin\dfrac{x}{2}\right)$

10. Compute $\arccos\left(-\dfrac{\sqrt{3}}{2}\right)$.

# GATEWAY TEST 3C
## Trigonometry

**Name** ____________________ **Date** ____________

**Class** ____________________ **Section** ____________

**1.** If $\cos\theta = \frac{12}{13}$ and $\theta$ is in the fourth quadrant, find $\csc\theta$.

**2.** Find all $\theta$ in the interval $[0, 2\pi)$ that satisfy the equation.

$$\sin 2\theta = -\frac{\sqrt{3}}{2}$$

**3.** Write the expression $\sqrt{x^2 - 9}$ in terms of $\theta$ when $x = 3\sec\theta$.

**4.** Simplify $(\sin\theta - \sin^3\theta)\csc\theta$.

**5.** Find $\sin 2A$ if $\sin A = -\frac{1}{3}$ and $\pi \le A \le \frac{3\pi}{2}$.

**6.** Find all $\theta$ in the interval $[0, 2\pi)$ that satisfy the equation.

$$\cos\theta\tan\theta + \sqrt{3}\cos\theta = 0$$

**7.** If $\cos 2\theta = -\frac{1}{4}$ and $\frac{\pi}{2} \le 2\theta \le \pi$, find $\cos\theta$.

**8.** Rewrite the given equation using the substitutions $x = r\cos\theta$ and $y = r\sin\theta$. Simplify your answer.

$$x^2 + y^2 - 4x = 0$$

**9.** Write the given expression in algebraic form.

$$\sin\left(\arctan\frac{x}{4}\right)$$

**10.** Compute $\arctan(-1)$.

# GATEWAY TEST 3D

## Trigonometry

**Name** ____________________ **Date** ____________

**Class** ____________________ **Section** ____________

**1.** If $\tan \theta = -\frac{7}{24}$ and $\theta$ is in the second quadrant, find $\sin \theta$.

**2.** Find all $\theta$ in the interval $[0, 2\pi)$ that satisfy the equation.

$$\sin 2\theta = \frac{1}{\sqrt{2}}$$

**3.** Write the expression $\sqrt{4x^2 + 25}$ in terms of $\theta$ when $2x = 5 \tan \theta$.

**4.** Simplify $\dfrac{\csc \theta}{\cot \theta}$.

**5.** Find $\cos 2A$ if $\sin A = -\dfrac{1}{3}$ and $\pi \le A \le \dfrac{3\pi}{2}$.

**6.** Find all $\theta$ in the interval $[0, 2\pi)$ that satisfy the equation.

$$2 \cot \theta \sin \theta - \cot \theta = 0$$

**7.** If $\cos 2\theta = -\dfrac{1}{4}$ and $\dfrac{\pi}{2} \le 2\theta \le \pi$, find $\sin \theta$.

**8.** Rewrite the given equation using the substitutions $x = r \cos \theta$ and $y = r \sin \theta$. Simplify your answer.

$$x^2 + y^2 + 5x = 0$$

**9.** Write the given expression in algebraic form.

$$\sin\left(\arctan \frac{5}{x}\right)$$

**10.** Compute $\operatorname{arccot}\left(-\dfrac{1}{\sqrt{3}}\right)$.

**GATEWAY TEST 4A** **Name** ____________ **Date** ____________

**Differentiation** **Class** ____________ **Section** ____________

In problems 1–9, find the derivative and simplify the answer. Write the answer without negative exponents.

1. $xe^x$

2. $\dfrac{x}{x+1}$

3. $\sqrt{2x-1}$

4. $\sin^2 2x$

5. $\arcsin \dfrac{x}{2}$

6. $e^{(x^2)}$

7. $\ln\left(\dfrac{x}{x+1}\right)$

8. $x^2\sqrt{2x+1}$

9. $x^4 - \dfrac{5}{x} + \sqrt{x} - x^e - \pi^3$

10. Find $\dfrac{dy}{dx}$: $y^3 + xy = 5$.

## GATEWAY TEST 4B

## Differentiation

**Name** ______________________ **Date** ______________

**Class** ______________________ **Section** ______________

In problems 1–9, find the derivative and simplify the answer. Write the answer without negative exponents.

**1.** $x \ln x$

**2.** $\dfrac{x}{1 - x}$

**3.** $\sqrt{4 + x^2}$

**4.** $\cos^3 2x$

**5.** $\arctan \dfrac{x}{3}$

**6.** $e^{(\sin x)}$

**7.** $\ln\left(\dfrac{x^2}{x^3 - 1}\right)$

**8.** $x^2\sqrt{1 - 3x}$

**9.** $x^3 - \dfrac{x}{2} + \sqrt{x} - x^{\pi} - e^3$

**10.** Find $\dfrac{dy}{dx}$: $y^2 - xy = 3$.

**Gateway Test 4C** **Name** ____________ **Date** ____________

**Differentiation** **Class** ____________ **Section** ____________

In problems 1–9, find the derivative and simplify the answer. Write the answer without negative exponents.

**1.** $x \sin x$

**2.** $\dfrac{x+2}{x-1}$

**3.** $(3-2x)^6$

**4.** $\tan^2 5x$

**5.** $\arcsin \sqrt{x}$

**6.** $e^{(5x)}$

**7.** $\ln\left[x\sqrt{2x+1}\right]$

**8.** $x^2\sqrt{x^2+1}$

**9.** $\dfrac{7}{x^2} + x^{3/2} - \sqrt{x} + e^{\pi}$

**10.** Find $\dfrac{dy}{dx}$: $x^2 + y^2 + xy = 2$.

## GATEWAY TEST 4D
## Differentiation

**Name** ____________________ **Date** ____________

**Class** ____________________ **Section** ____________

In problems 1–9, find the derivative and simplify the answer. Write the answer without negative exponents.

**1.** $x \cos x$

**2.** $\dfrac{x - 2}{x + 1}$

**3.** $(x^3 - 6)^7$

**4.** $\sec^2 3x$

**5.** $\arctan \sqrt{x}$

**6.** $e^{(1/x)}$

**7.** $\ln\left(\dfrac{x}{\sqrt{2x + 1}}\right)$

**8.** $x^2\sqrt{1 - x^2}$

**9.** $\dfrac{x^2}{5} + \dfrac{5}{x^2} - \sqrt[3]{x} + \pi^e$

**10.** Find $\dfrac{dy}{dx}$: $x^2 + y^2 - xy = 4$.

# GATEWAY TEST 5A

## Integration

Name ______________________ Date ______________

Class ______________________ Section ______________

Evaluate each of the integrals.

1. $\int \sqrt{2x - 1}\, dx$

2. $\int_1^2 (2x - 1)\, dx$

3. $\int \frac{1}{2x - 1}\, dx$

4. $\int \frac{x}{2x - 1}\, dx$

5. $\int \frac{3}{\sqrt{1 - x^2}}\, dx$

6. $\int \sqrt{1 - x^2}\, dx$

7. $\int e^{(2x-1)}\, dx$

8. $\int \ln(2x - 1)\, dx$

9. $\int \sec^2(2x - 1)\, dx$

10. $\int \frac{1}{(2x - 1)^2}\, dx$

# GATEWAY TEST 5B

## Integration

**Name** ______________________ **Date** ______________

**Class** ______________________ **Section** ______________

Evaluate each of the integrals.

1. $\int \sqrt{3x + 1}\, dx$

2. $\int_1^2 (3x + 1)\, dx$

3. $\int \frac{1}{3x + 1}\, dx$

4. $\int \frac{x}{3x + 1}\, dx$

5. $\int \frac{2}{1 + x^2}\, dx$

6. $\int \frac{1}{\sqrt{1 + x^2}}\, dx$

7. $\int e^{(3x+1)}\, dx$

8. $\int \ln(3x + 1)\, dx$

9. $\int \sin(3x + 1)\, dx$

10. $\int \frac{1}{(3x + 1)^3}\, dx$

## GATEWAY TEST 5C

## Integration

**Name** ____________________ **Date** ______________

**Class** ____________________ **Section** ______________

Evaluate each of the integrals.

**1.** $\int \sqrt{1 - 4x}\, dx$

**2.** $\int_1^2 (1 - 4x)\, dx$

**3.** $\int \frac{1}{1 - 4x}\, dx$

**4.** $\int \frac{x}{1 - 4x}\, dx$

**5.** $\int \frac{4}{\sqrt{1 - 4x^2}}\, dx$

**6.** $\int \frac{1}{\sqrt{x^2 - 1}}\, dx$

**7.** $\int e^{(1 - 4x)}\, dx$

**8.** $\int \ln(1 - 4x)\, dx$

**9.** $\int \cos(1 - 4x)\, dx$

**10.** $\int \frac{1}{(1 - 4x)^2}\, dx$

## GATEWAY TEST 5D
## Integration

**Name** ____________________ **Date** ____________

**Class** ____________________ **Section** ____________

Evaluate each of the integrals.

**1.** $\int \sqrt{2 - x}\, dx$

**2.** $\int_1^2 (2 - x)\, dx$

**3.** $\int \frac{1}{2 - x}\, dx$

**4.** $\int \frac{x}{2 - x}\, dx$

**5.** $\int \frac{1}{\sqrt{1 - 2x^2}}\, dx$

**6.** $\int \sqrt{4 - x^2}\, dx$

**7.** $\int e^{(2-x)}\, dx$

**8.** $\int \ln(2 - x)\, dx$

**9.** $\int \csc(2 - x)\cot(2 - x)\, dx$

**10.** $\int \frac{1}{(2 - x)^2}\, dx$

# GATEWAY TEST Answers

## Algebra—Tests 1A, 1B, 1C & 1D

### Test 1A

**1.** $\dfrac{3x-2}{\sqrt{x-1}}$ **2.** $\dfrac{1}{y(y-h)}$

**3.** $x^3+\dfrac{4}{\sqrt{3}}x^{(3/2)}+\dfrac{4}{3}$ **4.** $\dfrac{1\pm\sqrt{21}}{2}$

**5.** $-8$ **6.** $5x-2y+17=0$ **7.** $\dfrac{1-y}{x-1}$

**8.** $(x+1)^2+(y-3)^2=\dfrac{9}{2}$

center: $(-1, 3)$

radius: $\dfrac{3}{\sqrt{2}}$

**9.** $\frac{5}{3}$ **10.** $\left[-\frac{3}{2}, \infty\right)$

### Test 1B

**1.** $\dfrac{3(3x+5)}{\sqrt{2x+5}}$ **2.** $\dfrac{-3}{2x(x+h)}$

**3.** $x^5+3\sqrt{2}x^{5/2}+\dfrac{9}{2}$ **4.** $\dfrac{1\pm\sqrt{17}}{2}$

**5.** $-3$ **6.** $3x+4y-14=0$ **7.** $\dfrac{q+1}{h-k-6}$

**8.** $(x-2)^2+(y+1)^2=\dfrac{4}{5}$

center: $(2, -1)$

radius: $\dfrac{2}{\sqrt{5}}$

**9.** $-8$ **10.** $\left(-\infty, \frac{5}{3}\right]$

### Test 1C

**1.** $\dfrac{8(x-1)}{(3x-2)^{4/3}}$ **2.** $\dfrac{-2}{x(x+h)}$

**3.** $x^{8/3}-\sqrt{3}x^{4/3}+\dfrac{3}{4}$ **4.** $\dfrac{1\pm\sqrt{41}}{4}$

**5.** $-1$ **6.** $2x-3y+14=0$ **7.** $\dfrac{5r}{d-r}$

**8.** $(x+3)^2+(y+2)^2=\dfrac{9}{4}$

center: $(-3, -2)$

radius: $\dfrac{3}{2}$

**9.** $-\frac{4}{7}$ **10.** $\left[\frac{3}{4}, \infty\right)$

### Test 1D

**1.** $\dfrac{5(x+6)}{2(x+3)^{3/2}}$ **2.** $\dfrac{4}{x(x+h)}$

**3.** $x^{5/2}-\dfrac{4}{\sqrt{5}}x^{5/4}+\dfrac{4}{5}$ **4.** $1\pm\sqrt{6}$

**5.** $2$ **6.** $5x+3y-26=0$ **7.** $-\dfrac{2x+2y}{2x+3y^2}$

**8.** $\left(x-\dfrac{1}{2}\right)^2+(y-1)^2=\dfrac{3}{2}$

center: $\left(\dfrac{1}{2}, 1\right)$

radius: $\sqrt{\dfrac{3}{2}}$

**9.** $5$ **10.** $\left[-\frac{2}{7}, \infty\right)$

# GATEWAY TEST Answers

## The Exponential and Logarithmic Functions—Tests 2A, 2B, 2C & 2D

### Test 2A

1. $\frac{15}{7}$ 2. $\ln 3$ 3. $\frac{3}{4}$ 4. 3 5. 5

6.

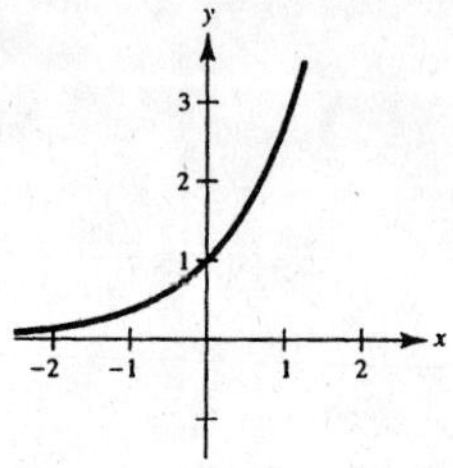

7. $(e^{-2}, 0)$

8. $\ln(4x^5 - x - 1) + \frac{1}{2}\ln(x - 7) - 3\ln(x^2 + 1)$

9. $\dfrac{e}{1 - e}$ 10. $\left(\dfrac{2}{3}, \infty\right)$

### Test 2B

1. $-3$ 2. $\ln\sqrt{5}$ 3. $\frac{5}{12}$ 4. $\frac{5}{2}$ 5. 4

6.

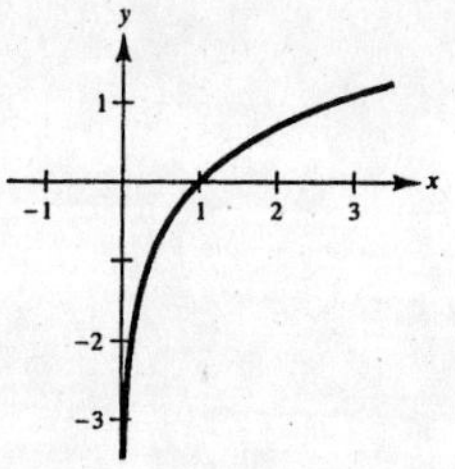

7. $(e^3, 0)$

8. $\ln(3x^6 + 2) + \frac{1}{2}\ln(x + 8) - 4\ln(x - 1)$

9. $\dfrac{e}{e - 1}$ 10. $\left(-\infty, \dfrac{5}{3}\right)$

### Test 2C

1. $\frac{1}{3}$ 2. $\ln 2$ 3. $\frac{2}{3}$ 4. 2 5. 3

6.

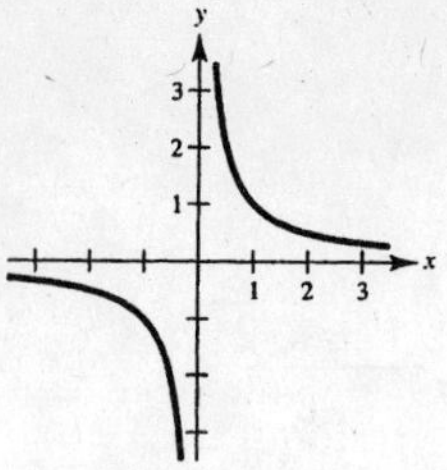

7. $(\sqrt{e}, 0)$

8. $\ln(2x - 1) - \frac{1}{2}\ln(3x + 1) - 9\ln(x^3 - 7)$

9. $\dfrac{2e}{1 - e}$ 10. $\left(-\dfrac{1}{2}, \infty\right)$

### Test 2D

1. $\frac{11}{3}$ 2. 0 3. $\frac{4}{9}$ 4. $\frac{1}{3}$ 5. 2

6.

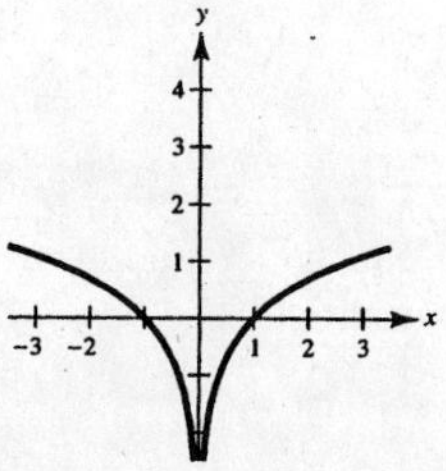

7. $(e^2, 0)$

8. $\ln(3x + 2) - \frac{1}{3}\ln(2x + 1) - 2\ln(x^4 + x + 1)$

9. $\dfrac{e}{1 - 2e}$ 10. $\left(-\infty, \dfrac{7}{4}\right)$

# GATEWAY TEST Answers

## Trigonometry—Tests 3A, 3B, 3C & 3D

### Test 3A

**1.** $-\frac{13}{12}$ **2.** $0, \frac{\pi}{2}, \pi, \frac{3\pi}{2}$ **3.** $2|\sec\theta|$

**4.** $\cos\theta$ **5.** $\frac{\sqrt{15}}{8}$ **6.** $0, \frac{2\pi}{3}, \pi, \frac{4\pi}{3}$

**7.** $\sqrt{\frac{2}{3}}$ **8.** $r = -3\cos\theta$

**9.** $\frac{\sqrt{9-x^2}}{x}$ **10.** $-\frac{\pi}{6}$

### Test 3B

**1.** $-\frac{4}{5}$ **2.** $\frac{\pi}{12}, \frac{5\pi}{12}, \frac{13\pi}{12}, \frac{17\pi}{12}$ **3.** $3|\cos\theta|$

**4.** $\csc\theta$ **5.** $-\frac{7}{8}$ **6.** $0, \frac{\pi}{4}, \pi, \frac{5\pi}{4}$ **7.** $\frac{1}{\sqrt{3}}$

**8.** $r = 2\sin\theta$ **9.** $\frac{\sqrt{4-x^2}}{x}$ **10.** $\frac{2\pi}{3}$

### Test 3C

**1.** $-\frac{13}{5}$ **2.** $\frac{2\pi}{3}, \frac{5\pi}{6}, \frac{5\pi}{3}, \frac{11\pi}{6}$ **3.** $3|\tan\theta|$

**4.** $\cos^2\theta$ **5.** $\frac{4\sqrt{2}}{9}$ **6.** $\frac{\pi}{2}, \frac{2\pi}{3}, \frac{3\pi}{2}, \frac{5\pi}{3}$

**7.** $\sqrt{\frac{3}{8}}$ **8.** $r = 4\cos\theta$

**9.** $\frac{x}{\sqrt{x^2+16}}$ **10.** $-\frac{\pi}{4}$

### Test 3D

**1.** $\frac{7}{25}$ **2.** $\frac{\pi}{8}, \frac{3\pi}{8}, \frac{9\pi}{8}, \frac{11\pi}{8}$ **3.** $5|\sec\theta|$

**4.** $\sec\theta$ **5.** $\frac{7}{9}$ **6.** $\frac{\pi}{6}, \frac{\pi}{2}, \frac{5\pi}{6}, \frac{3\pi}{2}$ **7.** $\sqrt{\frac{5}{8}}$

**8.** $r = -5\cos\theta$ **9.** $\frac{5}{\sqrt{x^2+25}}$ **10.** $\frac{5\pi}{6}$

# GATEWAY TEST Answers

## Differentiation— Tests 4A, 4B, 4C & 4D

### Test 4A

**1.** $e^x(x+1)$ **2.** $\dfrac{1}{(x+1)^2}$ **3.** $\dfrac{1}{\sqrt{2x-1}}$

**4.** $2\sin 4x$ **5.** $\dfrac{1}{\sqrt{4-x^2}}$ **6.** $2xe^{x^2}$

**7.** $\dfrac{1}{x(x+1)}$ **8.** $\dfrac{x(5x+2)}{\sqrt{2x+1}}$

**9.** $4x^3 - \dfrac{5}{x} + \dfrac{1}{2\sqrt{x}} - ex^{(e-1)}$ **10.** $\dfrac{-y}{3y^2+x}$

### Test 4B

**1.** $\ln x + 1$ **2.** $\dfrac{1}{(x-1)^2}$ **3.** $\dfrac{x}{\sqrt{x^2+4}}$

**4.** $-6\sin 2x\cos^2 2x$ **5.** $\dfrac{3}{x^2+9}$

**6.** $(\cos x)e^{(\sin x)}$ **7.** $-\dfrac{x^3+2}{x(x^3-1)}$ **8.** $\dfrac{x(4-15x)}{2\sqrt{1-3x}}$

**9.** $3x^2 - \dfrac{1}{2} + \dfrac{1}{2\sqrt{x}} - \pi x^{(\pi-1)}$ **10.** $\dfrac{y}{2y-x}$

### Test 4C

**1.** $x\cos x + \sin x$ **2.** $-\dfrac{3}{(x-1)^2}$

**3.** $12(2x-3)^5$ **4.** $10\tan 5x\sec^2 5x$

**5.** $\dfrac{1}{2\sqrt{x}\sqrt{1-x}}$ **6.** $5e^{5x}$ **7.** $\dfrac{3x+1}{x(2x+1)}$

**8.** $\dfrac{x(3x^2+2)}{\sqrt{x^2+1}}$ **9.** $-\dfrac{14}{x^3} + \dfrac{3}{2}x^{1/2} - \dfrac{1}{2\sqrt{x}}$

**10.** $-\dfrac{2x+y}{2y+x}$

### Test 4D

**1.** $\cos x - x\sin x$ **2.** $\dfrac{3}{(x+1)^2}$

**3.** $21x^2(x^3-6)^6$ **4.** $6\tan 3x\sec^2 3x$

**5.** $\dfrac{1}{2\sqrt{x}(x+1)}$ **6.** $-\dfrac{e^{(1/x)}}{x^2}$ **7.** $\dfrac{x+1}{x(2x+1)}$

**8.** $\dfrac{x(2-3x^2)}{\sqrt{1-x^2}}$ **9.** $\dfrac{2x}{5} - \dfrac{10}{x^3} - \dfrac{1}{3x^{2/3}}$

**10.** $\dfrac{y-2x}{2y-x}$

# GATEWAY TEST Answers

## Integration— Tests 5A, 5B, 5C & 5D

### Test 5A

**1.** $\frac{1}{3}(2x - 1)^{3/2} + C$ **2.** 2

**3.** $\frac{1}{2}\ln|2x - 1| + C$ **4.** $\frac{1}{4}\ln|2x - 1| + \frac{1}{2}x + C$

**5.** $3 \arcsin x + C$ **6.** $\frac{1}{2}\arcsin x + \frac{1}{2}x\sqrt{1 - x^2} + C$

**7.** $\frac{1}{2}e^{(2x-1)} + C$

**8.** $\frac{1}{2}(2x - 1)\ln(2x - 1) - x + C$

**9.** $\dfrac{1}{2}\tan(2x - 1) + C$ **10.** $\dfrac{1}{2(1 - 2x)} + C$

### Test 5B

**1.** $\frac{2}{9}(3x + 1)^{3/2} + C$ **2.** $\frac{11}{2}$

**3.** $\frac{1}{3}\ln|3x + 1| + C$ **4.** $\frac{1}{3}x - \frac{1}{9}\ln|3x + 1| + C$

**5.** $2 \arctan x + C$ **6.** $\ln\left[\sqrt{x^2 + 1} + x\right] + C$

**7.** $\frac{1}{3}e^{(3x+1)} + C$

**8.** $\frac{1}{3}(3x + 1)\ln(3x + 1) - x + C$

**9.** $-\dfrac{1}{3}\cos(3x + 1) + C$ **10.** $-\dfrac{1}{6(3x + 1)^2} + C$

### Test 5C

**1.** $-\frac{1}{6}(1 - 4x)^{3/2} + C$ **2.** $-5$

**3.** $-\frac{1}{4}\ln|1 - 4x| + C$

**4.** $-\frac{1}{4}x - \frac{1}{16}\ln|1 - 4x| + C$ **5.** $2 \arcsin(2x) + C$

**6.** $\ln\left[\sqrt{x^2 - 1} + x\right] + C$ **7.** $-\frac{1}{4}e^{(1-4x)} + C$

**8.** $\frac{1}{4}(4x - 1)\ln(1 - 4x) - x + C$

**9.** $-\dfrac{1}{4}\sin(1 - 4x) + C$ **10.** $\dfrac{1}{4(1 - 4x)} + C$

### Test 5D

**1.** $-\frac{2}{3}(2 - x)^{3/2} + C$ **2.** $\frac{1}{2}$

**3.** $-\ln|2 - x| + C$ **4.** $-2\ln|2 - x| - x + C$

**5.** $\dfrac{1}{\sqrt{2}}\arcsin\left(\sqrt{2}x\right) + C$

**6.** $2 \arcsin\dfrac{x}{2} + \dfrac{1}{2}x\sqrt{4 - x^2} + C$

**7.** $-e^{(2-x)} + C$ **8.** $(x - 2)\ln(2 - x) - x + C$

**9.** $\csc(2 - x) + C$ **10.** $\dfrac{1}{2 - x} + C$

# PART III SUGGESTED SOLUTIONS FOR SPECIAL FEATURES


Chapter Openers . . . . . . . . . . . . . . . . . . . . . . . . . . . . . . . . . . . . . 265

Explorations . . . . . . . . . . . . . . . . . . . . . . . . . . . . . . . . . . . . . . . 274

Technology Boxes . . . . . . . . . . . . . . . . . . . . . . . . . . . . . . . . . . . . 300

Section Projects . . . . . . . . . . . . . . . . . . . . . . . . . . . . . . . . . . . . . 307

# CHAPTER OPENERS

## Chapter P: *Eruptions of Old Faithful*

1. The graph of the data, which is called a scatter plot, suggests a relationship. The points on the scatter plot tend to cluster along a straight line. There is a correlation. Eruptions of longer duration are followed by longer intervals before the next eruption. Shorter eruptions are followed by shorter time intervals.

2. After an eruption that lasted for 2 minutes and 40 seconds, the next eruption would be expected to occur in about 67 minutes.

3. Using hand calculation: A line passing through the points $(1.80, 56)$ and $(4.60, 91)$, for example, looks like it fits the scatter plot fairly well. An equation for the line is

$$y = 12.5x + 33.5.$$

Using technology: The calculated regression line has equation $y = 11.8x + 35.6$.

## Chapter 1: *Swimming Speed: Taking It To The Limit*

1. The scatter plot for the men's 100m freestyle shows that the record time dropped by a "full second" from 1975 to 1976. Therefore, body shaving must have started in 1976.

2. Because the record time in the men's 800m freestyle dropped by about 15 seconds from 1957 to 1958, a technological advance may have occurred in 1958. Perhaps another advance occurred in 1967; for both scatter plots appear to change direction then.

3. The lower limit for men to swim 100m appears to be around 42 seconds. For 800m the limit appears to be 450 seconds. Each estimate was made by visualizing a gently bending curve that seems to fit the corresponding scatter plot and continues into the future years.

4. The scatter plot for the men's 100m freestyle brings to mind the hyperbolic curve $y = 1/x$. To retain the reciprocal relationship but gain flexibility to shift vertically and change scale, we postulate a model curve of the form

$$time = \frac{a}{year} + b,$$

where $a$ and $b$ are constants to be determined. The scatter plot suggests that $(year, time)$ points such as $(55.5, 55.5)$ and $(94, 48.2)$ might be on the curve. Substituting these values for *year* and *time* gives the two simultaneous equations

$$55.5 = \frac{a}{55.5} + b \text{ and } 42.2 = \frac{a}{94} + b.$$

The suitably rounded solution is $a = 989$, $b = 37.7$. This results in the equation

$$time = \frac{989}{year} + 37.7.$$

The scatter plot for the men's 800m freestyle seems to change direction more abruptly than the scatter plot for the men's 100m freestyle. Similarly, the graph of $y = x^4$ bends more rapidly than the graph of $y = x^2$. We postulate a model curve of the form

$$time = \frac{a}{(year)^4} + b,$$

where $a$ and $b$ are constants to be determined. By guessing that the curve passes through the $(year, time)$ points $(58, 555)$ and $(94, 467)$, we get the two simultaneous equations

$$555 = \frac{a}{58^4} + b \text{ and } 467 = \frac{a}{94^4} + b.$$

The suitably rounded solution is $a = 1165 \times 10^6$, $b = 452.1$. The resulting equation is

$$time = \frac{1{,}165{,}000{,}000}{(year)^4} + 452.1.$$

It is too much to hope that a simple formula can adequately model a process that goes through basic changes. Nonetheless, these two models are revealing.

5. The phrase "approach a limit asymptotically" suggests that in the distant future points on the scatter plots will always be just slightly above some horizontal line. On the scatter plot for the men's 800m freestyle, for example, the horizontal line with constant *time* = 450 seconds would correspond to the lower limit we gave in answer to question 3.

## Chapter 2: *Gravity: Finding It Experimentally*

1. The best-fitting model of position as a function of time is parabolic. Its quadratic equation is

$$s = -4.96784t^2 - 0.23743t + 0.29147,$$

where $s$ meters is the height and $t$ seconds is the time.

2. The scatter plot appears to be linear with one unusual observation. Experimental error is possible and the velocity at 0.12 seconds could be erroneous. Using linear regression on all 12 observations gives the equation

$$v = -9.7848t - 0.14696,$$

where $v$ m/sec is the velocity and $t$ sec is the time.

Removing the suspect observation $(0.12, -1.47655)$ and using linear regression on the remaining 11 pairs gives the alternative equation

$$v = -9.7551t - 0.13607.$$

Both of these equations for velocity have a coefficient of $t$ that is approximately twice the coefficient of $t^2$ in the quadratic equation for height.

3. The estimated value of $g$ is $-2(-4.96784) = 9.93568$. Based on only this data, there is no reason to think that either a greater or smaller estimate would be better.

## Chapter 3: *Packaging: The Optimal Form*

1. Formulae for the surface area $S$ and volume $V$ are given in the statement of question 2. In this case $V = 49.54$ cubic inches and so the height is $49.54/(\pi r^2)$. To make each line of the following table, we assign a value to the radius $r$, then compute $h$, and finally calculate $S$.

| Radius (in.) | Height (in.) | Surface Area (sq. in.) |
|---|---|---|
| 1.35 | 8.652 | 84.84 |
| 1.45 | 7.500 | 81.54 |
| 1.55 | 6.564 | 79.02 |

The cleanser container does not minimize the surface area.

**2.** First solve the equation $V = \pi r^2 h$ for $h = V/(\pi r^2)$. Then substitute this expression for $h$ into the equation for $S$ and simplify.

$$S = 2\pi r^2 + 2\pi r \frac{V}{\pi r^2} = 2\pi r^2 + \frac{2V}{r}.$$

If the volume is 48.42 cubic inches, then $S = 2\pi r^2 + \dfrac{96.84}{r}$.

**3.** Use the equation $S = 2\pi r^2 + 2V/r$, which was derived to answer question 2. For each volume in the table, the corresponding tabular radius is less than the optimal radius.

**4.** Even though the volume is fixed, you cannot maximize the surface area. By making the radius large enough, surface area can be made arbitrarily large.

## Chapter 4: *The Wankel Rotary Engine and Area*

**1.** The graph of $y = f(x)$ is an arc of the circle of radius 4 centered at the point $(0, -2\sqrt{3})$. To check this, move the constant term $-2\sqrt{3}$ in the equation for $f(x)$ to the left-hand side. Square both sides; then isolate the new constant term on the right-hand side. The result is

$$\left(f(x) + 2\sqrt{3}\right)^2 + (x - 2)^2 = 16.$$

(a) The region of interest has been graphed on a grid with a $\frac{1}{4}$-inch square mesh. Count the number of squares that lie entirely within the region and multiply by $\left(\frac{1}{4}\right)^2$ in.$^2$ to underestimate the area. Similarly, count the number of squares needed to completely cover the region and multiply by $\left(\frac{1}{4}\right)^2$ in.$^2$ to overestimate the area. Because we are familiar with circular arcs, only values of $f(2.25), f(2.5), f(2.75)$, and $f(3.5)$ must be calculated in order to know which squares to count. This method places the area between 0.875 and 1.875 in.$^2$. The average of these two values gives an approximate area $A \approx 1.375$ in.$^2$ that is accurate to within 0.5 in.$^2$.

(b) The region of interest is the difference between a sector of the circle having a central angle of $\pi/3$ and the largest equilateral triangle lying within the sector. Therefore, the area $A$ of the region is exactly equal to a difference of areas and the exact area is

$$A = \frac{\pi/3}{2\pi}\pi(4)^2 - \frac{1}{2}(4)\left(2\sqrt{3}\right) = \frac{8}{3}\pi - 4\sqrt{3}.$$

**2.** (a) We could change the mesh of the grid to improve accuracy. If the same strategy is used with a grid having a $1/n$-inch mesh, then the average of the under- and over-estimates will be in error by no more than $2/n$ in.$^2$. Values can be obtained that are arbitrarily close to the actual area.

(b) The accuracy with which $\pi$ and $\sqrt{3}$ are known determines the accuracy of this method. Thus, $A \approx 1.449377179$ and values that are arbitrarily close to the actual area can be obtained.

**3.** The "bulged triangle" consists of an equilateral triangle having sides of length 4 inches together with three copies of the region whose area has been approximated. The equilateral triangle has an area of $4\sqrt{3} \approx 6.92820323$ in.$^2$.

(a) The area of the "bulged triangle" is approximately $1.375 + 4\sqrt{3} = 11.053$ in.$^2$.

(b) The area of the "bulged triangle" is $8\pi - 8\sqrt{3} \approx 11.27633477$ in.$^2$.

## Chapter 5: *Plastics and Cooling*

1. Temperature decreases over time. There is a constant time increment $\Delta t$ between points on the scatter plot and the temperature change $\Delta T$ is always negative. Furthermore, $\Delta T$ gets closer to zero over time. Thus, $\Delta T/\Delta t$, the average rate at which temperature changes, is increasing to zero.

2. The water should continue to cool until its temperature is indistinguishable from the room temperature. Thus, the curve should remain above a horizontal asymptote, the line $T = 69.55$.

3. The temperature should be modeled by a decreasing function that is concave upward. That is, the derivative should be negative, but its derivative should be positive. Thus, the derivative of the modeling function will be increasing.

4. As the answer to question 2 explains, it is appropriate to assume that $c = 69.55$. We only need to find values for $a$ and $b$ that produce a reasonable model of the form $T - 69.55 = a \cdot b^t$. The scatter plot allows us to estimate two convenient and representative pairs of the form $(t, T - 69.55)$ such as $(14, 57)$ and $(28, 32)$. Plugging these numbers into the model gives the simultaneous equations

   $$57 = ab^{14} \text{ and } 32 = ab^{28}.$$

   Because $32 = a(b^{14})^2 = a(57/a)^2 = 57^2/a$, we find $a = 57^2/32 = 101.53$. Furthermore, division gives $32/57 = (ab^{28})/(ab^{14}) = b^{14}$. Therefore, $b = (32/57)^{1/14} = 0.9596$. Finally, we obtain the formula $T = 101.53(0.9596)^t + 69.55$.

## Chapter 6: *Constructing an Arch Dam*

1. Part of the cross section is a trapezoid with bases of 16 feet and $59 - (-16) = 75$ feet. Thus,

   $$\int_{-70}^{59} f(x)\,dx = \frac{16+75}{2}389 + \int_{-70}^{-16} (0.03x^2 + 7.1x + 350)\,dx$$

   $$= 17{,}699.5 + \left[0.01x^3 + 3.55x^2 + 350x\right]_{-70}^{-16} = 23{,}502.34$$

   Rounded to suitable precision, the cross section is about 23,500 square feet.

2. If the dam were straight instead of arched, we could multiply its length times the constant cross sectional area to get the volume. Now imagine that there is a flexible solid shaped like the arch dam. Bend it until all the circular arcs are straightened out. The shorter arcs will be stretched; the longer ones will be compressed. The base will become rectangular (instead of being shaped like a sector of a circular ring) and all the cross sections will be identical and perpendicular to all the straightened arcs. If we knew the length of the straightened dam, we could estimate its volume.

3. The base of the dam has an inner radius of 150 feet and inner arc of length $2\pi(150)\frac{150}{360} = 392.7$ feet. The outer radius is $150 + 59 - (-80) = 289$ feet and the outer arc has length $2\pi(289)\frac{150}{360} = 756.6$ feet.

   The average is 574.65 or about 575 feet. Using 574.65 feet for the length of the straightened dam gives an estimated volume of 13,505,619.68 cubic feet or approximately 500,000 cubic yards.

# Chapter 7: *Making a Mercator Map*

**1.** The expression $\sum_{i=1}^{n} R\Delta\phi \sec \phi_i$ approximates how far from the equator to draw the line representing the latitude $\phi_n$.

**2.** Let a positive latitude $\phi$ be given. If $n$ is a positive integer, let $\Delta\phi = \phi/n$. Then the answer to question 1 is a Riemann sum that approximates the integral $\int_0^{\phi} R \sec \lambda d\lambda = R \ln(\sec \phi + \tan \phi)$, which is the actual vertical distance between the lines representing the equator and the latitude $\phi$.

**3.**

| $\phi$ | Distance $6 \ln(\sec \phi + \tan \phi)$ |
|---|---|
| 10° | 1.053 |
| 20° | 2.138 |
| 30° | 3.296 |
| 40° | 4.577 |
| 50° | 6.064 |

**4.** Both sec $\phi$ and tan $\phi$ become infinitely large as $\phi$ increases to 90°.

# Chapter 8: *The Koch Snowflake: Infinite Perimeter?*

**1.** The triangles added in the $n$-th iteration have a side length of $\left(\frac{1}{3}\right)^n$.

**2.**

| $n$ | 0 | 1 | 2 | 3 |
|---|---|---|---|---|
| Perimeter | 3 | $3 \cdot 4 \cdot \frac{1}{3}$ | $3 \cdot 4^2 \left(\frac{1}{3}\right)^2$ | $3 \cdot 4^3 \left(\frac{1}{3}\right)^3$ |

As $n$ approaches infinity, so does the length of the perimeter, $3\left(\frac{4}{3}\right)^n$.

**3.**

| $n$ | 0 | 1 | 2 | 3 |
|---|---|---|---|---|
| Area | $\frac{\sqrt{3}}{4}$ | $\frac{\sqrt{3}}{4} + 3\frac{\sqrt{3}}{4}\left(\frac{1}{3}\right)^2$ | $\frac{\sqrt{3}}{4} + 3\frac{\sqrt{3}}{4}\left(\frac{1}{3}\right)^2 + 12\frac{\sqrt{3}}{4}\left(\frac{1}{9}\right)^2$ | $\frac{\sqrt{3}}{4} + 3\frac{\sqrt{3}}{4}\left(\frac{1}{3}\right)^2 + 12\frac{\sqrt{3}}{4}\left(\frac{1}{9}\right)^2 + 48\frac{\sqrt{3}}{4}\left(\frac{1}{27}\right)^2$ |

The area after the $n$-th iteration is $\frac{\sqrt{3}}{4}\left[1 + \frac{1}{3}\sum_{i=0}^{n-1}\left(\frac{4}{9}\right)^i\right]$, which converges as a geometric series to $\frac{2\sqrt{3}}{5}$ as $n$ approaches infinity. This is consistent with the observation that each iteration lies within the circle of radius $\frac{1}{\sqrt{3}}$ that is circumscribed about the original equilateral triangle.

**4.** The Koch Snowflake is a region of the plane with an area less than $\frac{\pi}{3}$. Hence, a region can have a finite area and also have infinite perimeter.

# Chapter 9: *Exploring New Planets*

1. The graph is elliptical with its major axis on the $x$-axis and the left focus at the origin. As $\theta$ varies from 0 to $\pi$, the upper half of the ellipse is traced counterclockwise from $x = 0.602$ to $x = -0.258$. As $\theta$ increases to $2\pi$, the counterclockwise trace continues until the bottom half of the ellipse has been drawn. As $\theta$ goes from $2\pi$ to $4\pi$, pixels are occasionally added to the display, indicating that the entire ellipse is retraced.

2. By symmetry, one-fourth of the orbit lies between the two points where $x$ is maximal ($\theta = 0$) and $y$ is maximal. The trace feature of the graphing utility can be used to see that $y$ is a maximum if $\theta$ is somewhere in the interval $(1, 1.4)$. At that point the derivative of $y$ must be zero. Because $y = r\sin\theta$, we can calculate

$$\frac{dy}{d\theta} = \frac{0.3612(\cos\theta - 0.4)}{(1 - 0.4\cos\theta)^2}.$$

In the interval $(1, 1.4)$, the solution of $\cos\theta = 0.04$ is $\theta = 1.16$ (about 66.4°). That is the angle corresponding to one-fourth of the orbit.

3. For the quarter of the orbit starting at $\theta = 0$. the time is

$$\frac{116.6}{2(0.5324)}\int_0^{\cos^{-1}0.4}\left(\frac{0.3612}{1 - 0.4\cos\theta}\right)^2 d\theta = 36.6 \text{ days.}$$

Because 70 Vir B completes an orbit every 116.6 days, the planet takes less time, namely $116.6/2 - 36.6 = 21.7$ days, to travel through the second quarter of its orbit.

The distance $r$ between the planet and the star 70 Virginis decreases as $\theta$ goes from 0 to $\pi$ because the denominator, $1 - 0.4\cos\theta$, increases. As $\theta$ goes from $\pi$ to $2\pi$, $r$ increases. The motion is periodic and follows an elliptical path. During a fixed interval of time, a shorter ray sweeps out the same area as a longer ray according to Kepler's second law of planetary motion. This is possible only if the planet moves faster when it is closer to 70 Virginis. Thus, the maximum speed occurs when the planet is nearest to its star and the minimum speed occurs when the planet is farthest away.

# Chapter 10: *Suspension Bridges*

1. The magnitudes or lengths of the vectors $\mathbf{T}$, $\mathbf{T}_0$, and $\mathbf{W}$ are denoted by $\|\mathbf{T}\|$, $\|\mathbf{T}_0\|$, and $\|\mathbf{W}\|$, respectively. Because the horizontal and vertical forces must be in equilibrium, the following trigonometric relationships hold: $\|\mathbf{T}_0\| = \|\mathbf{T}\|\cos\theta$ and $\|\mathbf{W}\| = \|\mathbf{T}\|\sin\theta$.

2. Introduce a coordinate system in the plane of the hanging cable with the $x$-axis running along the horizontal roadway and the upward-directed $y$-axis positioned midway between the towers. Let the unit of length be one foot. The points $(-2100, 520)$, $(0, 6)$, and $(2100, 520)$ are on a parabola $y = ax^2 + bx + c$, say. Therefore,

$$520 = (-2100)^2a - 2100b + c$$

$$520 = (2100)^2a + 2100b + c$$

$$6 = c$$

It follows that $b = 0$ and an equation describing the shape of the cable is $y = \dfrac{514}{(2100)^2}x^2 + 6$.

3. If $x = 400$, then $\tan\theta = \dfrac{dy}{dx} = 2\dfrac{514}{(2100)}(400)$ and $\theta = 0.0930$ radians (about 5.33°). Furthermore,

$\|\mathbf{T}\| = \sqrt{\|\mathbf{T}_0\|^2 + \|\mathbf{W}\|^2}$ and also $\|\mathbf{T}\| = \dfrac{\|\mathbf{W}\|}{\sin\theta} = 10.8\|\mathbf{W}\|$.

**4.** If $x = 200$, then $\theta = \tan^{-1}\left[2\frac{514}{(2100)}(200)\right] = 0.0466$ radians (about 2.67°). Also,

$$\|\mathbf{T}\| = \sqrt{\|\mathbf{T}_0\|^2 + \|\mathbf{W}\|^2} = 21.5\|\mathbf{W}\|.$$

**5.** Because $\|\mathbf{T}_0\|$ is constant, $\|\mathbf{W}\|$ increases from 0 as $D$ moves away from the center of the cable, and $\|\mathbf{T}\| = \sqrt{\|\mathbf{T}_0\|^2 + \|\mathbf{W}\|^2}$, we conclude that $\|\mathbf{T}\|$ increases from $\|\mathbf{T}_0\|$ as $D$ moves away from the center.

The least tension occurs at the center of the cable. The greatest tension is at the point where the cable attaches to the tower.

## Chapter 11: *Race Car Cornering*

**1.**

| $t$ | 0 | 0.5 | 1 | 1.5 | 2 | 2.5 | 3 | 3.5 | 4 | 4.5 |
|---|---|---|---|---|---|---|---|---|---|---|
| $x(t)$ | 840 | 827 | 789 | 728 | 644 | 540 | 420 | 287 | 146 | 0.249 |
| $y(t)$ | 0 | 146 | 287 | 420 | 540 | 643 | 727 | 789 | 827 | 840 |

The car is traveling counterclockwise in the first quadrant on a circle of radius 840 centered at the origin.

**2.** Yes.

**3.** The speed of the car appears to be constant, but the direction keeps changing. Therefore, the velocity, which comprises both speed and direction, is not constant.

**4.** The velocity vector for the car is

$$\begin{aligned}\mathbf{v}(t) &= -840(0.349)(\sin 0.349t)\mathbf{i} + 840(0.349)(\cos 0.349t)\mathbf{j}\\ &= -293(\sin 0.349t)\mathbf{i} + 293(\cos 0.349t)\mathbf{j}.\end{aligned}$$

The speed of the car is constant. More precisely, it is

$$\|\mathbf{v}(t)\| = \sqrt{(293 \sin 0.349t)^2 + (293 \cos 0.349t)^2} = 293.$$

**5.** The acceleration for the car is

$$\mathbf{a}(t) = -102(\cos 0.349t)\mathbf{i} - 102(\sin 0.349t)\mathbf{j} = -\tfrac{51}{420}\mathbf{r}(t).$$

The acceleration is directed from the car toward the center of the circle along $-r$. The driver feels thrown away from the center of the circle but does not actually move away because acceleration prevents it. The driver feels the effect of the acceleration.

## Chapter 12: *Satellite Receiving Dish*

**1.** The receiver should be located at the focus of the circular paraboloid. When the dish is pointed at a satellite, the incoming signals will be reflected to the receiver so that signal strength will increase.

**2.** Headlights, movie projectors, flashbulbs, and some telescopes use parabolic reflectors.

3. Suppose that the tangent plane has an equation of the form $ax + by + cz + d = 0$. Because the plane is not vertical, $c \neq 0$ and we may assume that $c = -1$. Because $(1, 1, 2)$ is a point on the plane, $a + b - 2 + d = 0$. Thus, the plane has equation $z = a(x - 1) + b(y - 1) + 2$.

   The vertical plane $y = 1$ intersects both the paraboloid and the tangent plane giving a parabola and one of its tangent lines. The parabola is also the intersection of the cylinder $z = x^2 + 1$ with the plane $y = 1$. Similarly, the tangent line is the intersection of the plane $z = a(x - 1) + 2$ and the plane $y = 1$.

   In the $xz$-plane ($y = 0$), the line $z = a(x - 1) + 2$ must be tangent to the parabola $z = x^2 + 1$ at the point $(1, 2)$. Also in the $xz$-plane, methods of single variable calculus show that the tangent line has equation $z = 2x$. Thus, $a = 2$.

   By symmetry (interchanging the roles of $x$ and $y$), $b = 2$. Thus, the tangent plane has equation $z = 2(x - 1) + 2(y - 1) + 2$.

4. You may talk about the normal vector for a plane, but it does not make sense to talk about the slope of a plane. Nonetheless, the tangent plane has a "slope in the $x$-direction" of $a = 2$ as derived (using differentiation) in the answer to question 3. Likewise, the "slope in the $y$-direction" is $b = 2$. These pieces of slope information are combined with $c = -1$ to give the normal vector for the plane: $a\mathbf{i} + b\mathbf{j} + c\mathbf{k} = 2\mathbf{i} + 2\mathbf{j} - \mathbf{k}$.

## Chapter 13: *Hyperthermia Treatments for Tumors*

1. Doubling the radius of a sphere increases its volume eightfold. Hence, $\dfrac{V_T}{V} = \dfrac{1}{8}$.

2. The method of Example 2 in Section 13.6 can be used to verify that $\frac{4}{3}\pi abc$ is the volume enclosed by the ellipsoid

$$\frac{x^2}{a^2} + \frac{y^2}{b^2} + \frac{z^2}{c^2} = 1,$$

   where $a$, $b$, and $c$ are positive.

| Equitherm | $\dfrac{V_T}{V}$ |
|---|---|
| $\dfrac{x^2}{0.1} + \dfrac{y^2}{0.26} + \dfrac{z^2}{0.1} = 1$ | $\dfrac{1}{125}$ |
| $\dfrac{x^2}{0.4} + \dfrac{y^2}{1.04} + \dfrac{z^2}{0.4} = 1$ | $\dfrac{8}{125}$ |
| $\dfrac{x^2}{0.9} + \dfrac{y^2}{2.34} + \dfrac{z^2}{0.9} = 1$ | $\dfrac{27}{125}$ |
| $\dfrac{x^2}{1.6} + \dfrac{y^2}{4.16} + \dfrac{z^2}{1.6} = 1$ | $\dfrac{64}{125}$ |
| $\dfrac{x^2}{2.5} + \dfrac{y^2}{6.50} + \dfrac{z^2}{2.5} = 1$ | 1 |

# Chapter 14: *Mathematical Sculpture*

1. Imagine a long flexible solid with a uniform cross section in the shape of an equilateral triangle. When the solid is straightened out, it has three identical sides, each lying in a distinct plane. Twist one end by 120° (one-third of a complete revolution). Then bring the twisted end around and butt it against the untwisted end. The alignment will be perfect and you will have made a torus with only one side. This is similar to the Möbius strip, which has only one side even though each small segment appears to have two sides. The umbilic torus is also similar to the twisted solid we described, but the cross sections of the umbilic torus are more interesting than equilateral triangles. In summary, the umbilic torus has only one side, but short segments seem to have three sides.

2. Use the equations given in terms of the two parameters $u$ and $v$.

3. Try multiplying the formula for $y$ by 4.

# EXPLORATIONS

## Chapter P, Section 1, page 3

**a.**

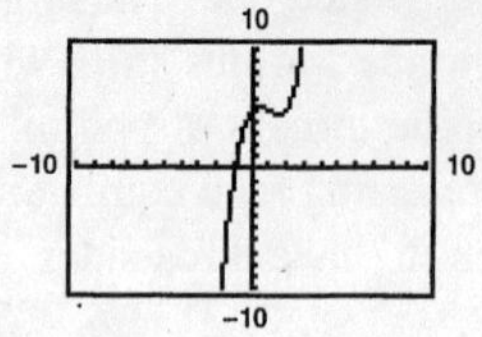


$y = x^3 - 3x^2 + 2x + 5$

**b.**

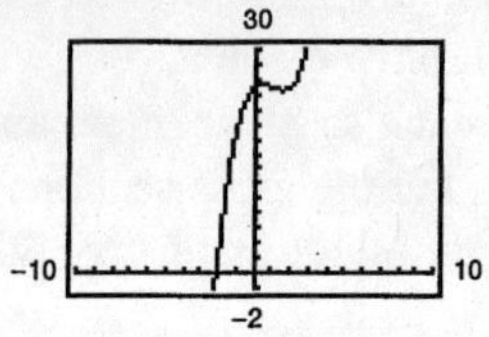


$y = x^3 - 3x^2 + 2x + 25$

**c.**

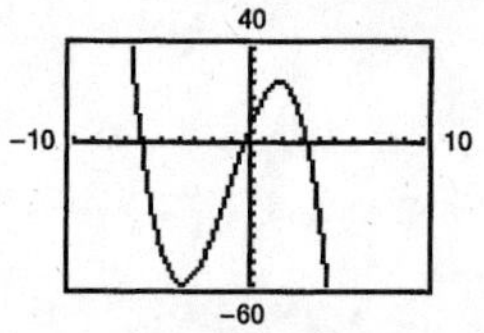


$y = x^3 - 3x^2 + 20x + 5$

**d.**

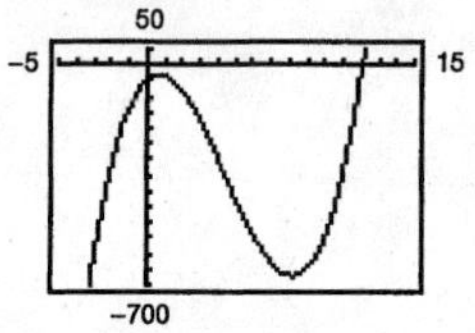


$y = 3x^3 - 40x^2 + 50x - 45$

**e.**

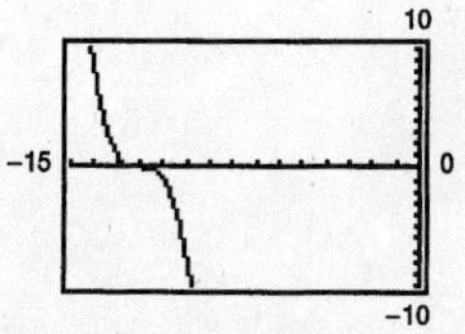


$y = -(x + 12)^3$

**f.**

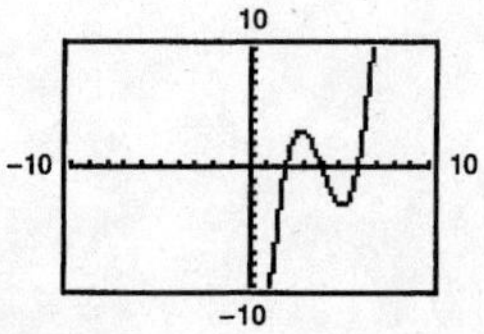


$y = (x - 2)(x - 4)(x - 6)$

With an analytic approach, some other things to consider are:

(1) How many turns does the graph have?

(2) What are the $x$- and $y$-intercepts?

(3) Does the graph begin (and end) by going up or down?

## Chapter P, Section 2, page 11

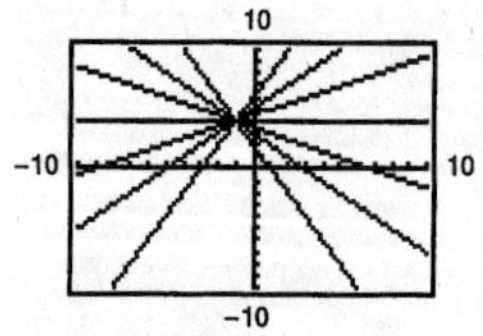


All seven lines have the point $(-1, 4)$ in common. The coefficient of the quantity $(x + 1)$ is the slope. An equation of the line passing through $(-1, 4)$ with a slope of $m$ would be $y - 4 = m(x + 1)$.

## Chapter P, Section 3, page 23

**a.** Horizontal shift 2 units to the left; $y = |x + 2|$

**b.** Vertical shift 1 unit upward; $y = \sin x + 1$

**c.** Horizontal shift 2 units to the right and vertical shift 1 unit downward; $y = (x - 2)^2 - 1$

**d.** Horizontal shift 1 unit to the left and vertical shift 2 units upward; $y = (x + 1)^3 + 2$

## Chapter P, Section 3, page 26

$f(x)$: even

$g(x)$: neither

$h(x)$: odd

$j(x)$: even

$k(x)$: neither

$p(x)$: odd

A function $f$ is even if $f(-x) = f(x)$; $f$ is odd if $f(-x) = -f(x)$.

## Chapter P, Section 5, page 36

**a.** Add 5. $\quad f^{-1}(x) = x + 5$

**b.** Divide by 6. $\quad f^{-1}(x) = \dfrac{x}{6}$

**c.** Multiply by 2. $\quad f^{-1}(x) = 2x$

**d.** Subtract 2, then divide by 3. $\quad f^{-1}(x) = \dfrac{x - 2}{3}$

**e.** Take the cube root. $\quad f^{-1}(x) = \sqrt[3]{x}$

**f.** Divide by 4, then add 2. $\quad f^{-1}(x) = \dfrac{x}{4} + 2$

The graphs of $f$ and $f^{-1}$ are symmetric with respect to the line $y = x$.

## Chapter P, Section 5, page 40

Many different domain restrictions yield an invertible function with the full range of the secant function. The textbook uses

$$\left[0, \frac{\pi}{2}\right) \cup \left(\frac{\pi}{2}, \pi\right].$$

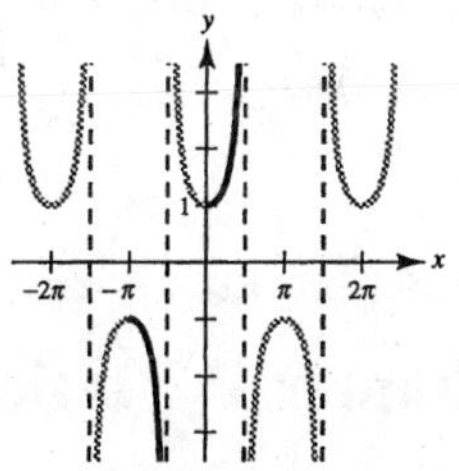


Another possibility is

$$\left[0, \frac{\pi}{2}\right) \cup \left[\pi, \frac{2\pi}{2}\right).$$

The graph at the right illustrates the choice

$$\left[-\pi, -\frac{\pi}{2}\right) \cup \left[0, \frac{\pi}{2}\right).$$

This choice makes sense for four reasons. First, why use big angles if small ones will do? Second, in applications it is convenient to have the formula $\tan\theta = \sqrt{\sec^2\theta - 1}$, without any $(\pm)$ sign ambiguity. Comparing the graphs at the right shows how to restrict the domain of the secant function to values for which the tangent function is not negative. Third, it helps if inverse trigonometric functions evaluate to acute angles whenever possible, especially in work with right triangles. Thus,

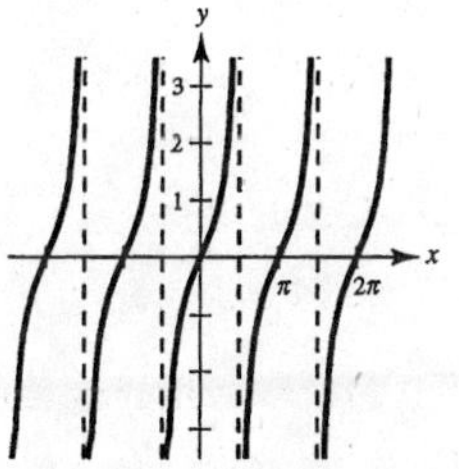


$$\left(0, \frac{\pi}{2}\right)$$

—CONTINUED—

## Chapter P, Section 5, page 40 —CONTINUED—

should be part of the domain of the restricted secant function. Fourth, it should be easy to compute values of the inverse secant function using available technology.

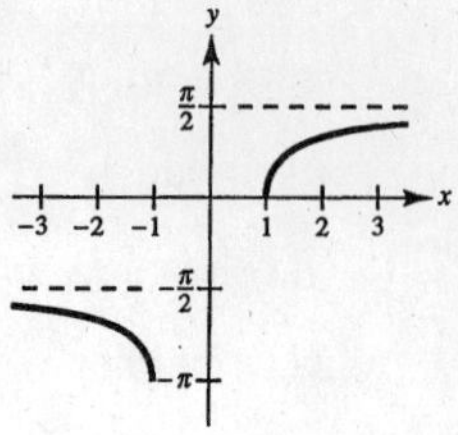


Choosing this restricted domain for the secant function would mean that the inverse function satisfies

$$y = \text{arcsec } x \text{ iff } x = \sec y, -\pi \le y < -\frac{\pi}{2} \text{ or } 0 \le y < \frac{\pi}{2}.$$

A graph is shown at the right. One benefit is the derivative formula

$$\frac{d}{dx} \text{arcsec } x = \frac{1}{x\sqrt{x^2 - 1}}$$

which is different from the one given in the textbook. Scientific and graphing calculators have a key for the inverse cosine function. Using the definition given in the textbook, it is easy to evaluate the inverse secant function on a calculator because arcsec $x$ = arccos $1/x$. By comparison, restricting the domain, as discussed, means that

$$\text{arcsec } x = \frac{x}{|x|} \arccos \frac{1}{x};$$

that is, after computing arccos $1/x$, adjust the sign to agree with $x$.

## Chapter P, Section 5, page 42

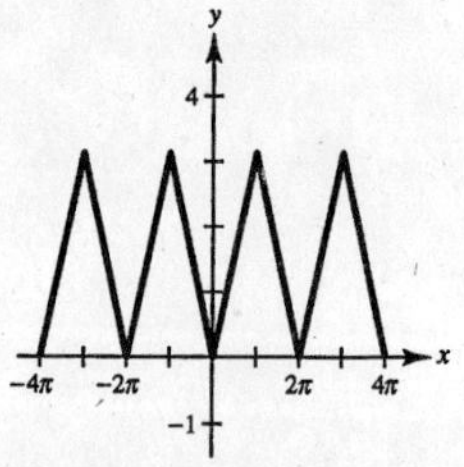


$y = \arccos(\cos x)$

As you can see on the graph, $\arccos(\cos x) = x$ only if $0 \le x \le \pi$.

## Chapter P, Section 6, page 50

The figure P.51 shows the graph of $y_2 = e^{\ln x}$. The domain of $y_1 = \ln e^x$ is all real numbers, whereas the domain of $y_2$ is all positive real numbers. The equation $\ln e^x = e^{\ln x}$ is true for all positive real numbers only because that is the domain of $\ln x$.

## Chapter 1, Section 1, page 63

$P(1, 1)$ and $Q_1(1.5, f(1.5))$: $m = \dfrac{2.25 - 1}{1.5 - 1} = \dfrac{1.25}{0.5} = 2.5$

$P(1, 1)$ and $Q_2(1.1, f(1.1))$: $m = \dfrac{1.21 - 1}{1.1 - 1} = \dfrac{0.21}{0.1} = 2.1$

$P(1, 1)$ and $Q_3(1.01, f(1.01))$: $m = \dfrac{1.0201 - 1}{1.01 - 1} = \dfrac{0.0201}{0.01} = 2.01$

$P(1, 1)$ and $Q_4(1.001, f(1.001))$: $m = \dfrac{1.002001}{1.001 - 1} = \dfrac{0.002001}{0.001} = 2.001$

$P(1, 1)$ and $Q_5(1.0001, f(1.0001))$: $m = \dfrac{1.00020001}{1.0001 - 1} = \dfrac{0.00020001}{0.0001} = 2.0001$

$m = 2$

## Chapter 1, Section 1, page 64

The width of each rectangle is $\frac{1}{5}$.

The area of the inscribed set of rectangles is:

$$\begin{aligned} A &= \tfrac{1}{5} \cdot f(0) + \tfrac{1}{5} \cdot f\left(\tfrac{1}{5}\right) + \tfrac{1}{5} \cdot f\left(\tfrac{2}{5}\right) + \tfrac{1}{5} \cdot f\left(\tfrac{3}{5}\right) + \tfrac{1}{5} \cdot f\left(\tfrac{4}{5}\right) \\ &= \tfrac{1}{5}\left[0 + \tfrac{1}{25} + \tfrac{4}{25} + \tfrac{9}{25} + \tfrac{16}{25}\right] \\ &= \tfrac{1}{5}\left[\tfrac{30}{25}\right] \\ &= \tfrac{6}{25} = 0.24. \end{aligned}$$

The area of the circumscribed set of rectangles is:

$$\begin{aligned} A &= \tfrac{1}{5} \cdot f\left(\tfrac{1}{5}\right) + \tfrac{1}{5} \cdot f\left(\tfrac{2}{5}\right) + \tfrac{1}{5} \cdot f\left(\tfrac{3}{5}\right) + \tfrac{1}{5} \cdot f\left(\tfrac{4}{5}\right) + \tfrac{1}{5} \cdot f(1) \\ &= \tfrac{1}{5}\left[\tfrac{1}{25} + \tfrac{4}{25} + \tfrac{9}{25} + \tfrac{16}{25} + 1\right] \\ &= \tfrac{1}{5}\left[\tfrac{55}{25}\right] \\ &= \tfrac{11}{25} = 0.44. \end{aligned}$$

The area of the region is approximately $\frac{1}{2}\left(\frac{6}{25} + \frac{11}{25}\right) = \frac{17}{50} = 0.34$.

## Chapter 1, Section 2, page 66

| $x$ | 1.75 | 1.9 | 1.99 | 1.999 | 2 | 2.001 | 2.01 | 2.1 | 2.25 |
|---|---|---|---|---|---|---|---|---|---|
| $f(x)$ | 0.75 | 0.9 | 0.99 | 0.999 | 1 | 1.001 | 1.01 | 1.1 | 1.25 |

The graph of $f(x) = \dfrac{x^2 - 3x + 2}{x - 2}$ agrees with the graph of $g(x) = x - 1$ at all points but one, where $x = 2$. If you trace along $f$ getting closer and closer to $x = 2$, the value of $y$ will get closer and closer to 1.

# Chapter 1, Section 4, page 86

**a.**

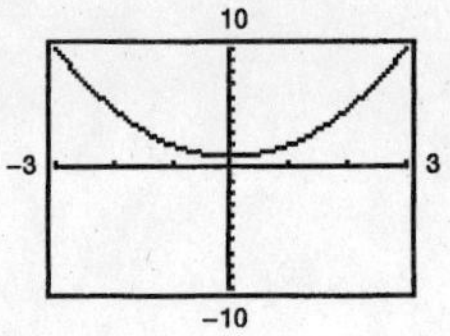


$y = x^2 + 1$ looks continuous on $(-3, 3)$.

**b.**

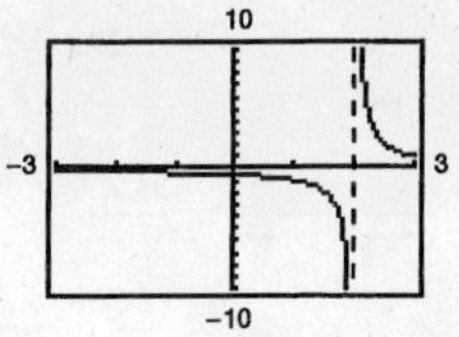


$y = \dfrac{1}{x-2}$ does not look continuous on $(-3, 3)$.

**c.**

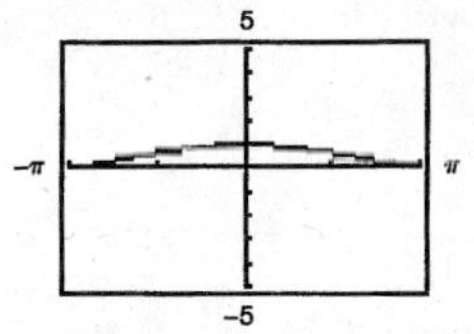


$y = \dfrac{\sin x}{x}$ looks continuous on $(-\pi, \pi)$.

**d.**

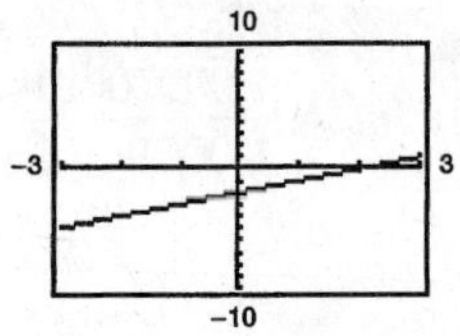


$y = \dfrac{x^2 - 4}{x + 2}$ looks continuous on $(-3, 3)$.

**e.**

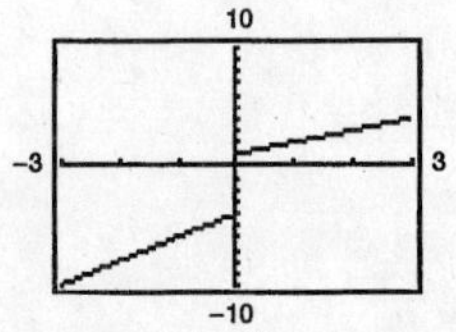


$y = \begin{cases} 2x - 4, & x \le 0 \\ x + 1, & x > 0 \end{cases}$ looks continuous on $(-3, 3)$.

You cannot trust the results you obtain graphically. In these examples, only part **a** is continuous. Part **b** is discontinuous at $x = 2$; part **c** is discontinuous at $x = 0$; part **d** is discontinuous at $x = -2$; part **e** is discontinuous at $x = 0$.

# Chapter 1, Section 5, page 99

**a.**

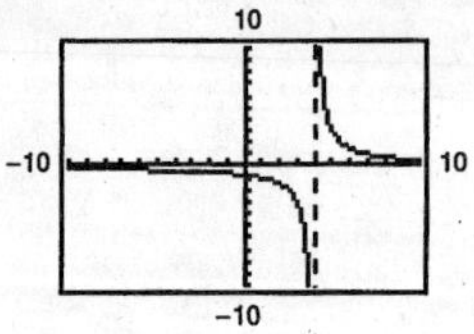


$c = 4;\ \lim_{x\to 4^-} f(x) = -\infty;\ \lim_{x\to 4^+} f(x) = +\infty$

**b.**

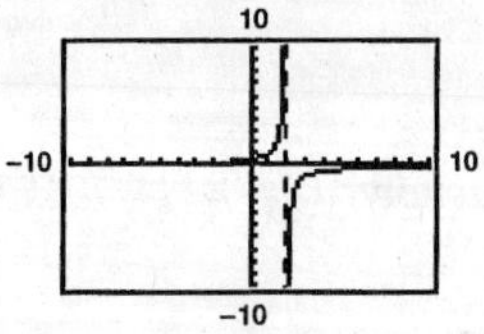


$c = 2;\ \lim_{x\to 2^-} f(x) = +\infty;\ \lim_{x\to 2^+} f(x) = -\infty$

**c.**

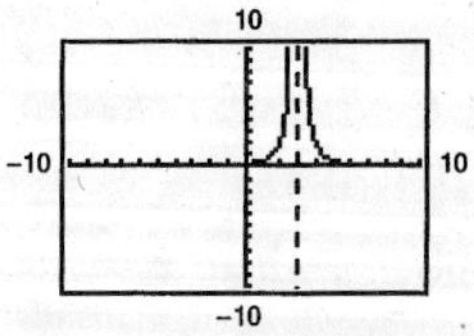


$c = 3;\ \lim_{x\to 3^-} f(x) = +\infty;\ \lim_{x\to 3^+} f(x) = +\infty$

**d.**

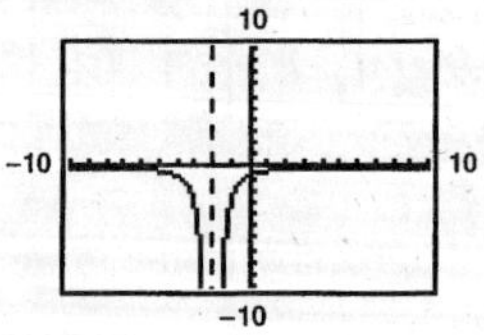


$c = -2;\ \lim_{x\to 2^-} f(x) = -\infty;\ \lim_{x\to -2^+} f(x) = -\infty$

## Chapter 2, Section 1, page 112

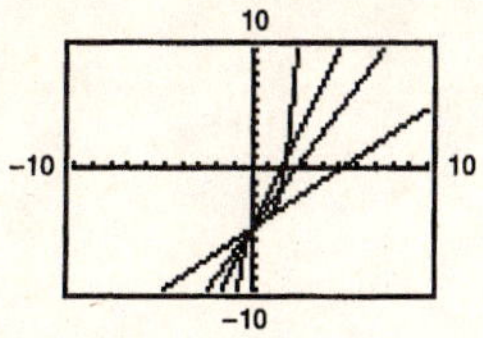


The line $y = x - 5$ appears to be tangent to the graph of $f$ at the point $(0, -5)$ because it seems to intersect the graph at only that point.

## Chapter 2, Section 2, page 123

**a.** $f(x) = x^1$

$$f'(x) = \lim_{\Delta x \to 0} \frac{f(x + \Delta x) - f(x)}{\Delta x}$$
$$= \lim_{\Delta x \to 0} \frac{x + \Delta x - x}{\Delta x}$$
$$= \lim_{\Delta x \to 0} \frac{\cancel{\Delta x}}{\cancel{\Delta x}}$$
$$= \lim_{\Delta x \to 0} 1$$
$$= 1$$

**b.** $f(x) = x^2$

$$f'(x) = \lim_{\Delta x \to 0} \frac{f(x + \Delta x) - f(x)}{\Delta x}$$
$$= \lim_{\Delta x \to 0} \frac{(x + \Delta x)^2 - x^2}{\Delta x}$$
$$= \lim_{\Delta x \to 0} \frac{x^2 + 2x\Delta x + (\Delta x)^2 - x^2}{\Delta x}$$
$$= \lim_{\Delta x \to 0} \frac{2x\Delta x + (\Delta x)^2}{\Delta x}$$
$$= \lim_{\Delta x \to 0} \frac{\cancel{\Delta x}(2x + \Delta x)}{\cancel{\Delta x}}$$
$$= \lim_{\Delta x \to 0} (2x + \Delta x)$$
$$= 2x$$

**c.** $f(x) = x^3$

$$f'(x) = \lim_{\Delta x \to 0} \frac{f(x + \Delta x) - f(x)}{\Delta x}$$
$$= \lim_{\Delta x \to 0} \frac{(x + \Delta x)^3 - x^3}{\Delta x}$$
$$= \lim_{\Delta x \to 0} \frac{x^3 + 3x^2\Delta x + 3x(\Delta x)^2 + (\Delta x)^3 - x^3}{\Delta x}$$
$$= \lim_{\Delta x \to 0} \frac{3x^2\Delta x + 3x(\Delta x)^2 + (\Delta x)^3}{\Delta x}$$
$$= \lim_{\Delta x \to 0} \frac{\cancel{\Delta x}[3x^2 + 3x\Delta x + (\Delta x)^2]}{\cancel{\Delta x}}$$
$$= \lim_{\Delta x \to 0} [3x^2 + 3x\Delta x + (\Delta x)^2]$$
$$= 3x^2$$

**—CONTINUED—**

## Chapter 2, Section 2, page 123 —CONTINUED—

**d.** $f(x) = x^4$

$$
\begin{aligned}
f'(x) &= \lim_{\Delta x \to 0} \frac{f(x+\Delta x) - f(x)}{\Delta x} \\
&= \lim_{\Delta x \to 0} \frac{(x+\Delta x)^4 - x^4}{\Delta x} \\
&= \lim_{\Delta x \to 0} \frac{x^4 + 4x^3\Delta x + 6x^2(\Delta x)^2 + 4x(\Delta x)^3 + (\Delta x)^4 - x^4}{\Delta x} \\
&= \lim_{\Delta x \to 0} \frac{4x^3\Delta x + 6x^2(\Delta x)^2 + 4x(\Delta x)^3 + (\Delta x)^4}{\Delta x} \\
&= \lim_{\Delta x \to 0} \frac{\cancel{\Delta x}\left[4x^3 + 6x^2\Delta x + 4x(\Delta x)^2 + (\Delta x)^3\right]}{\cancel{\Delta x}} \\
&= \lim_{\Delta x \to 0} \left[4x^3 + 6x^2\Delta x + 4x(\Delta x)^2 + (\Delta x)^3\right] \\
&= 4x^3
\end{aligned}
$$

**e.** $f(x) = x^{1/2} = \sqrt{x}$

$$
\begin{aligned}
f'(x) &= \lim_{\Delta x \to 0} \frac{f(x+\Delta x) - f(x)}{\Delta x} \\
&= \lim_{\Delta x \to 0} \frac{\sqrt{x+\Delta x} - \sqrt{x}}{\Delta x} \\
&= \lim_{\Delta x \to 0} \left(\frac{\sqrt{x+\Delta x} - \sqrt{x}}{\Delta x}\right)\left(\frac{\sqrt{x+\Delta x} + \sqrt{x}}{\sqrt{x+\Delta x} + \sqrt{x}}\right) \\
&= \lim_{\Delta x \to 0} \frac{(x+\Delta x) - x}{\Delta x\left(\sqrt{x+\Delta x} + \sqrt{x}\right)} \\
&= \lim_{\Delta x \to 0} \frac{\cancel{\Delta x}}{\cancel{\Delta x}\left(\sqrt{x+\Delta x} + \sqrt{x}\right)} \\
&= \lim_{\Delta x \to 0} \frac{1}{\sqrt{x+\Delta x} + \sqrt{x}} \\
&= \frac{1}{2\sqrt{x}} \\
&= \frac{1}{2}x^{-1/2}
\end{aligned}
$$

**f.** $f(x) = x^{-1} = \frac{1}{x}$

$$
\begin{aligned}
f'(x) &= \lim_{\Delta x \to 0} \frac{f(x+\Delta x) - f(x)}{\Delta x} \\
&= \lim_{\Delta x \to 0} \frac{\frac{1}{x+\Delta x} - \frac{1}{x}}{\Delta x} \\
&= \lim_{\Delta x \to 0} \frac{\frac{x - (x+\Delta x)}{x(x+\Delta x)}}{\Delta x} \\
&= \lim_{\Delta x \to 0} \frac{-\cancel{\Delta x}}{\cancel{\Delta x}\,(x)(x+\Delta x)} \\
&= \lim_{\Delta x \to 0} \frac{-1}{x(x+\Delta x)} \\
&= -\frac{1}{x^2} \\
&= -x^{-2}
\end{aligned}
$$

The exponent of $f$ becomes the coefficient of $f'$ and the power of $x$ decreases by 1.

$$(x^n)' = n\left(x^{n-1}\right)$$

## Chapter 2, Section 2, page 127

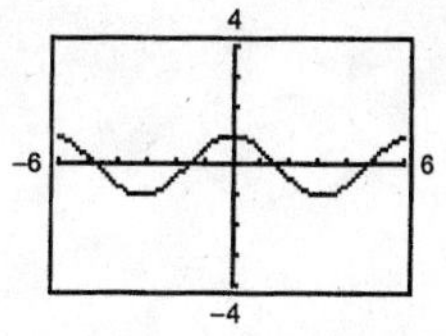


This function represents the graph of $y = \cos x$. The derivative of the sine function equals the cosine function.

## Chapter 2, Section 2, page 129

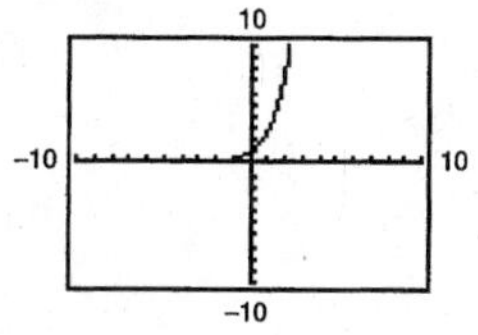


This function represents the graph of $y = e^x$. The derivative of the exponential function is the exponential function.

## Chapter 2, Section 3, page 142

**a.** $y = e^x$

**b.** $y = e^x, y = \dfrac{1}{e^x}$

**c.** $y = e^x$

**d.** $y = e^x, y = \dfrac{1}{e^x}, y = \sin x, y = \cos x$

$y = y^{(8)}$ is true for $y = \sin x$. If $y = \sin x$, $y = y^{(4n)}$ where $n$ is a natural number.

## Chapter 2, Section 4, page 147

**a.** Using the Quotient Rule: $\left(\dfrac{2}{3x+1}\right)' = \dfrac{(3x+1)(0) - 2(3)}{(3x+1)^2} = -\dfrac{6}{(3x+1)^2}$

Using the Chain Rule: $\left(\dfrac{2}{3x+1}\right)' = [2(3x+1)^{-1}]'$

$$= 2(-1)(3x+1)^{-2}(3)$$

$$= -\frac{6}{(3x+1)^2}$$

**b.** Using algebra before differentiating: $[(x+2)^3]' = [x^3 + 6x^2 + 12x + 8]' = 3x^2 + 12x + 12$

Using the Chain Rule: $[(x+2)^3]' = 3(x+2)^2(1) = 3x^2 + 12x + 12$

**c.** Using a trigonometric identity and the Product Rule:

$[\sin 2x]' = [2 \sin x \cos x]' = 2[(\sin x)(-\sin x) + (\cos x)(\cos x)]$

$$= 2[\cos^2 x - \sin^2 x]$$

$$= 2 \cos 2x$$

Using the Chain Rule: $[\sin 2x]' = (\cos 2x)(2)$

$$= 2 \cos 2x$$

In general, the Chain Rule is simpler.

## Chapter 2, Section 4, page 152

| $x$ | $y_1$ | $y_2$ |
|---|---|---|
| 0 | ERROR | ERROR |
| 1 | 0 | 1 |
| 2 | .69315 | .5 |
| 3 | 1.0986 | .33333 |
| 4 | 1.3863 | .25 |
| 5 | 1.6094 | .2 |
| 6 | 1.7918 | .16667 |

The derivative of $\ln x$ is $1, \frac{1}{2}, \frac{1}{3}, \frac{1}{4}, \frac{1}{5}, \ldots$ as $x = 1, 2, 3, 4, \ldots$. It appears to be $\frac{1}{x}$.

| $x$ | $y_1$ | $y_2$ |
|---|---|---|
| 0 | 1 | 0 |
| 1 | $-1$ | 0 |
| 2 | 1 | 0 |
| 3 | $-1$ | 0 |
| 4 | 1 | 0 |
| 5 | $-1$ | 0 |
| 6 | 1 | 0 |

The derivative of $\cos \pi x$ is 0 where $x = 0, 1, 2, 3, \ldots$.

## Chapter 2, Section 4, page 161

No, the graph does not have a tangent line at the point (0, 1) because it has a sharp turn there.

## Chapter 2, Section 6, page 171

Because $y' = 2\sqrt{1 - x^2}$, at the point (0, 0), $m = 2\sqrt{1 - 0} = 2$ and the equation of the tangent line at the origin is $y = 2x$. On the interval $(-0.266, 0.266)$ the tangent line is within 0.01 units of the graph of the function. To say that a function is "locally linear" means that it behaves like a line near $x = 0$.

## Chapter 2, Section 7, page 175

$$V = \frac{\pi}{3} r^2 h \qquad \text{Given: } \frac{dh}{dt} = -0.2 \text{ ft/min}$$

$$\frac{dV}{dt} = \frac{\pi}{3}\left(r^2 \frac{dh}{dt} + 2rh \frac{dr}{dt}\right) \qquad \frac{dr}{dt} = -0.1 \text{ ft/min}$$

$$= \frac{\pi}{3}[(1 \text{ ft})^2 (-0.2 \text{ ft/min}) + 2(1 \text{ ft})(2 \text{ ft})(-0.1 \text{ ft/min})] \qquad r = 1 \text{ ft}$$

$$= \frac{\pi}{3}\left(-0.2 \text{ ft}^3/\text{min} - 0.4 \text{ ft}^3/\text{min}\right) \qquad h = 2 \text{ ft}$$

$$= \frac{\pi}{3}\left(-0.6 \text{ ft}^3/\text{min}\right)$$

$$= -\frac{\pi}{5} \text{ ft}^3/\text{min}$$

The rate of change in the volume does depend on the values of $r$ and $h$ because both variables are in the function $\frac{dV}{dt}$.

## Chapter 3, Section 1, page 198

**a.** Minimum: $x = 1.9999998,\ y = 1$
Maximum: $x = -0.9999934,\ y = 9.9999606$

**b.** Minimum: $x = 1.8685161,\ y = -8.064605$
Maximum: $x = -0.5351851,\ y = -1.12058$

In each case the $x$-values are approximate. For the graphing utility to come up with the exact answer, the approximation must be rational.

## Chapter 3, Section 2, page 206

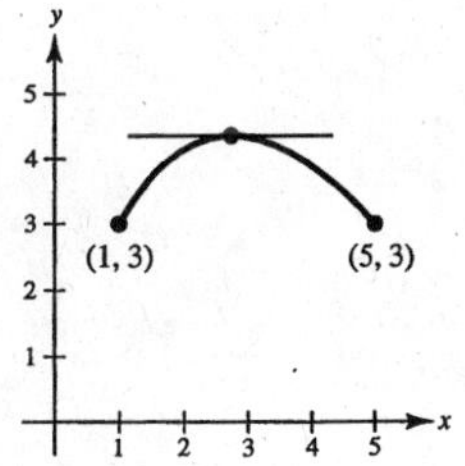


There is at least one point on the graph for which the derivative is zero. It would not be possible to draw the graph otherwise, because to be differentiable, the graph must be continuous.

## Chapter 3, Section 3, page 215

Relative Minimum: $x = 1.0471976,\ y = -0.3424266$
Relative Maximum $x = 5.2359878,\ y = 3.4840193$

Both approximations are close to the actual $x$ values. The relative minimum occurs at $x = \frac{\pi}{3} \approx 1.047197551$ and the relative maximum ocurs at $x = \frac{5\pi}{3} \approx 5.235987756.$

## Chapter 3, Section 4, page 225

The value of $c$ has a bearing on the shape of the graph, but has no bearing on the value of the second derivative at given values of $x$. Graphically, this is true because changes in the value of $c$ change the location of the extrema but do not affect the location of the point of inflection.

## Chapter 3, Section 5, page 231

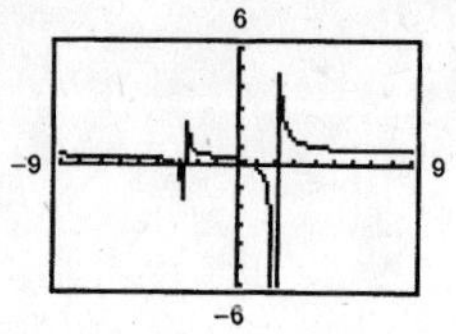


There are vertical asymptotes at $x = 2$ and $x = -\frac{8}{3}$.
There are no extrema.
There is a point of inflection at $(1, 0)$.
The graph is decreasing on the intervals $\left(-\infty, -\frac{8}{3}\right)$, $\left(-\frac{8}{3}, 2\right)$, and $(2, \infty)$.

No one viewing window shows all features. For example, you need to zoom in to see the point of inflection. The horizontal asymptote is $y = \frac{2}{3}$. If $x = 890$, $y = 0.6676664 < 0.667\overline{6}$ and, therefore, within 0.001 units of its horizontal asymptote.

## Chapter 3, Section 7, page 254

The minimum area occurs when $x = \dfrac{4}{\pi + 4} \approx 0.56$. Then, $r = \dfrac{2(1 - x)}{\pi} = \dfrac{2}{\pi + 4}$.

## Chapter 3, Section 8, page 262

The TI-83 distinguishes between the two graphs after zooming in twice. As the $x$-values get closer to 1, the $y$-values do not change until the 5th decimal place.

## Chapter 4, Section 1, page 276

**a.** $F(x) = x^2$

**b.** $F(x) = \dfrac{1}{2}x^2$

**c.** $F(x) = \dfrac{1}{3}x^3$

**d.** $F(x) = -\dfrac{1}{x}$

**e.** $F(x) = -\dfrac{1}{2x^2}$

**f.** $F(x) = \sin x$

One way to find $F$ is to guess, check, and revise.

## Chapter 4, Section 2, page 293

| $n$ | 10 | 100 | 1000 |
|---|---|---|---|
| $s(n)$ | $\frac{684}{300} = 2.28$ | $\frac{78{,}804}{30{,}000} = 2.6268$ | $\frac{7{,}988{,}004}{3{,}000{,}000} = 2.662668$ |
| $S(n)$ | $\frac{924}{300} = 3.08$ | $\frac{81{,}204}{30{,}000} = 2.7068$ | $\frac{8{,}012{,}004}{3{,}000{,}000} = 2.670668$ |

Because $s(n) \le \text{area} \le S(n)$, area $= \frac{8}{3}$.

## Chapter 4, Section 3, page 301

One example to show that the converse of Theorem 4.4 is false is

$$f(x) = \begin{cases} 0 & \text{if } x \neq 3 \\ 7 & \text{if } x = 3 \end{cases}.$$

Although $f$ is not continuous at 3, $f$ is integrable on the closed interval $[2, 4]$ and $\int_2^4 f(x)dx = 0$.

To see this, let $\varepsilon > 0$ and choose $\delta = \varepsilon/8$. Any Riemann sum $\sum_{i=1}^{n} f(c_i)\Delta x_i$ of $f$ for a partition $\Delta$ of $[2, 4]$ can have at most one nonzero term and its value is between 0 and $7\|\Delta\|$.

Hence, if $\|\Delta\| < \delta$ then

$$\left|0 - \sum_{i=1}^{n} f(c_i)\Delta x_i\right| = \sum_{i=1}^{n} f(c_i)\Delta x_i \leq 7\|\Delta\| < 7 \cdot \frac{\varepsilon}{8} < \varepsilon.$$

A second example is $g(x) = \begin{cases} 4 & \text{if } x \leq 1 \\ 5 & \text{if } x > 1 \end{cases}$. As intuition suggests, $\int_0^2 g(x)dx = 9$.

If a function is differentiable on a closed interval $[a, b]$ as defined in Section 2.1, then the function is continuous on $[a, b]$. (The proof of Theorem 2.1 is easily extended to include the one-sided limits that occur at the endpoints $a$ and $b$.) By Theorem 4.4, if a function is differentiable on $[a, b]$, then the function is integrable there. That is, differentiability on $[a, b]$ is stronger than continuity on $[a, b]$, which is stronger than integrability on $[a, b]$.

The examples given here show that integrability is weaker than continuity on $[a, b]$. Examples in Section 2.1 show that a continuous function can have a sharp turn or a vertical tangent line and not be differentiable. Thus, continuity is weaker than differentiability.

## Chapter 4, Section 4, page 309

The symbol $\int$ was first applied to the definite integral. Leibniz calculated area as an infinite sum, hence, the letter *S*.

## Chapter 4, Section 4, page 315

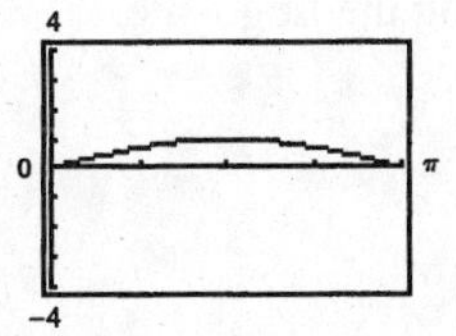


The graph of *F* is the graph of the sine function. The antiderivative of the cos *t* is the sin *t* and when evaluated on the interval $[0, x]$ using the Fundamental Theorem of Calculus, the result is sin *x*.

## Chapter 4, Section 5, page 322

**a.**

$$f(u) = u^4$$
$$u = g(x) = x^2 + 1$$
$$g'(x) = 2x$$
$$\int 2x(x^2 + 1)^4\,dx = \frac{1}{5}(x^2 + 1)^5 + C$$

**b.**

$$f(u) = \sqrt{u}$$
$$u = g(x) = x^3 + 1$$
$$g'(x) = 3x^2$$
$$\int 3x^2\sqrt{x^3 + 1}\,dx = \frac{2}{3}(x^3 + 1)^{3/2} + C$$

—CONTINUED—

## Chapter 4, Section 5, page 322 —CONTINUED—

**c.**
$$f(u) = u^1$$
$$u = g(x) = \tan x + 3$$
$$g'(x) = \sec^2 x$$
$$\int \sec^2 x(\tan x + 3)dx = \frac{1}{2}(\tan x + 3)^2 + C$$

**d.**
$$\int x(x^2 + 1)^4 dx = \frac{1}{2}\int 2x(x^2 + 1)^4 dx$$
$$= \frac{1}{2}\left[\frac{1}{5}(x^2 + 1)^5\right] + C$$
$$= \frac{1}{10}(x^2 + 1)^5 + C$$

**e.**
$$\int x^2\sqrt{x^3 + 1}\, dx = \frac{1}{3}\int 3x^2\sqrt{x^3 + 1}\, dx$$
$$= \frac{1}{3}\left[\frac{2}{3}(x^3 + 1)^{3/2}\right] + C$$
$$= \frac{2}{9}(x^3 + 1)^{3/2} + C$$

**f.**
$$\int 2\sec^2 x(\tan x + 3)\, dx = 2\int\left(\frac{1}{2}\right)(2\sec^2 x)(\tan x + 3)\, dx$$
$$= 2\left[\frac{1}{2}(\tan x + 3)^2\right] + C = (\tan x + 3)^2 + C$$

## Chapter 4, Section 5, page 327

(a) Choose $\int x^2\sqrt{x^3 + 1}\, dx$ because with $u = x^3 + 1$, this integral can easily by evaluated using the General Power Rule for Integration, Theorem 4.13.

(b) Choose $\int \tan 3x \sec^2 3x$ because with $u = \tan 3x$, this integral can easily be evaluated using the General Power Rule for Integration, Theorem 4.13.

## Chapter 4, Section 7, page 342

$\dfrac{1}{x^2 + 1}, \dfrac{x^2}{3x^2 + 5}, \dfrac{x^3}{x^2 + 1}$ are examples of rational functions that cannot be integrated using the Log Rule.

For each of these, if you let $u$ equal the denominator, the numerator is not $u'$.

## Chapter 5, Section 1, page 374

The solutions are hyperbolas.

Yes, $y' = 2\left(\dfrac{x}{y}\right)$.

To see this, use implicit differentiation.

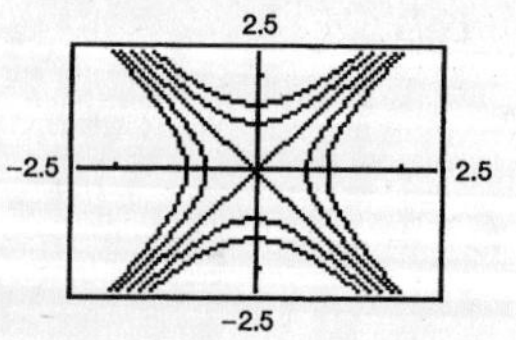


$$y^2 - 2x^2 = C$$
$$2yy' - 4x = 0$$
$$2yy' = 4x$$
$$y' = \frac{2x}{y}$$

Yes, all curves for which this statement is true are represented by the general solution.

## Chapter 6, Section 5, page 450

No. For work to be done an object must move in the direction of an applied force. There is no force in the direction of horizontal movement.

## Chapter 6, Section 5, page 452

Less work is done in compressing the spring from $x = 0$ to $x = 3$.

$$\begin{aligned} W &= \int_0^3 250x\,dx \\ &= \left[125x^2\right]_0^3 \\ &= 1125 \text{ inch-pounds} \end{aligned}$$

## Chapter 6, Section 6, page 466

For the shell method $V = 2\pi \int_a^b ph\,dx$, where

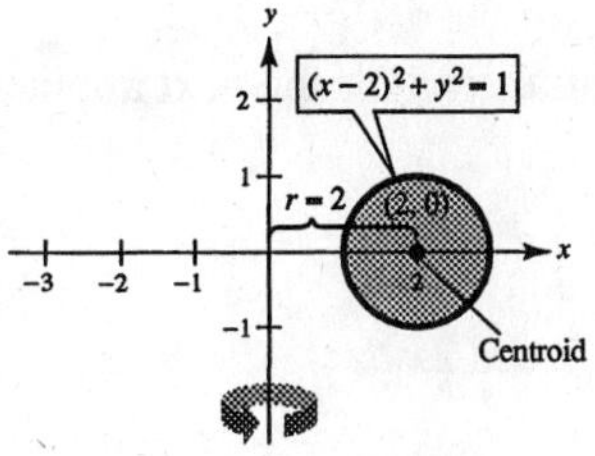


$$a = 1$$

$$b = 3$$

$$p = x$$

$$\begin{aligned} h &= \sqrt{1-(x-2)^2} - \left[-\sqrt{1-(x-2)^2}\right] \\ &= 2\sqrt{1-(x-2)^2}. \end{aligned}$$

Hence, $V = 2\pi \int_1^3 x\left[2\sqrt{1-(x-2)^2}\right] dx = 4\pi \int_1^3 x\sqrt{1-(x-2)^2}\,dx.$

The TI-83 gives 39.47842289 as a value for this integral.

## Chapter 7, Section 1, page 482

Parts **a.** and **b.** can be evaluated using the basic integration rules.

**a.** $$\begin{aligned} \int \frac{3}{\sqrt{1-x^2}}\,dx &= 3\int \frac{1}{\sqrt{1-x^2}}\,dx \\ &= 3 \arcsin x + C \end{aligned}$$

**b.** $$\begin{aligned} \int \frac{3x}{\sqrt{1-x^2}}\,dx &= 3\int x(1-x^2)^{-1/2}\,dx \\ &= 3\left(-\frac{1}{2}\right)\int (-2x)(1-x^2)^{-1/2}\,dx \\ &= \frac{-\frac{3}{2}(1-x^2)^{1/2}}{\frac{1}{2}} + C \\ &= -3\sqrt{1-x^2} + C \end{aligned}$$

Part **c** cannot be evaluated using the basic integration rules.

## Chapter 7, Section 2, page 488

In the special case when $f$ and $g$ are both increasing functions, this graph gives the definite integral formulation of integration by parts.

$$\Big[uv\Big]_{(p,\,r)}^{(q,\,s)}$$

is unfamiliar notation. $(p, r)$ and $(q, s)$ are ordered pairs of numbers where the first components are values of $u$ and the second components are values of $v$.

## Chapter 7, Section 2, page 491

$$dv = e^x\,dx \quad \rightarrow \quad v = \int e^x\,dx = e^x$$

$$u = \cos 2x \rightarrow du = -2\sin 2x\,dx$$

Integration by parts produces the following.

$$\int e^x \cos 2x\,dx = e^x \cos 2x + 2\int e^x \sin 2x\,dx$$ First use of parts

Make the same type of substitutions for the next application of integration by parts.

$$dv = e^x\,dx \quad \rightarrow \quad v = \int e^x\,dx = e^x$$

$$u = \sin 2x \rightarrow du = 2\cos 2x\,dx$$

Integration by parts now produces

$$\int e^x \sin 2x\,dx = e^x \sin 2x - 2\int e^x \cos 2x\,dx.$$ Second use of parts

Therefore, you have

$$\int e^x \cos 2x\,dx = e^x \cos 2x + 2e^x \sin 2x - 4\int e^x \cos 2x\,dx.$$

Because the right-hand integral is a constant multiple of the original integral, you can add it to both sides of the equation to obtain

$$5\int e^x \cos 2x\,dx = e^x \cos 2x + 2e^x \sin 2x$$

$$\int e^x \cos 2x\,dx = \frac{1}{5}e^x \cos 2x + \frac{2}{5}e^x \sin 2x + C.$$ Divide by 5

## Chapter 7, Section 4, page 506

The exact value of the integral $\int_{-1}^{1} \sqrt{1 - x^2}\, dx$ is $\frac{1}{2}\pi$. From geometry, you know that $y = \sqrt{1 - x^2}$ is a semicircle with radius 1. Therefore, Area $= \frac{1}{2}\pi r^2 = \frac{1}{2}\pi$.

To determine the accuracy of an approximation using Simpson's Rule or the Trapezoidal Rule, the derivatives of the function are needed. For this function, all derivatives are undefined at the limits of integration.

Let $x = \sin \theta$

$dx = \cos \theta \, d\theta$

At $x = 1,\ 1 = \sin \theta \Rightarrow \theta = \frac{\pi}{2}$.

At $x = -1,\ -1 = \sin \theta \Rightarrow \theta = \frac{\pi}{2}$.

$$\int_{-1}^{1} \sqrt{1 - x^2}\, dx = \int_{-\pi/2}^{\pi/2} \sqrt{1 - \sin^2 \theta} \cos \theta \, d\theta$$

$$= \int_{-\pi/2}^{\pi/2} \cos^2 \theta \, d\theta$$

$$= \frac{1}{2}\int_{-\pi/2}^{\pi/2} [1 + \cos 2\theta]\, d\theta$$

$$= \frac{1}{2}\left[\theta + \frac{1}{2}\sin 2\theta\right]_{-\pi/2}^{\pi/2}$$

$$= \frac{1}{2}\left[\frac{\pi}{2} + \frac{1}{2}\sin \pi\right] - \frac{1}{2}\left[-\frac{\pi}{2} + \frac{1}{2}\sin(-\pi)\right]$$

$$= \frac{\pi}{4} + \frac{\pi}{4}$$

$$= \frac{\pi}{2}$$

## Chapter 7, Section 6, page 525

Yes, because $\displaystyle\int \frac{2\,du}{u^2 + 1} = 2 \arctan u + C$

$$= 2 \arctan \sqrt{x - 1} + C$$

## Chapter 7, Section 7, page 532

A graphical approach gave the following results.

**a.** $\displaystyle\lim_{x \to 0} \frac{2^{2x} - 1}{x} \approx 1.39$

**b.** $\displaystyle\lim_{x \to 0} \frac{3^{2x} - 1}{x} \approx 2.19$

**c.** $\displaystyle\lim_{x \to 0} \frac{4^{2x} - 1}{x} \approx 2.77$

**d.** $\displaystyle\lim_{x \to 0} \frac{5^{2x} - 1}{x} \approx 3.22$

**—CONTINUED—**

## Chapter 7, Section 7, page 532 —CONTINUED—

It is difficult to observe a pattern here. But the analytic approach leads you to notice that

$1.39 \approx 2 \ln 2$

$2.19 \approx 2 \ln 3$

$2.77 \approx 2 \ln 4$

$3.22 \approx 2 \ln 5.$

Thus, $\lim_{x \to 0} \dfrac{a^{2x} - 1}{x} = 2 \ln a.$

## Chapter 8, Section 1, page 556

**a.** The denominators are powers of 2.

$$a_n = \frac{1}{2^{n-1}}, \quad n \geq 1$$

As $n \to \infty$, $\dfrac{1}{2^{n-1}} \to 0$ because the numerator stays constant as the denominator increases.

**b.** The denominators are factorials.

$$a_n = \frac{1}{n!}, \quad n \geq 1$$

As $n \to \infty$, $\dfrac{1}{n!} \to 0$ because the numerator stays constant as the denominator increases.

**c.** The denominator is the sum $1 + 2 + 3 + \cdot \cdot \cdot + n$.

$$a_n = \frac{10}{\frac{1}{2}(n)(n+1)}, \quad n \geq 1$$

As $n \to \infty$, $\dfrac{10}{\frac{1}{2}(n)(n+1)} \to 0$ because the numerator stays constant as the denominator increases.

**d.** The numerator is an integer squared and the corresponding denominator is that integer plus one squared.

$$a_n = \frac{n^2}{(n+1)^2}, \quad n \geq 1$$

As $n \to \infty$, $\dfrac{n^2}{(n+1)^2} \to 1$ because the numerator and the denominator are equal degrees.

**e.** The numerator is an odd integer starting with 3 and each denominator is three more than the previous one.

$$a_n = \frac{2n+1}{4+3n}, \quad n \geq 1$$

As $n \to \infty$, $\dfrac{2n+1}{4+3n} \to \dfrac{2}{3}$ because the numerator and the denominator are equal degrees.

## Chapter 8, Section 2, page 567

**a.** $0.1 + 0.01 + 0.001 + 0.0001 + \cdots = \frac{1}{9}$

This sum is $\frac{1}{3}$ of part (b), whose sum is $\frac{1}{3}$.

**b.** $\frac{3}{10} + \frac{3}{100} + \frac{3}{1000} + \frac{3}{10{,}000} + \cdots = 0.333\overline{3} = \frac{1}{3}$

This sum is common knowledge.

**c.** $1 + \frac{1}{2} + \frac{1}{4} + \frac{1}{8} + \frac{1}{16} + \cdots = 2$

Example 1 of this section shows that $\frac{1}{2} + \frac{1}{4} + \frac{1}{8} + \frac{1}{16} + \cdots = 1$.

Therefore, $1 + \left(\frac{1}{2} + \frac{1}{4} + \frac{1}{8} + \frac{1}{16} + \cdots\right) = 1 + 1 = 2$.

**d.** $\frac{15}{100} + \frac{15}{10{,}000} + \frac{15}{1{,}000{,}000} + \cdots = \frac{5}{33}$

$$\begin{aligned}\frac{15}{100} + \frac{15}{10{,}000} + \frac{15}{1{,}000{,}000} + \cdots &= 0.151515\overline{15} \\ &= 15(0.010101\overline{01}) \\ &= 15\left(\frac{1}{99}\right) \\ &= \frac{5}{33}\end{aligned}$$

## Chapter 8, Section 2, page 569

Because $\triangle PQR$ and $\triangle TSP$ are similar triangles, the ratios of corresponding sides are equal.

In $\triangle TSP$, the ratio of the vertical side to the horizontal side is $\dfrac{1 + r + r^2 + r^3 + \cdots}{1}$.

In $\triangle PQR$, the ratio of the vertical side to the horizontal side is $\dfrac{1}{1 - r}$.

Therefore, $1 + r + r^2 + r^3 + \cdots = \dfrac{1}{1 - r}$.

This result is a special case of Theorem 8.6 where $r > 0$ and $a = 1$.

## Chapter 8, Section 6, page 597

Parts **a.** and **b.** converge; part **c.** has no guarantee and part **d.** diverges. These conclusions are all a result of Theorem 8.17.

## Chapter 8, Section 8, page 616

**a.** $e^{-x}$

**b.** $\cos x$

**c.** $\sin x$

**d.** $\arctan x$

**e.** $e^{2x}$

## Chapter 9, Section 6, page 702

If $a \neq 0$ and $b = 1$, the graph is a parabola.
If $a \neq 0$ and $0 < b < 1$, the graph is an ellipse.
If $a \neq 0$ and $b > 1$, the graph is a hyperbola.

## Chapter 10, Section 3, page 735

At 30°, the vector is

$$\frac{\sqrt{3}}{2}\mathbf{i} + \frac{1}{2}\mathbf{j}.$$

At 330° the vector is

$$\frac{\sqrt{3}}{2}\mathbf{i} - \frac{1}{2}\mathbf{j}.$$

The angle between these two vectors is 60°.

$$\left(\frac{\sqrt{3}}{2}\mathbf{i} + \frac{1}{2}\mathbf{j}\right) \cdot \left(\frac{\sqrt{3}}{2}\mathbf{i} - \frac{1}{2}\mathbf{j}\right) = \frac{3}{4} - \frac{1}{4} = \frac{1}{2} \text{ and } \cos 60° = \frac{1}{2}.$$

At 60° the vector is

$$\frac{1}{2}\mathbf{i} + \frac{\sqrt{3}}{2}\mathbf{j}.$$

At 180° the vector is $-\mathbf{i}$. The angle between these two vectors is 120°.

$$\left(\frac{1}{2}\mathbf{i} + \frac{\sqrt{3}}{2}\mathbf{j}\right) \cdot (-\mathbf{i}) = -\frac{1}{2} \text{ and } \cos 120° = -\frac{1}{2}.$$

Conjecture: the dot product of two unit vectors is equal to the cosine of the angle between the vectors.

## Chapter 10, Section 4, page 744

**a.** $\begin{vmatrix} \mathbf{i} & \mathbf{j} & \mathbf{k} \\ 3 & 0 & 3 \\ 3 & 0 & -3 \end{vmatrix} = -\mathbf{j}(-9 - 9) = 18\mathbf{j}$

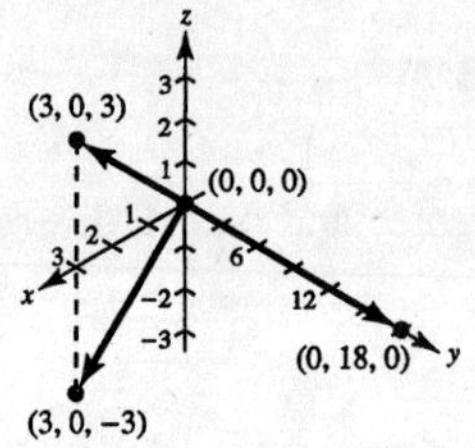


**b.** $\begin{vmatrix} \mathbf{i} & \mathbf{j} & \mathbf{k} \\ 0 & 3 & 3 \\ 0 & -3 & 3 \end{vmatrix} = \mathbf{i}(9 + 9) = 18\mathbf{i}$

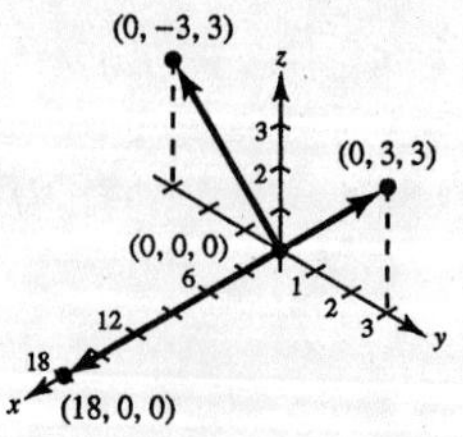


—CONTINUED—

## Chapter 10, Section 4, page 744 —CONTINUED—

c. $\begin{vmatrix} \mathbf{i} & \mathbf{j} & \mathbf{k} \\ 3 & 3 & 0 \\ 3 & -3 & 0 \end{vmatrix} = \mathbf{k}(-9-9) = -18\mathbf{k}$

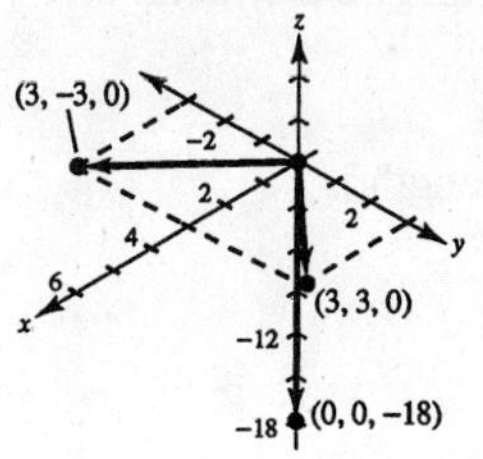


In each of the above cases, the vectors **u** and **v** lie on a plane parallel to one of the three coordinate planes and $\mathbf{u} \times \mathbf{v}$ is parallel to the axis perpendicular to the plane.

Conjecture: $\mathbf{u} \times \mathbf{v}$ is perpendicular to both **u** and **v**.

## Chapter 11, Section 2, page 797

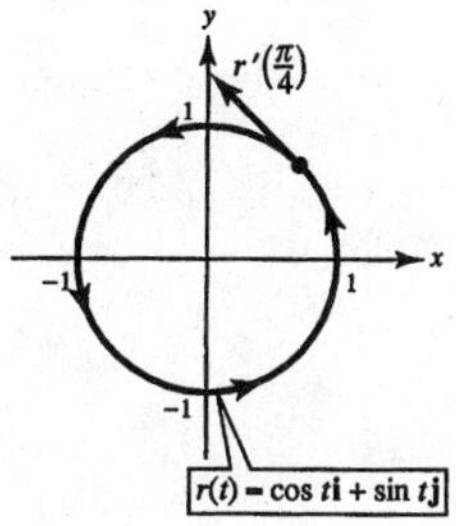


$\mathbf{r}(t) = \cos t\mathbf{i} + \sin t\mathbf{j}$

Let $x = \cos t$ and $y = \sin t$. Because $\cos^2 t + \sin^2 t = 1$ you obtain $x^2 + y^2 = 1$, the unit circle centered at the origin.

$$\mathbf{r}(t) = \cos t\mathbf{i} \quad + \sin t\mathbf{j} \Rightarrow \mathbf{r}\left(\frac{\pi}{4}\right) = \frac{1}{\sqrt{2}}\mathbf{i} \quad + \frac{1}{\sqrt{2}}\mathbf{j}$$

$$\mathbf{r}'(t) = -\sin t\mathbf{i} + \cos t\mathbf{j} \Rightarrow \mathbf{r}'\left(\frac{\pi}{4}\right) = -\frac{1}{\sqrt{2}}\mathbf{i} + \frac{1}{\sqrt{2}}\mathbf{j}$$

$\mathbf{r}\left(\frac{\pi}{4}\right)$ and $\mathbf{r}'\left(\frac{\pi}{4}\right)$ are perpendicular.

$$\begin{aligned} \mathbf{r}(t) \cdot \mathbf{r}(t) &= (\cos t\mathbf{i} + \sin t\mathbf{j}) \cdot (\cos t\mathbf{i} + \sin t\mathbf{j}) \\ &= \cos^2 t + \sin^2 t \\ &= 1, \text{ a constant} \end{aligned}$$

$$\begin{aligned} \mathbf{r}(t) \cdot \mathbf{r}'(t) &= (\cos t\mathbf{i} + \sin t\mathbf{j}) \cdot (-\sin t\mathbf{i} + \cos t\mathbf{j}) \\ &= -\sin t \cos t + \sin t \cos t \\ &= 0 \end{aligned}$$

This example is a direct application of Property 7 of Theorem 11.2 because $\mathbf{r}(t) \cdot \mathbf{r}(t) = 1$, a constant, and $\mathbf{r}(t) \cdot \mathbf{r}'(t) = 0$.

## Chapter 11, Section 3, page 802

$\mathbf{r}(t) = (\cos \omega t)\mathbf{i} + (\sin \omega t)\mathbf{j}$

$\mathbf{v}(t) = -(\omega \sin \omega t)\mathbf{i} + (\omega \cos \omega t)\mathbf{j}$

$$\begin{aligned} \text{speed} = |\mathbf{v}(t)| &= \sqrt{\omega^2 \sin^2 \omega t + \omega^2 \cos^2 \omega t} \\ &= \sqrt{\omega^2 (\sin^2 \omega t + \cos^2 \omega t)} \\ &= \sqrt{\omega^2(1)} \\ &= \omega, \text{ a constant} \end{aligned}$$

Acceleration $= \mathbf{v}'(t) = \mathbf{r}''(t) = -\omega^2 \cos \omega t\mathbf{i} - \omega^2 \sin \omega t\mathbf{j}$ is not a constant.

## Chapter 11, Section 5, page 820

No, the arc length of a curve does not depend on the parameter being used. If it did, the arc length would not be the same, no matter which parameter was used.

With $\mathbf{r}(t) = t^2\mathbf{i} + \frac{4}{3}t^3\mathbf{j} + \frac{1}{2}t^4\mathbf{k}$, the arc length is

$$s = \int_0^{\sqrt{2}} \sqrt{(2t)^2 + (4t^2)^2 + (2t^3)^2}\, dt$$

$$= \int_0^{\sqrt{2}} \sqrt{4t^2 + 16t^4 + 4t^6}\, dt$$

$$= \int_0^{\sqrt{2}} \sqrt{4t^2(1 + 4t^2 + t^4)}\, dt$$

$$= \int_0^{\sqrt{2}} 2t\sqrt{(t^2 + 2) - 3}\, dt$$

$$= \left[\frac{t^2 + 2}{2}\sqrt{(t^2 + 2)^2 - 3} - \frac{3}{2}\ln\left|(t^2 + 2) + \sqrt{(t^2 + 2)^2 - 3}\right|\right]_0^{\sqrt{2}}$$

$$\approx 4.816$$

The results are the same.

## Chapter 12, Section 1, page 838

**a.** Paraboloid

**b.** Plane

**c.** Cylinder

**d.** Cone (the half above the $xy$-plane)

**e.** Hyperboloid of One Sheet (the half above the $xy$-plane)

## Chapter 12, Section 7, page 896

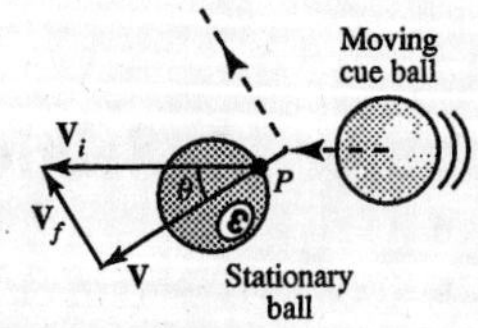


The diagram at the right depicts what happens if both balls have the same mass and the collision is elastic (absorbs no energy). The angle $\theta$ between the line of impact (the normal line to the stationary ball at the point $P$ of collision) and the line of motion of the cue ball before impact can vary from 0 to $\pi/2$. The pre impact velocity $\mathbf{v}_i$ of the cue ball is resolved into orthogonal components. The collision imparts a speed of $\|\mathbf{v}\| = \|\mathbf{v}_i\| \cos\theta$ to the stationary ball and its direction is away from $P$ along the line of impact.
The post impact speed of the cue ball is $\|\mathbf{v}_f\| = \|\mathbf{v}_i\| \sin\theta$, directed along a line that is perpendicular to the line of impact.
The topmost figure in the statement of the exploration illustrates the case $\theta = 0$, where the stationary ball acquires the greatest possible speed—the same speed that the cue ball has before impact. As $\theta$ increases, the speed acquired decreases.
The stationary ball in the bottom figure would acquire the least speed.
Two physical principles are needed to justify this explanation: conservation of linear momentum and conservation of kinetic energy.

## Chapter 12, Section 8, page 906

There appears to be an absolute minimum on the surface $z = f(x, y) = x^3 - 3xy + y^3$ at $f(1, 1) = -1$. However, this point is not an absolute minimum for the surface because there are points on the surface where $z < -1$. For example, $f(-1, -1) = -5$.

## Chapter 13, Section 2, page 946

Using the depth at the center of each square as the height, the volume of each section can be approximated by multiplying the length (10) by the width (10) by the height.

$$V \approx (10)(10)(100) + (10)(10)(90) + (10)(10)(70) + (10)(10)(70) + (10)(10)(70) + (10)(10)(40) + (10)(10)(50) + (10)(10)(50) + (10)(10)(40) + (10)(10)(40) + (10)(10)(50) + (10)(10)(30)$$

$$= 7000 \text{ cubic yards}$$

## Chapter 13, Section 2, page 949

The volume of the circular paraboloid can be found in three ways.

(1) Using a double integral:

$$V\int_{-a}^{a}\int_{-\sqrt{a^2-x^2}}^{\sqrt{a^2-x^2}} (a - x^2 - y^2)\, dy\, dx = \frac{\pi a^4}{2}$$

(2) Using the disc method for the volume of a solid of revolution:

$$V = \pi \int_0^{a^2} R^2 - r^2\, dz \qquad \text{where } R = \sqrt{a^2 - z}$$

$$r = 0$$

$$= \pi \int_0^{a^2} (a^2 - z)\, dz$$

$$= \frac{\pi a^4}{2}$$

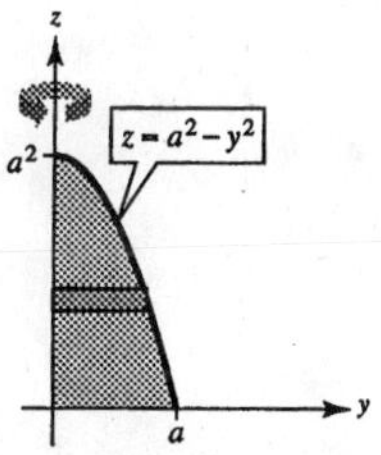


(3) Using the shell method for the volume of a solid of revolution

$$V = 2\pi \int_0^{a} ph\, dy \qquad \text{where } p = y$$

$$h = a^2 - y^2$$

$$= 2\pi \int_0^{a} y(a^2 - y^2)\, dy$$

$$= \frac{\pi a^4}{2}$$

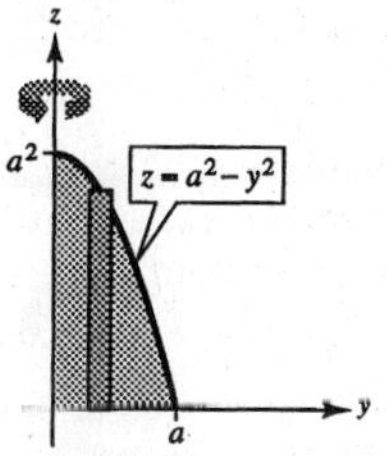

## Chapter 13, Section 3, page 957

The volume of the circular paraboloid can be found using polar coordinates as follows.

$$V = \int\!\!\int_R z\,dA$$

where $z = a^2 - (x^2 + y^2)$

$= a^2 - r^2$

and $R$ is as shown

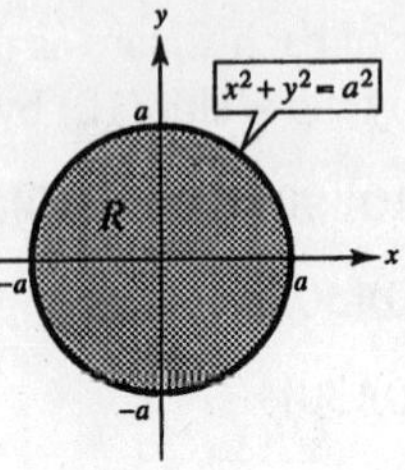


$$= \int_0^{2\pi}\int_0^a (a^2 - r^2)r\,dr\,d\theta$$

$$= \int_0^{2\pi}\int_0^a (a^2 r - r^3)\,dr\,d\theta$$

$$= \int_0^{2\pi}\left[\frac{a^2r^2}{2} - \frac{r^4}{4}\right]_0^a d\theta$$

$$= \int_0^{2\pi}\left[\frac{a^4}{2} - \frac{a^4}{4}\right]d\theta$$

$$= \int_0^{2\pi}\frac{a^4}{4}\,d\theta$$

$$= \frac{a^4}{4}\Big[\theta\Big]_0^{2\pi}$$

$$= \frac{a^4}{4}(2\pi)$$

$$= \frac{\pi a^4}{2}$$

## Chapter 13, Section 6, page 979

The volume of the circular paraboloid can be found using a triple integral.

$$V = \int_{-a}^{a}\int_{-\sqrt{a^2-x^2}}^{\sqrt{a^2-x^2}}\int_{0}^{a^2-x^2-y^2} dz\,dy\,dx$$

$$= \int_{-a}^{a}\int_{-\sqrt{a^2-x^2}}^{\sqrt{a^2-x^2}} (a^2 - x^2 - y^2)\,dy\,dx$$

$$= \int_{-a}^{a}\left[(a^2 - x^2)y - \frac{y^3}{3}\right]_{-\sqrt{a^2-x^2}}^{\sqrt{a^2-x^2}} dy\,dx$$

$$= \int_{-a}^{a} \frac{4}{3}(a^2 - x^2)^{3/2}\,dx \qquad \text{let } x = a\sin\theta,\; dx = a\cos\theta\,d\theta$$

$$= \frac{4}{3}\int_{-\pi/2}^{\pi/2} (a^2\cos^2\theta)^{3/2} \cdot a\cos\theta\,d\theta$$

$$= \frac{8}{3}a^4\int_{0}^{\pi/2} \cos^4\theta\,d\theta$$

$$= \frac{8}{3}a^4\left(\frac{1}{2}\right)\left(\frac{3}{4}\right)\left(\frac{\pi}{2}\right) \qquad \text{Wallis's Formula}$$

$$= \frac{\pi a^4}{2}$$

## Chapter 13, Section 6, page 983

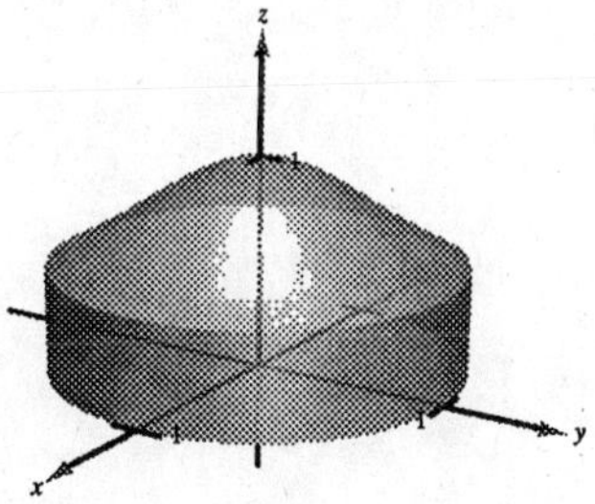


Estimate: $\left(0, 0, \frac{1}{3}\right)$

$$m = \int_0^{2\pi}\int_0^1\int_0^{1/(1+r^2)} kr\,dz\,dr\,d\theta = k\pi\ln 2$$

$$\bar{x} = \frac{\int_0^{2\pi}\int_0^1\int_0^{1/(1+r^2)} kxr\,dz\,dr\,d\theta}{m} = 0$$

$$\bar{y} = \frac{\int_0^{2\pi}\int_0^1\int_0^{1/(1+r^2)} kyr\,dz\,dr\,d\theta}{m} = 0$$

$$\bar{z} = \frac{\int_0^{2\pi}\int_0^1\int_0^{1/(1+r^2)} kzr\,dz\,dr\,d\theta}{m} = \frac{k\pi/4}{k\pi\ln 2} = \frac{1}{4\ln 2} \approx 0.3607$$

## Chapter 13, Section 7, page 989

The volume of the circular paraboloid can be found using cylindrical coordinates.

Because $z = a^2 - x^2 - y^2 = a^2 - r^2$, the bounds on $z$ are $0 \leq z \leq a^2 - r^2$.

$$V = \int_0^{2\pi} \int_0^a \int_0^{a^2-r^2} r\,dz\,dr\,d\theta$$

$$= \int_0^{2\pi} \int_0^a \Big[rz\Big]_0^{a^2-r^2} dr\,d\theta$$

$$= \int_0^{2\pi} \int_0^a r(a^2 - r^2)\,dr\,d\theta$$

$$= \frac{\pi a^4}{2}$$

(For details, refer to the Exploration for Chapter 13, Section 3, page 957.)

## Chapter 14, Section 2, page 1028

You can use a line integral to find the surface area of the piece of tin by integrating the height function over the path of the base of the cylinder. Begin by writing a parametric form for the circular base, $x^2 + y^2 = 9$.

$x = 3 \cos t$ and $y = 3 \sin t$

$$\mathbf{r}(t) = 3 \cos t\mathbf{i} + 3 \sin t\mathbf{j}$$

$$\mathbf{r}'(t) = -3 \sin t\mathbf{i} + 3 \cos t\mathbf{j}$$

$$\|\mathbf{r}'(t)\| = 3$$

With this parametrization,

$ds = \|\mathbf{r}'(t)\|\,dt = 3\,dt$ and $f(x, y) = 1 + \cos \dfrac{\pi x}{4} = 1 + \cos \dfrac{\pi}{4}(3 \cos t)$.

Therefore,

$$\text{lateral surface area} = \int_C f(x, y)\,ds$$

$$= \int_0^{2\pi} \left[1 + \cos \frac{\pi}{4}(3 \cos t)\right](3\,dt).$$

Using the integration capabilities of a graphing utility gives a value of approximately 19.330.

## Chapter 14, Section 5, page 1057

If $u$ is fixed, then $\mathbf{r}(u, v) = C_1 \cos v\mathbf{i} + C_2 \sin v\mathbf{j} + C_2\mathbf{k}$ where $C_1 = 2 + \cos u$ and $C_2 = \sin u$. Then $z = C_2$ and, because $x = C_1 \cos v$ and $y = C_1 \sin v$, $x^2 + y^2 = C_1^2$. This is a circle with center at $(0, 0, C_2)$ on a plane parallel to the $xy$-plane and $C_2$ units from it.

If $v$ is fixed, then $\mathbf{r}(u, v) = C_1(2 + \cos u)\mathbf{i} + C_2(2 + \cos u)\mathbf{j} + \sin u\mathbf{k}$ where

$$\frac{x}{C_1} = 2 + \cos u = \frac{y}{C_2} \text{ and } z = \sin u.$$

Hence,

$$\cos u = \frac{x}{C_1} - 2 = \frac{y}{C_2} - 2$$

and you obtain the equations

$$\left(\frac{x}{C_1} - 2\right)^2 + z^2 = 1 \text{ and } \left(\frac{y}{C_2} - 2\right)^2 + z^2 = 1.$$

The first equation is the generating curve of an elliptic cylinder with rulings parallel to the $y$-axis. The second equation is also the generating curve of an elliptic cylinder, but with rulings parallel to the $x$-axis. Thus, the surface is two elliptic cylinders that intersect each other at right angles.

# TECHNOLOGY BOXES

## Chapter P, Section 6, page 51

The domain $f(x) = \ln x^2$ is all real numbers except $x = 0$ and the domain of $g(x) = 2 \ln x$ is all positive real numbers. Therefore, the graph on top is $f$ and the bottom graph is $g$.

## Chapter 2, Section 1, page 119

For $f(x) = |x|$, $\dfrac{f(x + \Delta x) - f(x - \Delta x)}{2\,\Delta x} = \dfrac{|x + \Delta x| - |x - \Delta x|}{2\,\Delta x}$.

When $x = 0$, $= \dfrac{|\Delta x| - |-\Delta x|}{2\,\Delta x}$

$$= \frac{0}{2\,\Delta x}$$

$$= 0.$$

## Chapter 2, Section 2, page 128

Estimates:

For $y = \frac{1}{2} \sin x$, the slope at $(0, 0)$ appears to be $\frac{1}{2}$.
For $y = \sin x$, the slope at $(0, 0)$ appears to be 1.
For $y = \frac{3}{2} \sin x$, the slope at $(0, 0)$ appears to be $\frac{3}{2}$.
For $y = 2 \sin x$, the slope at $(0, 0)$ appears to be 2.

Verification:

$f(x) = \frac{1}{2} \sin x \quad f'(x) = \frac{1}{2} \cos x \quad f'(0) = \frac{1}{2} \cos 0 = \frac{1}{2}(1) = \frac{1}{2}$

$f(x) = \sin x \quad f'(x) = \cos x \quad f'(0) = \cos 0 = 1$

$f(x) = \frac{3}{2} \sin x \quad f'(x) = \frac{3}{2} \cos x \quad f'(0) = \frac{3}{2} \cos 0 = \frac{3}{2}(1) = \frac{3}{2}$

$f(x) = 2 \sin x \quad f'(x) = 2 \cos x \quad f'(0) = 2 \cos 0 = 2(1) = 2$

## Chapter 2, Section 3, page 138

The value of $y'$ is 0 at the two points that have horizontal tangent lines.

## Chapter 2, Section 5, page 162

If the window is set for $-2\pi < t < 2\pi$, the graph appears the same but not connected at $x = 0$.

## Chapter 3, Section 1, page 200

The function $f(x) = \sin x$ has critical numbers.

## Chapter 3, Section 3, page 216

The graph on the left is incorrect. This graph is produced by a calculator that can only graph powers of negative numbers that have integer exponents.

## Chapter 3, Section 5, page 234

A viewing window in which $-4000 \leq x \leq 2000$, $-2 \leq y \leq 3$ shows a vertical asymptote at $x = -1000$ and a horizontal asymptote at $y = 2$.

## Chapter 3, Section 6, page 241

A viewing window in which $-10 \leq x \leq 25$, $-10 \leq y \leq 10$ will show the zero at $x = 20$ and the vertical asymptote at $x = 21$ for the function $g$.

## Chapter 3, Section 7, page 250

The TI-86 indicates a maximum value of $V$ that is approximately 107.997.

## Chapter 4, Section 6, page 336

The TI-83 gave an answer of 2 for $\int_0^{\pi} \sin x\, dx$. This answer differs from the text answer of 1.974 by 0.026. The TI-83 gave an error message for $\int_{-1}^{2} \frac{1}{x}\, dx$.

## Chapter 4, Section 7, page 344

When we used MATHCAD, the antiderivative for Example 5 was the same as that given in the text. For Example 6 the antiderivative was given as $\frac{2}{(x+1)} + 2\ln(x+1)$.

## Chapter 4, Section 8, page 351

You can illustrate $y = \arctan \sqrt{e^{2x} - 1}$ as shown on the right. The length of side $s$ can be calculated as follows

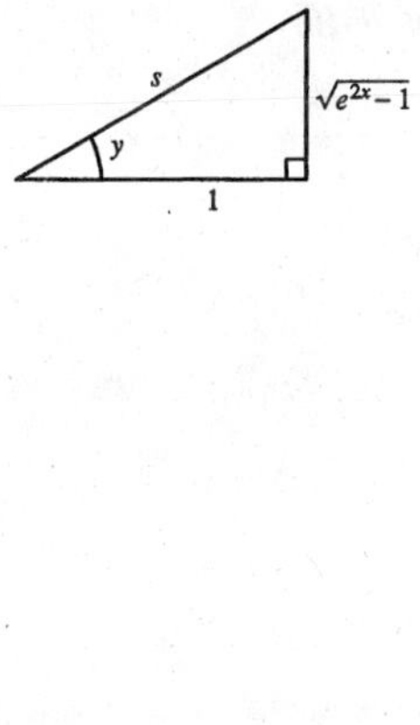


$$s^2 = (1)^2 + \left(\sqrt{e^{2x} - 1}\right)^2$$

$$s^2 = 1 + e^{2x} - 1$$

$$s^2 = e^{2x}$$

$$s = e^x$$

Hence, $y = \arctan \sqrt{e^{2x} - 1} = \text{arcsec}\, \frac{s}{1} = \text{arcsec}\, e^x$.

## Chapter 6, Section 7, page 473

The TI-83 gives 1608.495487 as a value for this integral.

## Chapter 7, Section 1, page 485

$$\begin{aligned} \ln 2x + C &= \ln 2 + \ln x + C \\ &= \ln x + (\ln 2 + C) \\ &= \ln x + C_1, \qquad \text{where } C_1 = \ln 2 + C \end{aligned}$$

## Chapter 7, Section 2, page 489

The graphs are the same.

## Chapter 7, Section 2, page 490

To evaluate $\int_0^1 \arctan x\, dx$, find the antiderivative and then apply the Fundamental Theorem of Calculus.
To evaluate $\int_0^1 \arctan x^2\, dx$, use numerical approximation.

## Chapter 7, Section 3, page 498

The results are equivalent.

$$\begin{aligned}
-\cos^5 x\left(\tfrac{1}{7}\sin^2 x + \tfrac{2}{35}\right) + C &= -\cos^5 x\left[\tfrac{1}{7}(1 - \cos^2 x) + \tfrac{2}{35}\right] + C \\
&= -\tfrac{1}{7}\cos^5 x + \tfrac{1}{7}\cos^7 x - \tfrac{2}{35}\cos^5 x + C \\
&= -\tfrac{7}{35}\cos^5 x + \tfrac{1}{7}\cos^7 x + C \\
&= -\tfrac{1}{5}\cos^5 x + \tfrac{1}{7}\cos^7 x + C
\end{aligned}$$

## Chapter 7, Section 4, page 507

DERIVE: $\displaystyle\int \frac{dx}{\sqrt{9 - x^2}} = \text{asin}\,\frac{x}{3}$

Trigonometric substitution:

let $x = 3\sin\theta$

$dx = 3\cos\theta\, d\theta$

$$\begin{aligned}
\int \frac{dx}{\sqrt{9 - x^2}} &= \int \frac{3\cos\theta\, d\theta}{\sqrt{9 - 9\sin^2\theta}} \\
&= \int \frac{3\cos\theta\, d\theta}{3\cos\theta} \\
&= \int d\theta \\
&= \theta + C \\
&= \arcsin\frac{x}{3} + C
\end{aligned}$$

—CONTINUED—

## Chapter 7, Section 4, page 507 —CONTINUED—

DERIVE: $\displaystyle\int \frac{dx}{x\sqrt{9-x^2}} = \frac{\ln\left[\frac{\sqrt{9-x^2}-3}{x}\right]}{3}$

Trigonometric substitution:

let $x = 3\sin\theta$

$dx = 3\cos\theta\,d\theta$

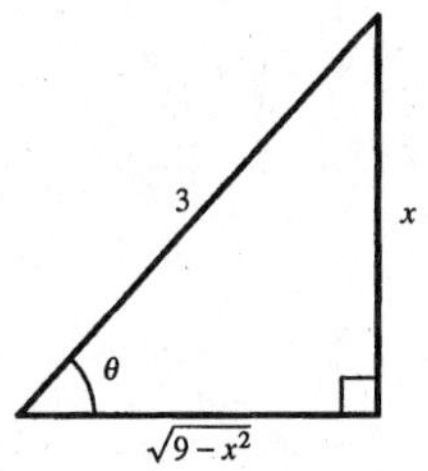


$$\int \frac{dx}{x\sqrt{9-x^2}} = \int \frac{3\cos\theta\,d\theta}{3\sin\theta\sqrt{9-9\sin^2\theta}}$$
$$= \int \frac{3\cos\theta\,d\theta}{(3\sin\theta)(3\cos\theta)}$$
$$= \int \frac{d\theta}{3\sin\theta}$$
$$= \frac{1}{3}\int \csc\theta\,d\theta$$
$$= \frac{1}{3}\ln|\csc\theta - \cot\theta| + C$$
$$= \frac{1}{3}\ln\left|\frac{3}{x} - \frac{\sqrt{9-x^2}}{x}\right| + C$$
$$= \frac{1}{3}\ln\left|\frac{3-\sqrt{9-x^2}}{x}\right| + C$$
$$= \frac{1}{3}\ln\left|\frac{\sqrt{9-x^2}-3}{x}\right| + C$$

DERIVE: $\displaystyle\int \frac{dx}{x^2\sqrt{9-x^2}} = -\frac{\sqrt{9-x^2}}{9x}$

Trigonometric substitution:

let $x = 3\sin\theta$

$dx = 3\cos\theta\,d\theta$

$$\int \frac{dx}{x\sqrt{9-x^2}} = \int \frac{3\cos\theta\,d\theta}{(3\sin\theta)^2\sqrt{9-9\sin^2\theta}}$$
$$= \int \frac{3\cos\theta\,d\theta}{(9\sin^2\theta)(3\cos\theta)}$$
$$= \int \frac{d\theta}{9\sin^2\theta}$$
$$= \frac{1}{9}\int \csc^2\theta\,d\theta$$
$$= -\frac{1}{9}\cot\theta + C$$
$$= -\frac{1}{9}\cdot\frac{\sqrt{9-x^2}}{x} + C$$
$$= -\frac{\sqrt{9-x^2}}{9x} + C$$

—CONTINUED—

## Chapter 7, Section 4, page 507 —CONTINUED—

DERIVE: $\displaystyle\int \frac{dx}{x^3\sqrt{9-x^2}} = \frac{1}{54}\ln\left[\frac{\sqrt{9-x^2}-3}{x}\right] - \frac{\sqrt{9-x^2}}{18x^2}$

Trigonometric substitution:

let $x = 3\sin\theta$

$dx = 3\cos\theta\, d\theta$

$$\int \frac{dx}{x^3\sqrt{9-x^2}} = \frac{3\cos\theta\, d\theta}{(3\sin\theta)^3\sqrt{9-9\sin^2\theta}}$$

$$= \int \frac{3\cos\theta\, d\theta}{27\sin^3\theta\,(3\cos\theta)}$$

$$= \int \frac{d\theta}{27\sin^3\theta}$$

$$= \frac{1}{27}\int \csc^3\theta\, d\theta$$

$$= \frac{1}{27}\left[\frac{1}{2}\ln|\cot\theta - \csc\theta| - \frac{1}{2}\cot\theta\csc\theta\right] + C$$

$$= \frac{1}{27}\left[\frac{1}{2}\ln\left|\frac{\sqrt{9-x^2}}{x} - \frac{3}{x}\right| - \frac{1}{2}\left(\frac{\sqrt{9-x^2}}{x}\right)\left(\frac{3}{x}\right)\right] + C$$

$$= \frac{1}{54}\ln\left|\frac{\sqrt{9-x^2}-3}{x}\right| - \frac{\sqrt{9-x^2}}{18x^2} + C$$

## Chapter 7, Section 5, page 520

Yes, the results are equivalent.

$$4\ln(x^2+2) = \ln(x^2+2)^4$$

$$= \ln[(x^2)^4 + 4(x^2)^3(2) + 6(x^2)^2(2)^2 + 4(x^2)(2)^3 + (2)^4]$$

$$= \ln[x^8 + 8x^6 + 24x^4 + 32x^2 + 16]$$

# Chapter 7, Section 7, page 535

$$y = \lim_{x\to 0} (1 - \cos x)^x$$

$$\ln y = \ln\left[\lim_{x\to 0} (1 - \cos x)^x\right]$$

$$= \lim_{x\to 0} [\ln(1 - \cos x)^x]$$

$$= \lim_{x\to 0} [x \ln(1 - \cos x)]$$

$$= \lim_{x\to 0} \frac{\ln(1 - \cos x)}{1/x}$$

$$= \lim_{x\to 0} \frac{\frac{\sin x}{1 - \cos x}}{-1/x^2}$$

$$= \lim_{x\to 0} \left[-\frac{x^2 \sin x}{1 - \cos x}\right]$$

$$= \lim_{x\to 0} \left[-\frac{x^2 \cos x + 2x \sin x}{\sin x}\right]$$

$$= \lim_{x\to 0} \left[-\frac{x^2 \sin x + 2x \cos x + 2x \cos x + 2 \sin x}{\cos x}\right]$$

$$= -\frac{2(0)}{1}$$

$$= 0$$

$$\ln y = 0$$

$$y = e^0 = 1$$

Hence, $\lim_{x\to 0} (1 - \cos x)^x = 1$

$$y = \lim_{x\to 0^+} (\tan x)^x$$

$$\ln y = \ln \lim_{x\to 0^+} (\tan x)^x$$

$$= \lim_{x\to 0^+} [\ln(\tan x)^x]$$

$$= \lim_{x\to 0^+} [x \ln(\tan x)]$$

$$= \lim_{x\to 0^+} \left[\frac{\ln(\tan x)}{1/x}\right]$$

$$= \lim_{x\to 0^+} \frac{\frac{\sec^2 x}{\tan x}}{-1/x^2}$$

$$= \lim_{x\to 0^+} \frac{-x^2}{\sin x \cos x}$$

$$= \lim_{x\to 0^+} \frac{-x^2}{1/2 \cos 2x}$$

$$= \frac{0}{1/2}$$

$$= 0$$

$$\ln y = 0$$

$$y = e^0 = 1$$

Hence, $\lim_{x\to 0^+} (\tan x)^x = 1$

# Chapter 8, Section 1, page 558

The first 20 terms of the sequence $a_n = \dfrac{n^2}{2^n - 1}$ are:

$a_1 = 1$

$a_2 = \frac{4}{3} \approx 1.33333$

$a_3 = \frac{9}{7} \approx 1.28571$

$a_4 = \frac{16}{15} \approx 1.06667$

$a_5 = \frac{25}{31} \approx 0.80645$

$a_6 = \frac{4}{7} \approx 0.57143$

$a_7 = \frac{49}{127} \approx 0.38583$

$a_8 = \frac{64}{255} \approx 0.25098$

$a_9 = \frac{81}{511} \approx 0.15851$

$a_{10} = \frac{100}{1023} \approx 0.09775$

$a_{11} = \frac{121}{2047} \approx 0.05911$

$a_{12} = \frac{16}{455} \approx 0.03516$

$a_{13} = \frac{169}{8191} \approx 0.02063$

$a_{14} = \frac{196}{16,383} \approx 0.01196$

$a_{15} = \frac{225}{32,767} \approx 0.00687$

$a_{16} = \frac{256}{65,535} \approx 0.00391$

$a_{17} = \frac{289}{131,071} \approx 0.00220$

$a_{18} = \frac{12}{9709} \approx 0.00124$

$a_{19} = \frac{361}{524,287} \approx 0.00069$

$a_{20} = \frac{400}{1,048,575} \approx 0.00038$

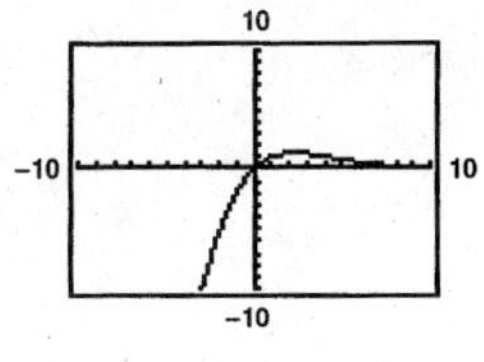

## Chapter 8, Section 2, page 570

You can compute the sum of the first 20 terms of the series $\sum_{n=0}^{\infty} \frac{3}{2^n}$ using the TI-83 as follows.

sum seq $(3/2 \wedge x, x, 0, 19, 1)$

You should obtain 5.999994278.

## Chapter 8, Section 5, page 592

When $n = 8$, $|R_8| \le a_9 = \frac{1}{9!} \approx 0.00000276 < 0.00001$. Therefore, you need 8 terms for the desired accuracy.

## Chapter 8, Section 7, page 606

As $n$ increases, the graph of $P_n$ becomes a better and better approximation of the graph of $f$ near $x = 0$.

## Chapter 9, Section 2, page 666

Yes, the curve represents the same graph as that shown in Figures 9.19 and 9.20; however, it has the opposite orientation.

## Chapter 9, Section 4, page 690

The graphs of $r = a \cos n\theta$ and $r = a \sin n\theta$ for noninteger values of $n$ are not rose curves because the petals overlap.

## Chapter 14, Section 6, page 1063

When we used DERIVE we got the value $12\pi + 36$, the same result as in Example 2.

## Chapter 14, Section 7, page 1075

$$\int_0^3 \int_0^{6-2x} \int_0^{3-x-z/2} (2 + 2y)\, dy\, dz\, dx = \frac{63}{2}$$

$$\int_0^6 \int_0^{3-z/2} \int_0^{3-y-z/2} (2 + 2y)\, dx\, dy\, dz = \frac{63}{2}$$

# SECTION PROJECTS

## Chapter 1, Section 5, page 105

(a) On the graph of $f$ it appears that the $y$-coordinates of points lie as close to 1 as desired as long as we consider only those points with an $x$-coordinate near to but not equal to 0.

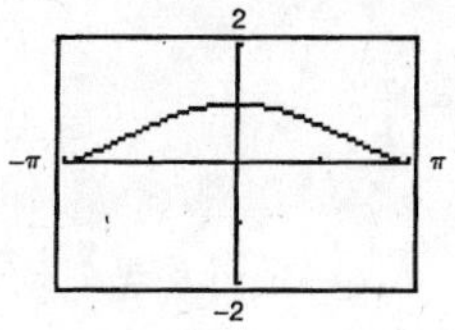


(b) Use a table of values of $x$ and $f(x)$ that includes several values of $x$ near 0. Check to see if the corresponding values of $f(x)$ are close to 1. In this case, because $f$ is an even function, only positive values of $x$ are needed.

(c) The slope of the sine function at the origin appears to be 1. (It is necessary to use radian measure and have the same unit of length on both axes.)

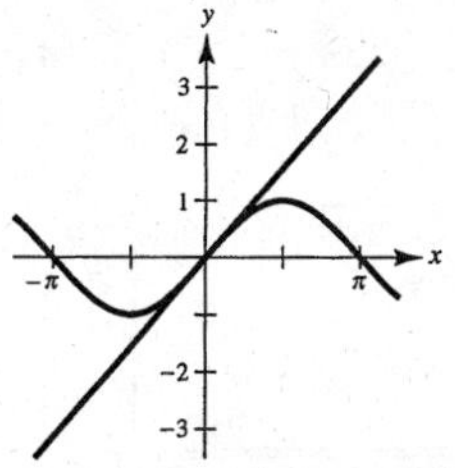


(d) In the notation of Section 1.1, $c = 0$ and $c + \Delta x = x$. Thus, $m_{sec} = \dfrac{\sin x - 0}{x - 0}$.

This formula has a value of 0.998334 if $x = 0.1$; $m_{sec} = 0.999983$ if $x = 0.01$.

The exact slope of the tangent line to $g$ at $(0, 0)$ is $\lim\limits_{x \to 0} m_{sec} = \lim\limits_{x \to 0} \dfrac{\sin x}{x} = 1$.

(e) The slope of the tangent line to the cosine function at the point $(0, 1)$ is 0. The analytical proof is as follows:

$$\lim_{\Delta x \to 0} m_{sec} = \lim_{\Delta x \to 0} \frac{\cos(0 + \Delta x) - 1}{\Delta x} = -\lim_{\Delta x \to 0} \frac{1 - \cos(\Delta x)}{\Delta x} = 0.$$

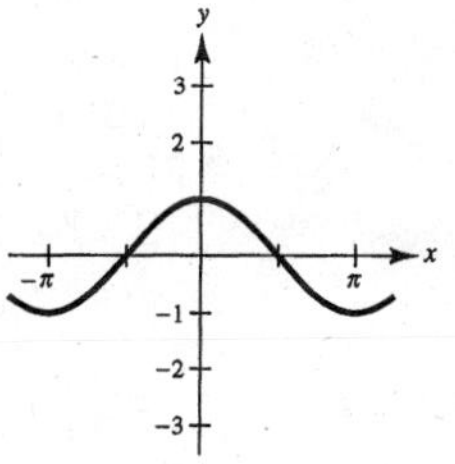


(f) The slope of the tangent line to the graph of the tan function at $(0, 0)$ is:

$$\lim_{\Delta x \to 0} m_{sec} = \lim_{\Delta x \to 0} \frac{\tan(0 + \Delta x) - 0}{\Delta x} = \lim_{\Delta x \to 0} \frac{\sin(\Delta x)}{\Delta x} \cdot \frac{1}{\cos \Delta x} = 1 \cdot \frac{1}{1} = 1.$$

## Chapter 2, Section 5, page 168

(a) $x^2 + y^2 = C^2$

$2x + 2yy' = 0$

$y' = -\dfrac{x}{y}$

at the point $(3, 4)$, $y' = -\dfrac{3}{4}$

(b) $xy = C$

$xy' + y = 0$

$y' = -\dfrac{y}{x}$

at the point $(1, 4)$, $y' = -4$

—CONTINUED—

## Chapter 2, Section 5, page 168 —CONTINUED—

(c) $ax = by$

$a = by'$

$y' = \frac{a}{b}$

at $a = \sqrt{3}$ and $b = 1, y' = \sqrt{3}$

(d) $y = C \cos x$

$y' = -C \sin x$

at $x = \frac{\pi}{3}$ and $C = \frac{2}{3}, y' = -\frac{2}{3} \sin \frac{\pi}{3}$

$= -\frac{1}{\sqrt{3}}$

## Chapter 3, Section 3, page 221

(a) $D = \pi + 2\alpha - 4 \sin^{-1}\left(\frac{1}{k} \sin \alpha\right)$ where $k \approx 1.33$

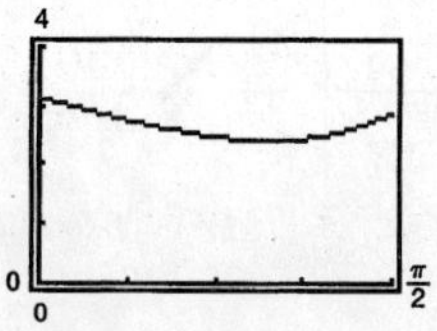

(b) $D'(\alpha) = 2 - 4\frac{d}{d\alpha}\left[\sin^{-1}\left(\frac{1}{k} \sin \alpha\right)\right]$

Let $y = \sin^{-1}[(1/k) \sin \alpha]$, then $\sin y = (1/k) \sin \alpha$ and

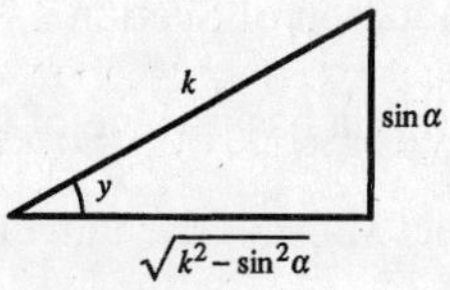

$$\cos y \frac{dy}{d\alpha} = \frac{1}{k} \cos \alpha$$

$$\frac{dy}{d\alpha} = \frac{\frac{1}{k} \cos \alpha}{\cos y} = \frac{\frac{1}{k} \cos \alpha}{\frac{\sqrt{k^2 - \sin^2 \alpha}}{k}} = \frac{\cos \alpha}{\sqrt{k^2 - \sin^2 \alpha}}.$$

Thus, $D'(\alpha) = 2 - 4\left(\frac{\cos \alpha}{\sqrt{k^2 - \sin^2 \alpha}}\right)$. The critical number occurs when $D'(\alpha) = 0$.

$$2 - 4\left(\frac{\cos \alpha}{\sqrt{k^2 - \sin^2 \alpha}}\right) = 0$$

$$\frac{1}{2} = \frac{\cos \alpha}{\sqrt{k^2 - (1 - \cos^2 \alpha)}}$$

$$\frac{1}{4} = \frac{\cos^2 \alpha}{k^2 - 1 + \cos^2 \alpha}$$

$$k^2 - 1 + \cos^2 \alpha = 4 \cos^2 \alpha$$

$$\frac{k^2 - 1}{3} = \cos^2 \alpha$$

Thus, $\cos \alpha = \sqrt{(k^2 - 1)/3}$.

—CONTINUED—

## Chapter 3, Section 3, page 221 —CONTINUED—

**Note:** Because $0 \le \alpha \le \pi/2$, $\cos\alpha \ge 0$. Then, $D_{min}$ occurs when

$$\alpha = \cos^{-1}\sqrt{\frac{k^2-1}{3}} = \cos^{-1}\sqrt{\frac{1.33^2-1}{3}} \approx 1.04 \text{ radians.}$$

$$\text{So, } D_{min} = D\left(\cos^{-1}\sqrt{\frac{1.33^2-1}{3}}\right) \approx 2.4 \text{ radians}$$

and $\pi - D_{min} \approx 0.74$ radian.

## Chapter 3, Section 7, page 261

(a) On April 14, the river was rising most rapidly because on that day the rate of change of the level of the river attained a positive extreme value.

(b) On April 23, the river was falling most rapidly because on that day the rate of change of the level of the river was a negative extreme value.

(c) The river rose, then fell, then rose again on April 17 and 18. The graph shows that the rate of change of the river level was positive, then negative, then positive again on those dates.

(d) Because the rate of change of the level of the river appears to have averaged about 2.8 feet per day on April 14, the river must have risen to about 113.8 feet by the end of the day. The next day, April 15, the rate of change of the level of the river seems to have averaged about $-0.6$ feet per day. Therefore, the estimated level of the river at 12:01 AM on April 16 is 113.2 feet.

(e) It appears that the river crested on April 19.

## Chapter 4, Section 4, page 321

(a)

| $x$ | 0 | $\frac{\pi}{6}$ | $\frac{\pi}{3}$ | $\frac{\pi}{2}$ | $\frac{2\pi}{3}$ | $\frac{5\pi}{6}$ | $\pi$ |
|---|---|---|---|---|---|---|---|
| $F(x)$ | 0 | 0.0453 | 0.3071 | 0.7854 | 1.2637 | 1.5255 | 1.5708 |

According to the Second Fundamental Theorem of Calculus, $F'(x) = \sin^2 x$, which is positive for all $x$ in the interval $(0, \pi)$. By Theorem 3.5, $F$ is increasing.

(b)

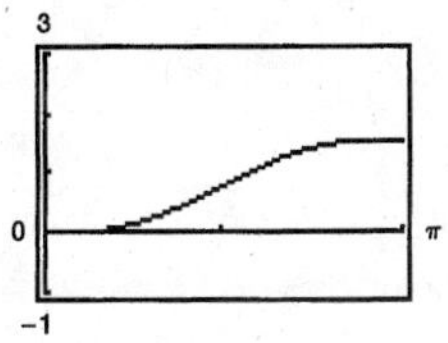


$$y = \int_0^x \sin^2 t\, dt$$



—CONTINUED—

## Chapter 4, Section 4, page 321 —CONTINUED—

(c) 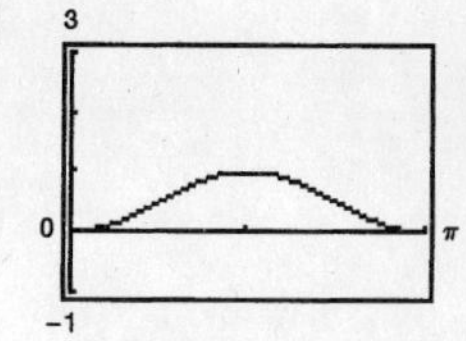


$$y = \frac{d}{dx}\int_0^x \sin^2 t\, dt$$

The graph in part (b) is increasing because this graph of $F'$ is always positive (except at the endpoints). The graphs also have corresponding symmetries.

(d) 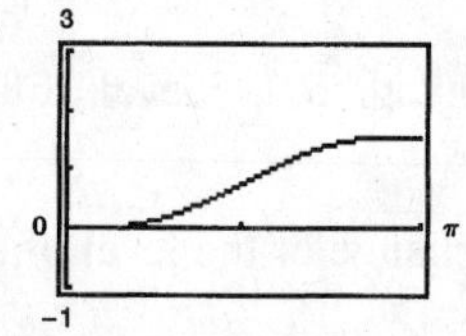


$$y = \frac{1}{2}x - \frac{\sin 2x}{4}$$

$$\frac{dy}{dt} = \frac{1}{2} - \frac{2\cos 2t}{4} = \frac{1 - \cos 2t}{2} = \sin^2 t$$

This graph is identical to the graph in part (b) and therefore, its derivative, $\sin^2 t$, has the same graph that appears in part (c).

## Chapter 4, Section 9, page 366

(a) $y = 693.8597 - 68.7672 \cosh 0.0100333x$

$y' = -68.7672(0.0100333) \sinh 0.0100333x = 0$ when $x = 0$.

Maximum height ($x = 0$): $y = 693.8597 - 68.7672 \cosh 0 \approx 625$ feet

(b) The height of the arch equals $y$ (when $x = 0$) plus the distance from the center of mass of the triangle to the top of the highest triangle.

$$A = \frac{1}{2}(2c)\left(\sqrt{3}c\right) = \sqrt{3}c^2$$

When $x = 0$, $A \approx 125$.

$$125 = \sqrt{3}c^2$$

$$c = \sqrt{\frac{125}{\sqrt{3}}} \approx 8.5$$

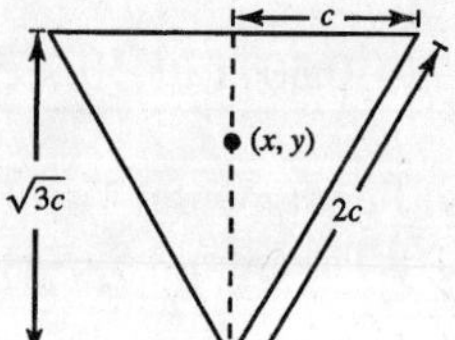


Height of the triangle: $8.5\sqrt{3}$

Distance from the center of mass of the triangle to the top of the highest triangle: $\frac{1}{3} \cdot 8.5\sqrt{3} \approx 5$ feet

Height of the arch: $625 + 5 = 630$ feet

—CONTINUED—

## Chapter 4, Section 9, page 366 —CONTINUED—

(c) The width of the arch equals $2 \cdot 299.2239 + 2$ times the distance from the center of mass of the triangle to the outer side of the triangle at the arch's base.

$$A = \sqrt{3}c^2$$

When $x \approx 299.2239$, $A \approx 1263$.

$$1263 = \sqrt{3}c^2$$

$$c = \sqrt{\frac{1263}{\sqrt{3}}} \approx 27.0$$

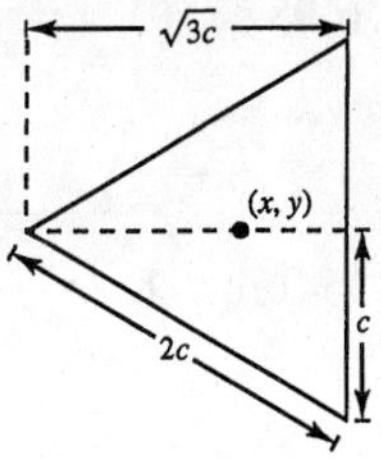


Height of the triangle: $27\sqrt{3}$

Distance from the center of mass of the triangle to the outer side of the triangle at the arch's base:

$$\frac{1}{3} \cdot 27\sqrt{3} \approx 16 \text{ feet}$$

Width of the arch: $598 + 32 = 630$ feet

## Chapter 5, Section 3, page 404

(a) Separate variables and integrate, using a convenient form of the constant term.

$$\frac{\ln|7.5w - C|}{17.5} = \int \frac{dw}{17.5w - C} = \int -\frac{dt}{3500} = \frac{t}{3500} + \frac{\ln|K|}{17.5}$$

We may assume that $K$ has the same sign as the expression $17.5w - C$ to solve for it.

$$17.5w - C = Ke^{(-17.5t/3500)}$$

If a person weights $w_0$ pounds at a time $t = 0$, then $K = 17.5w_0 - C$. Therefore,

$$w = \frac{C}{17.5} + \left(w_0 - \frac{C}{17.5}\right)e^{-t/200}$$

(b) Solving for $t$ in the answer to part (a) gives $t = 200 \ln \dfrac{17.5w_0 - C}{17.5w - C}$.

To lose 10 pounds will take $t = 200 \ln \dfrac{17.5(180) - 2500}{17.5(170) - 2500} \approx 63$ days.

To lose 35 pounds will take approximately 571 days.

(c) The "limiting" weight is $\lim_{t \to \infty} w = \dfrac{2500}{17.5} \approx 143$ pounds.

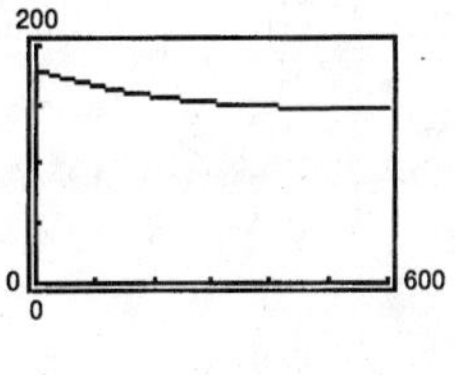


(d) To lose 10 pounds will take approximately 38 days.
To lose 35 pounds will take about 190 days.
The "limiting"weight is still 143 pounds.

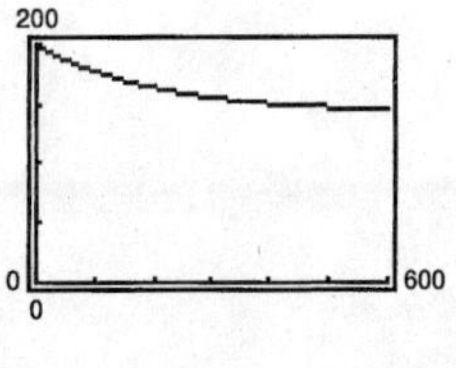

## Chapter 6, Section 3, page 439

(a) Volume of sphere:

$$V = \frac{4}{3}\pi(60{,}268)^3 \approx 9.16957 \times 10^{14}$$

Volume of oblate ellipsoid:

$$V = 2\pi\int_0^{60{,}268} 2(54{,}364)x\sqrt{1 - \frac{x^2}{60{,}268^2}}\,dx$$

$$= {}^2 17{,}456\pi\int_0^{60{,}268}\left(1 - \frac{x^2}{60{,}268^2}\right)^{1/2} x\,dx$$

$$= \left[(217{,}456\pi)\left(-\frac{60{,}268^2}{2}\right)\left(\frac{2}{3}\right)\left(1 - \frac{x^2}{60{,}268^2}\right)^{3/2}\right]_0^{60{,}268} \approx 8.27130 \times 10^{14}$$

Ratio: $\dfrac{8.27130 \times 10^{14}}{9.16957 \times 10^{14}} \approx 0.902$

(b) $$V = \frac{4}{3}\pi r^3$$

$$8.2713 \times 10^{14} = \frac{4}{3}\pi r^3$$

$$8.2713 \times 10^{14} \cdot \frac{3}{4\pi} = r^3$$

$$r \approx 58{,}232.002 \text{ kilometers}$$

## Chapter 6, Section 5, page 458

(a) $$V = 500\int_0^{1000}\left(25 - \frac{x^2}{40{,}000}\right)dx$$

$$= 500\left[25x - \frac{x^3}{120{,}000}\right]_0^{1000}$$

$$\approx 8.33 \times 10^6 \text{ ft}^3$$

(b) $$W = 500(64)\int_0^{25} y\left(200\sqrt{y}\right)dy$$

$$= \left[6{,}400{,}000\left(\frac{2}{5}y^{5/2}\right)\right]_0^{25}$$

$$= 8.00 \times 10^9 \text{ ft} \cdot \text{lb}$$

## Chapter 7, Section 3, page 505

(a) $\text{Length} = 2\int_0^{L/2} \sqrt{1 + \left[\frac{T}{ug}\left(\sinh\frac{ugx}{T}\right)\left(\frac{ug}{T}\right)\right]^2}\,dx$

$$= 2\int_0^{L/2} \sqrt{1 + \sinh^2\left(\frac{ugx}{T}\right)}\,dx = 2\int_0^{L/2} \cosh\frac{ugx}{T}\,dx = \left[\frac{2T}{ug}\sinh\frac{ugx}{T}\right]_0^{L/2} = \frac{2T}{ug}\sinh\frac{ugL}{2T}\text{ feet}$$

(b) $t = \dfrac{2(\text{length})}{v} = \dfrac{2\left(\frac{2T}{ug}\sinh\frac{ugL}{2T}\right)}{\sqrt{\frac{T}{u}}} = \dfrac{4}{g}\sqrt{\dfrac{T}{u}}\sinh\dfrac{ugL}{2T} = \dfrac{4}{g}v\sinh\dfrac{ugL}{2T}$

(c) $s = 12\left(\dfrac{T}{ug}\right)\left(\cosh\dfrac{ugL}{2T} - 1\right) = \dfrac{12}{g}\left(\dfrac{T}{ug}\right)\left(\cosh\dfrac{ugL}{2T} - 1\right)$

$$= \frac{12}{g}\left(\frac{tg}{4\sinh\frac{ugL}{2T}}\right)\left(\cosh\frac{ugL}{2T} - 1\right) \text{ from part (b).}$$

$$= \frac{12}{g}\left(\frac{t^2g^2}{16\sinh^2\frac{ugL}{2T}}\right)\left(\cosh\frac{ugL}{2T} - 1\right)$$

$$= \frac{12t^2g\left(\cosh\frac{ugL}{2T} - 1\right)}{16\left(\cosh^2\frac{ugL}{2T} - 1\right)} = \frac{3t^2g}{4\left(\cosh\frac{ugL}{2T} + 1\right)} \approx \frac{3t^2(32.2)}{4(2)} = 12.075t^2$$

## Chapter 8, Section 2, page 576

$$L - \frac{1}{4}L - 2\left(\frac{1}{16}\right)L - 4\left(\frac{1}{64}\right)L - 8\left(\frac{1}{256}\right)L - \cdots = L - L\left(\frac{1}{4} + \frac{1}{8} + \frac{1}{16} + \frac{1}{32} + \cdots\right)$$

$$= L - L\sum_{n=0}^{\infty}\frac{1}{4}\left(\frac{1}{2}\right)^n = L\left(1 - \frac{1/4}{1 - (1/2)}\right) = \frac{1}{2}L$$

The remaining pieces are getting smaller and smaller, thus making the table appear to disappear.

## Chapter 8, Section 3, page 582

(a) Grouping shows that the sum of the first $2^r$ terms of the harmonic series exceeds $1 + \frac{r}{2}$. It follows that no number $L$ can be the limit of the partial sums. In fact, if you pick any positive integer $r > 2L$ then $S_n > 1 + L$ for all $n > 2^r$.

—CONTINUED—

## Chapter 8, Section 3, page 582 —CONTINUED—

(b) Since $f(x) = 1/x$, the inequality

$$\sum_{i=2}^{n} f(i) \le \int_1^n f(x)\,dx$$

appearing in the proof of Theorem 8.10 means

$$\frac{1}{2} + \frac{1}{3} + \cdot \cdot \cdot + \frac{1}{n} \le \ln n.$$

Also, the inequality

$$\int_1^n f(x)\,dx \le \sum_{i-1}^{n-1} f(i),$$

with $n$ replaced by $n + 1$, means

$$\ln(n+1) = \int_1^{n+1} f(x)\,dx \le \sum_{i=1}^{n} f(i) = 1 + \frac{1}{2} + \frac{1}{3} + \cdot \cdot \cdot + \frac{1}{n}.$$

The combination means $\ln(n+1) \le 1 + \frac{1}{2} + \frac{1}{3} + \cdot \cdot \cdot + \frac{1}{n} \le 1 + \ln n.$

(c) According to part (b), $\sum_{n=1}^{M} \frac{1}{n} > 50$ if $\ln(M+1) > 50$. Because logarithm and exponential functions are increasing, we want

$$M + 1 = e^{\ln(M+1)} > e^{50} \approx 5.1847 \times 10^{21}.$$

Therefore, $M = 5.185 \times 10^{21}$, for example, will do.

(d) Part (b) implies that $1 + \frac{1}{2} + \cdot \cdot \cdot + \frac{1}{1{,}000{,}000} \le 1 + \ln 1{,}000{,}000 < 14.816 < 15.$

(e) Suppose that $1 < r \le s + 1$. The exact area of the plane region that lies above the interval $[r - 1, s]$ and below the graph of $y = 1/x$ is $\ln[s/(r-1)]$. The total area of $s - r + 1$ rectangles of width 1 inscribed in the region is

$$\frac{1}{r} + \cdot \cdot \cdot + \frac{1}{s}.$$

The exact area of the plane region over the interval $[r, s + 1]$ and under the curve $y = 1/x$ is $\ln[(s+1)/r]$. The total area of $s - r + 1$ rectangles of width 1 circumscribing the region is

$$\frac{1}{r} + \cdot \cdot \cdot + \frac{1}{s}.$$

Therefore,

$$\ln\frac{s+1}{r} < \frac{1}{r} + \cdot \cdot \cdot + \frac{1}{s} < \ln\frac{s}{r-1}.$$

Take $r = 10$ and $s = 20$ to get the first inequality. For the second inequality, use $r = 100$ and $s = 200$.

(f) We know from the solution to part (e) that

$$\ln\frac{2m+1}{m} \le \sum_{n=m}^{2m} \frac{1}{n} \le \ln\frac{2m}{m-1}.$$

By the Squeeze Theorem for Sequences, $\lim_{m\to\infty} \sum_{n=m}^{2m} \frac{1}{n} = \ln 2.$

# Chapter 8, Section 4, page 589

(a) $f(n, 5) = \binom{n-1}{4}\left(\frac{1}{2}\right)^{n+1}, \; n \geq 5$

Recall that $\binom{n}{k} = \dfrac{n!}{k!(n-k)!}$ and thus, $\binom{n-1}{4} = \dfrac{(n-1)!}{4!(n-5)!}$.

1981 $n = 1$: $f(1, 5) = 0$ since $n \not\geq 5$

1982 $n = 2$: $f(2, 5) = 0$ since $n \not\geq 5$

1983 $n = 3$: $f(3, 5) = 0$ since $n \not\geq 5$

1984 $n = 4$: $f(4, 5) = 0$ since $n \not\geq 5$

1985 $n = 5$: $f(5, 5) = \binom{4}{4}\left(\frac{1}{2}\right)^6 = \frac{1}{64} \approx 0.1563$

1986 $n = 6$: $f(6, 5) = \binom{5}{4}\left(\frac{1}{2}\right)^7 = 5\left(\frac{1}{128}\right) \approx 0.03906$

1987 $n = 7$: $f(7, 5) = \binom{6}{4}\left(\frac{1}{2}\right)^8 = 15\left(\frac{1}{256}\right) \approx 0.05859$

1988 $n = 8$: $f(8, 5) = \binom{7}{4}\left(\frac{1}{2}\right)^9 \approx 0.06836$

1989 $n = 9$: $f(9, 5) = \binom{8}{4}\left(\frac{1}{2}\right)^{10} \approx 0.06836$

1990 $n = 10$: $f(10, 5) \approx 0.06152$

1991 $n = 11$: $f(11, 5) \approx 0.05127$

1992 $n = 12$: $f(12, 5) \approx 0.04028$

1993 $n = 13$: $f(13, 5) \approx 0.03021$

1994 $n = 14$: $f(14, 5) \approx 0.02182$

1995 $n = 15$: $f(15, 5) \approx 0.01527$

1988 and 1989 were the years of greatest removal of 1980 wine.

(b) $\displaystyle\sum_{n=0}^{\infty} a_n = \sum_{n=5}^{\infty} \binom{n-1}{4}\left(\frac{1}{2}\right)^{n+1} = 1$

All the 1980 wine will be ultimately removed. This can also be verified by using math induction and the property of binomial coefficients:

$$\binom{n}{k} = \binom{n-1}{k} + \binom{n-1}{k-1}.$$

## Chapter 9, Section 2, page 674

I. $H(8, 3)$ matches (d).

II. $E(8, 3)$ matches (c).

III. $H(8, 7)$ matches (a).

IV. $E(24, 3)$ matches (e).

V. $H(24, 7)$ matches (f).

VI. $E(24, 7)$ matches (b).

## Chapter 9, Section 4, page 693

(a) $y = 3$

$r = y + 16 = 19$

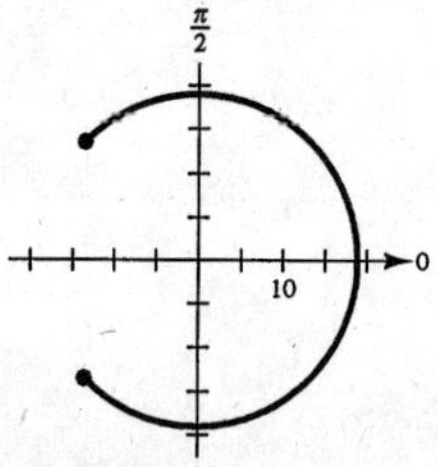


(b) $x = 2$

$\theta = -\frac{\pi}{8}x = -\frac{\pi}{4}$

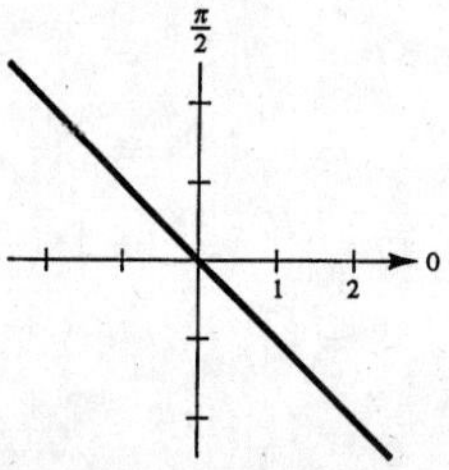


(c) $y = x + 5$

$r = y + 16 = x + 21$

$= -\frac{8}{\pi}\theta + 21$

$-\frac{3\pi}{4} \le \theta \le \frac{3\pi}{4}$

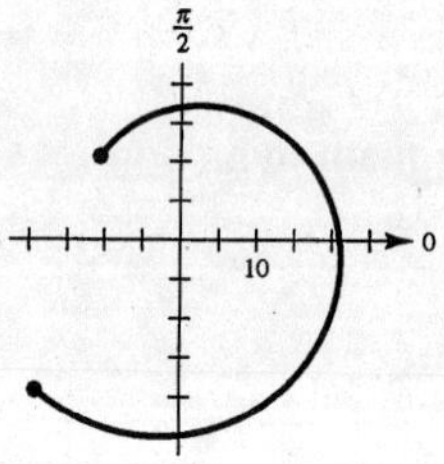


(d) $x^2 + (y - 5)^2 = 5^2$

$\left(-\frac{80}{\pi}\right)^2 + [(r - 16) - 5]^2 = 25$

$(r - 21)^2 = 25 - \frac{64\,\theta^2}{\pi^2}$

$r = 21 \pm \sqrt{25 - \frac{64\,\theta^2}{\pi^2}}$

$-\frac{3\pi}{4} \le \theta \le \frac{3\pi}{4}$

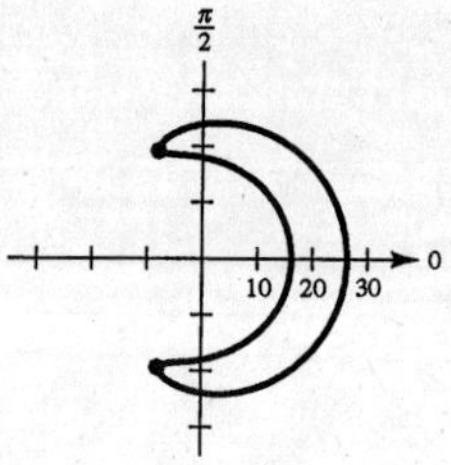

# Chapter 10, Section 5, page 762

(a) (i) The lines are not parallel because the direction vector $\langle 5, 5, -4\rangle$ for the line $L_1$ is not a scalar multiple of the direction vector $\langle 1, 8, -3\rangle$ for $L_2$.

(ii) At a point $(x, y, z)$ of intersection, the parameters $s$ and $t$ would satisfy the system of equations $4 + 5t = x = 4 + s$, $5 + 5t = y = -6 + 8s$, and $1 - 4t = z = 7 - 3s$. The lines do not intersect because, as the following computation shows, the system is inconsistent.

$$\begin{cases} s - 5t = 0 \\ 8s - 5t = 11 \\ -3s + 4t = -6 \end{cases}$$

$$\begin{cases} s - 5t = 0 & \\ 35t = 11 & \text{(equation 2)} - 8\text{(equation 1)} \\ -11t = -6 & \text{(equation 3)} + 3\text{(equation 1)} \end{cases}$$

$$\begin{cases} s - 5t = 0 & \\ 35t = 11 & \\ 0 = -\frac{89}{35} & \text{(equation 3)} + \frac{11}{35}\text{(equation 2)} \end{cases}$$

(iii) The vector $\langle 5, 5, -4\rangle \times \langle 1, 8, -3\rangle = \langle 17, 11, 35\rangle$ is orthogonal to the direction vectors for $L_1$ and $L_2$. Clearly $(4, 5, 1)$ is a point on $L_1$ and $(4, -6, 7)$ is on $L_2$. Therefore, $L_1$ lies in the plane

$$17(x - 4) + 11(y - 5) + 35(z - 1) = 0$$

and $L_2$ lies in the parallel plane

$$17(x - 4) + 11(y + 6) + 35(z - 7) = 0.$$

(iv) The distance between $L_1$ and $L_2$ is the same as the distance $D$ from a point in one of the parallel planes to the other plane.

$$D = \frac{|17 \cdot (4 - 4) + 11 \cdot (-6 - 5) + 35 \cdot (7 - 1)|}{\sqrt{17^2 + 11^2 + 35^2}} = \frac{89}{\sqrt{1635}} \approx 2.2.$$

(b) The vector $\langle 2, 4, 6\rangle \times \langle -1, 1, 1\rangle = \langle -2, -8, 6\rangle$ is normal to the parallel planes containing $L_1$ and $L_2$. The origin is a point on $L_1$. Therefore, $L_1$ lies in the plane $-2x - 8y + 6z = 0$. Because the point $(1, 4, -1)$ is on $L_2$, the distance between $L_1$ and $L_2$ is

$$\frac{|-2 \cdot 1 - 8 \cdot 4 + 6 \cdot (-1)|}{\sqrt{4 + 64 + 36}} = \frac{20}{\sqrt{26}} \approx 3.9.$$

(c) The vector $\langle 3, -1, 1\rangle \times \langle 4, 1, -3\rangle = \langle 2, 13, 7\rangle$ is normal to the parallel planes containing $L_1$ and $L_2$. The point $(0, 2, -1)$ is on $L_1$. Therefore, $L_1$ lies in the plane $2x + 13(y - 2) + 7(z + 1) = 0$. Because the point $(1, -2, -3)$ is on $L_2$, the distance between the lines is

$$\frac{|2 + 13(-4) + 7(-2)|}{\sqrt{4 + 169 + 49}} = \frac{64}{\sqrt{222}} \approx 4.3.$$

(d) Let $\mathbf{n} = \langle a_1, b_1, c_1\rangle \times \langle a_2, b_2, c_2\rangle$. The vector $\mathbf{n}$ is normal to the parallel planes containing $L_1$ and $L_2$. The point $(x_1, y_1, z_1)$ is on $L_1$. Therefore, $L_1$ lies in the plane $\mathbf{n} \cdot \langle x - x_1, y - y_1, z - z_1\rangle = 0$. Because the point $(x_2, y_2, z_2)$ is on $L_2$, the distance $D$ between $L_1$ and $L_2$ is

$$D = \frac{|\mathbf{n} \cdot \langle x_2 - x_1, y_2 - y_1, z_2 - z_1\rangle|}{\|\mathbf{n}\|}.$$

# Chapter 11, Section 1, page 793

(a) The figure shows that if $\mathbf{r}_A(\theta) = x\mathbf{i} + y\mathbf{j}$, then $y = 2a$ and, by right triangle trigonometry, $\cot\theta = \dfrac{x}{2a}$.

Therefore, $\mathbf{r}_A(\theta) = 2a\cot\theta\,\mathbf{i} + 2a\mathbf{j}$.

(b) The triangle with vertices at $O$, $B$, and the center of the circle $(0, a)$, is isosceles. The interior angles at $O$ and $B$ both equal $\pi/2 - \theta$. Because the angles of a triangle sum to $\pi$ radians, the interior angle at the vertex $(0, a)$ equals $2\theta$. Consequently, the angle between the positive $y$-axis and the radius through $B$, equals $\pi - 2\theta$. If $\mathbf{r}_B(\theta) = x\mathbf{i} + y\mathbf{j}$, then by circular function trigonometry,

$$x = a\sin(\pi - 2\theta) = a(\sin\pi\cos 2\theta - \cos\pi\sin 2\theta) = a\sin 2\theta$$

and

$$y = a + a\cos(\pi - 2\theta) = a + a(\cos\pi\cos 2\theta + \sin\pi\sin 2\theta) = a - a\cos 2\theta.$$

That is, $\mathbf{r}_B(\theta) = a\sin 2\theta\,\mathbf{i} + a(1 - \cos 2\theta)\mathbf{j}$.

(c) Because $P$ and $A$ have the same $x$-coordinate while $P$ and $B$ have the same $y$-coordinate,

$$\mathbf{r}(\theta) = 2a\cot\theta\mathbf{i} + a(1 - \cos 2\theta)\mathbf{j}.$$

To simplify, use a double angle formula.

$$\mathbf{r}(\theta) = 2a\cot\theta\mathbf{i} + 2a\sin^2\theta\mathbf{j}.$$

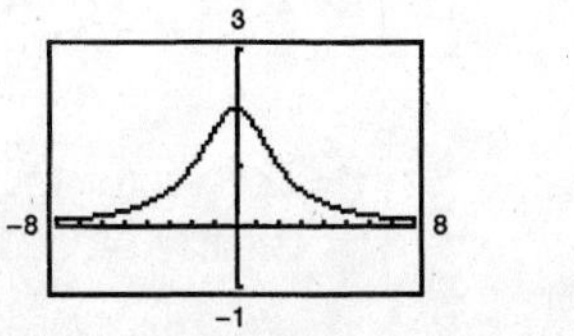


$\mathbf{r}(\theta) = 2\cot\theta\mathbf{i} + 2\sin^2\theta\mathbf{j}$

(d) Both limits would seem to be infinitely long vectors; therefore, neither limit exists.

(e) Using the answer to part (c), write the coordinates of $P$ as $x = 2a\cot\theta$ and $y = 2a\sin^2\theta$. Because

$$1 + \cot^2\theta = \csc^2\theta$$

and the cosecant is the reciprocal of the sine, we have

$$1 + \left(\frac{x}{2a}\right)^2 = \frac{2a}{y}.$$

Solve for $y$ to get $y = \dfrac{8a^3}{x^2 + 4a^2}$. If $a = 1$, the graph of this curve is the same as the graph obtained in part (c).

# Chapter 12, Section 3, page 868

The essay should explain the terms "grating" and "pitch." It should indicate how the sign of the expression $\nabla f \cdot \nabla g$ corresponds to how the eye identifies a moiré pattern. The pattern appearing on the left in this Project appears in the article as Figure 1. The pattern appearing on the right in this Project is a reflection (in the $y$-axis) of Figure 9 in the article and comes from superimposing three gratings, not just two.

## Chapter 12, Section 7, page 904

(a) $H = -\sum_{i=1}^{n} \rho_i \log_2 \rho_i$

May: $H = 3\left(-\frac{5}{16}\right)\log_2 \frac{5}{16} - \frac{1}{16}\log_2 \frac{1}{16} \approx 1.8232$

June: $H = 4\left(-\frac{1}{4}\right)\log_2 \frac{1}{4} = 2.0000$

August: $H = 2\left(-\frac{1}{4}\right)\log_2 \frac{1}{4} - \frac{1}{2}\log_2 \frac{1}{2} = 1.5000$

September: $H = 0 + 1\log_2 1 = 0$

September had no diversity of wildflowers. The greatest diversity occurred in June.

(b) $H = 10\left(-\frac{1}{10}\right)\log_2 \frac{1}{10} \approx 3.3219 > 2.0000$

The diversity is greater with 10 types of wildflowers. An equal proportion of each type of wildflower would produce a maximum diversity.

(c) $H_n = n\left(-\frac{1}{n}\right)\log_2 \frac{1}{n} = \log_2 n$

$\lim_{x \to \infty} H_n \to \infty$

## Chapter 12, Section 9, page 920

**Summary.** It will cost at least $16.4 million to build a pipeline to supply refinery $B$ with oil from the offshore facility $A$. The actual cost will depend on where the pipeline crosses the shoreline. The most expensive plausible route, costing almost $22.5 million, goes from $A$ to the nearest point on the shore, then to $B$. An alternative route through the point on the shoreline closest to $B$ would cost about $17.4 million. A straight line direct route from $A$ to $B$ would cost approximately $16.7 million.

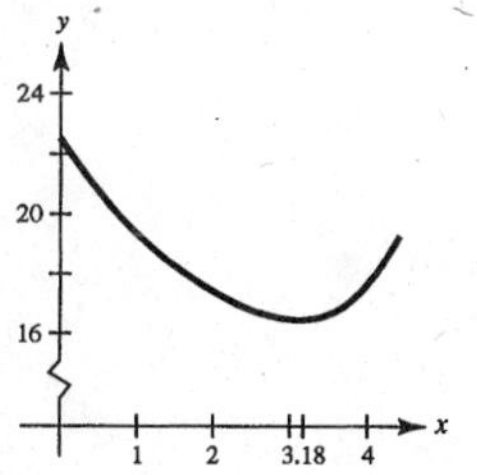


At the left is a graph of the cost $C$ in millions of dollars as a function of $x$, which is the number of miles between the point on the shore that is nearest to $A$ and the point where the pipeline crosses the shore. The least cost occurs at $x \approx 3.18$ miles.

**Analysis.** It is convenient to introduce a rectangular coordinate system as shown at the right with a unit of length representing one mile. The point $P$ where the pipeline crosses the shore is at $(x, 0)$ and $A$ is located at $(0, 2)$. The point $(0, -1)$ is one vertex of a right triangle with other vertices at $A$ and $B$. By the Pythagorean Theorem, the coordinates of $B$ are $(4, -1)$. The line from $A$ to $B$ meets the shore at $\left(\frac{8}{3}, 0\right)$.

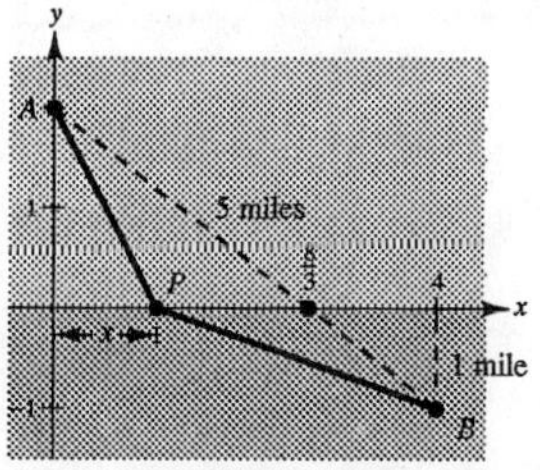


—CONTINUED—

## Chapter 12, Section 9, page 920 —CONTINUED—

The cost per mile to build the pipeline is \$3 million in water and \$4 million on land. Thus, in millions of dollars, building the pipeline from $A$ to $P$ costs $3\sqrt{x^2 + 4}$ and building the pipeline from $P$ to $B$ costs $4\sqrt{(4 - x)^2 + 1}$. The total cost is

$$C(x) = 3\sqrt{x^2 + 4} + 4\sqrt{(4 - x)^2 + 1}.$$

The above summary cites the costs $C(0) = 6 + 4\sqrt{17} \approx 22.49242250$, $C(4) = 6\sqrt{5} + 4 \approx 17.41640787$, and $C\left(\frac{8}{3}\right) = 16.\overline{6}$. The minimum cost occurs at a zero of the derivative

$$C'(x) = \frac{3x}{\sqrt{x^2 + 4}} - \frac{4(4 - x)}{\sqrt{(4 - x)^2 + 1}}.$$

A graphing utility can be used to find that $x = 3.1785$ is an approximate solution of $C'(x) = 0$ on the interval $[0, 4]$. (Newton's method or other methods of numeric approximation can also be used.) Consequently, the minimum cost is $C(3.1785) = 16.4427921$.

## Chapter 13, Section 4, page 970

$$\int_R\int y\,dA = \int_0^2\int_{x/2}^{-2x+5} y\,dy\,dx$$

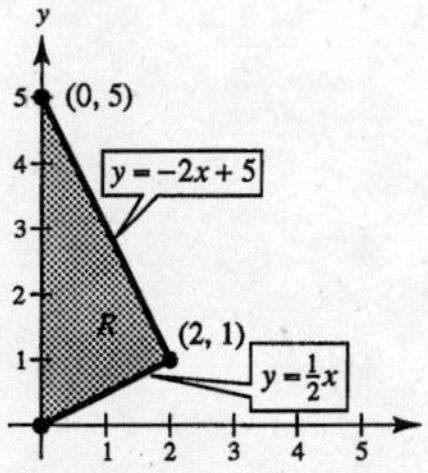


$$= \int_0^2 \left[\frac{y^2}{2}\right]_{x/2}^{-2x+5} dx$$

$$= \frac{1}{2}\int_0^2 \left[4x^2 - 20x + 25 - \frac{x^2}{4}\right] dx$$

$$= \frac{1}{2}\int_0^2 \left[\frac{15}{4}x^2 - 20x + 25\right] dx$$

$$= \frac{1}{2}\left[\frac{5}{4}x^3 - 10x^2 + 25x\right]_0^2$$

$$= \frac{1}{2}\left[\left(\frac{5}{4}\right)(8) - 40 + 50\right]$$

$$= 10$$

—CONTINUED—

## Chapter 13, Section 4, page 970 —CONTINUED—

$$\int\!\!\int_R xy\,dA = \int_0^2 \int_{x/2}^{-2x+5} xy\,dy\,dx$$

$$= \int_0^2 \left[\frac{xy^2}{2}\right]_{x/2}^{-2x+5} dx$$

$$= \frac{1}{2}\int_0^2 x\left[\frac{15}{4}x^2 - 20x + 25\right] dx$$

$$= \frac{1}{2}\int_0^2 \left[\frac{15}{4}x^3 - 20x^2 + 25x\right] dx$$

$$= \frac{1}{2}\left[\frac{15}{16}x^4 - \frac{20}{3}x^3 + \frac{25}{2}x^2\right]_0^2$$

$$= \frac{1}{2}\left[\frac{15}{16}(16) - \frac{20}{3}(8) + \frac{25}{2}(4)\right]$$

$$= \frac{35}{6}$$

$$\int\!\!\int_R y^2\,dA = \int_0^2 \int_{x/2}^{-2x+5} y^2\,dy\,dx$$

$$= \int_0^2 \left[\frac{y^3}{3}\right]_{x/2}^{-2x+5} dx$$

$$= \frac{1}{3}\int_0^2 \left[(-2x+5)^3 - \left(\frac{1}{2}x\right)^3\right] dx$$

$$= \frac{1}{3}\int_0^2 \left[-\frac{65}{8}x^3 + 60x^2 - 150x + 125\right] dx$$

$$= \frac{1}{3}\left[-\frac{65}{8}\left(\frac{x^4}{4}\right) + 20x^3 - 75x^2 + 125x\right]_0^2$$

$$= \frac{1}{3}\left[-\frac{65}{32}(16) + 20(8) - 75(4) + 125(2)\right]$$

$$= \frac{155}{6}$$

$$x_p = \frac{\int\!\!\int_R xy\,dA}{\int\!\!\int_R y\,dA} = \frac{35/6}{10} = \frac{7}{12} \qquad y_p = \frac{\int\!\!\int_R y^2\,dA}{\int\!\!\int_R y\,dA} = \frac{155/6}{10} = \frac{31}{12}$$

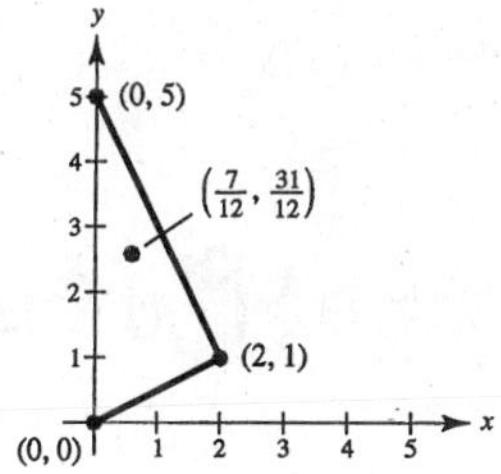


## Chapter 13, Section 5, page 977

(a) $V = \int\!\!\int_R \frac{1}{\sqrt{x^2+y^2}}\,dy\,dx$

$$= \int\!\!\int_R \frac{1}{r}\,r\,dr\,d\theta$$

$$= \int_{2\arctan 0.01}^{\pi/2} \int_0^{9\csc\theta} dr\,d\theta$$

$$= \Big[9\ln|\csc\theta - \cot\theta|\Big]_{2\arctan 0.01}^{\pi/2} \approx 41.45 \text{ in.}^3$$

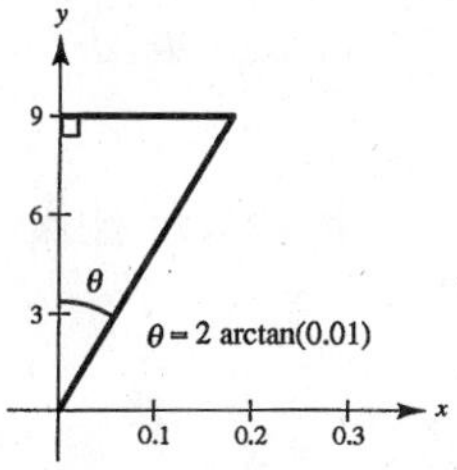


—CONTINUED—

## Chapter 13, Section 5, page 977 —CONTINUED—

(b) $f(x, y) = \dfrac{1}{\sqrt{x^2 + y^2}}$

$$\sqrt{1 + (f_x)^2 + (f_y)^2} = \sqrt{1 + \frac{x^2}{(x^2 + y^2)^3} + \frac{y^2}{(x^2 + y^2)^3}} = \frac{\sqrt{(x^2 + y^2)^2 + 1}}{x^2 + y^2}$$

$$S = \iint_R \frac{\sqrt{(x^2 + y^2)^2 + 1}}{x^2 + y^2}\, dy\, dx = \iint_R \frac{\sqrt{r^4 + 1}}{r^2}\, r\, dr\, d\theta = \int_{2 \arctan 0.01}^{\pi/2} \int_0^{9 \csc \theta} \frac{\sqrt{r^4 + 1}}{r}\, dr\, d\theta$$

$$\approx 501.69 \text{ in.}^2$$

## Chapter 13, Section 7, page 994

(a) $\displaystyle\int_0^{2\pi} \int_0^{\pi} \int_0^{1 + 0.2 \sin 8\theta \sin \phi} \rho^2 \sin \phi\, d\rho\, d\phi\, d\theta$

$$= \frac{1}{3} \int_0^{2\pi} \int_0^{\pi} (1 + 0.2 \sin 8\theta \sin \phi)^3 \sin \phi\, d\phi\, d\theta$$

$$= \frac{1}{3} \int_0^{2\pi} \int_0^{\pi} [\sin \phi + 0.6 \sin 8\theta \sin^2\phi + 0.12 \sin^2 8\theta \sin^3 \phi + 0.008 \sin^3 8\theta \sin^4 \phi]\, d\phi\, d\theta$$

$$= \frac{1}{3} \int_0^{2\pi} \Bigg[-\cos \phi + 0.6 \sin 8\theta\left(\frac{1}{2}\right)(\phi - \sin \phi \cos \phi) + 0.12 \sin^2 8\theta\left(-\frac{\sin^2 \phi \cos \phi}{3} - \frac{2}{3} \cos \phi\right)$$

$$+ 0.008 \sin^3 8\theta\left(-\frac{\sin^3 \phi \cos \phi}{4} + \frac{3}{8}(\phi - \sin \phi \cos \phi)\right)\Bigg]_0^{\pi} d\theta$$

$$= \frac{1}{3} \int_0^{2\pi} (2 + 0.3\pi \sin 8\theta + 0.16 \sin^2 8\theta + 0.003\pi \sin^3 8\theta)\, d\theta$$

$$= \frac{1}{3}\left[2\theta - \frac{0.3\pi}{8} \cos 8\theta + \frac{0.16}{8}\left(\frac{1}{2}\right)(8\theta - \sin 8\theta \cos 8\theta) + \frac{0.003\pi}{8}\left(-\frac{\sin^2 8\theta \cos 8\theta}{3} - \frac{2}{3} \cos 8\theta\right)\right]_0^{2\pi}$$

$$= \frac{104\pi}{75}$$

(b) $\displaystyle\int_0^{2\pi} \int_0^{\pi} \int_0^{1 + 0.2 \sin 8\theta \sin 4\phi} \rho^2 \sin \phi\, d\rho\, d\phi\, d\theta = \frac{1}{3} \int_0^{2\pi} \int_0^{\pi} (1 + 0.2 \sin 8\theta \sin 4\phi)^3 \sin \phi\, d\phi\, d\theta \approx 4.316$

## Chapter 14, Section 4, page 1050

(a) If $x = \cosh t$ and $y = \sinh t$, then $x^2 - y^2 = \cosh^2 t - \sinh^2 t = 1$.

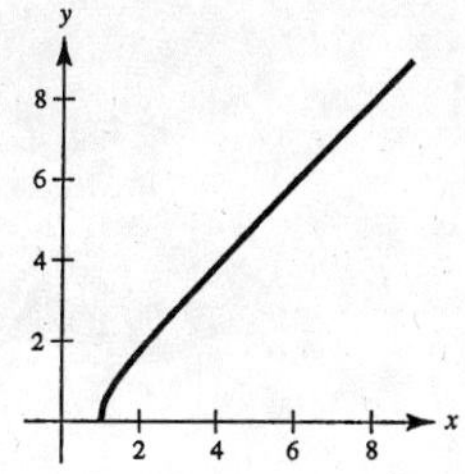


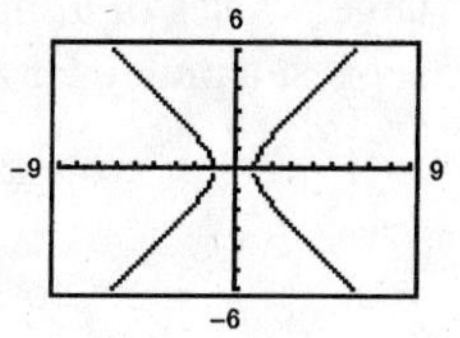


$x^2 - y^2 = 1$

$\mathbf{r}(t) = \cosh t\mathbf{i} + \sinh t\mathbf{j},\ 0 \le t \le 5$

(b) Divide the boundary $C$ into three curves as shown at the right and introduce the parameterization

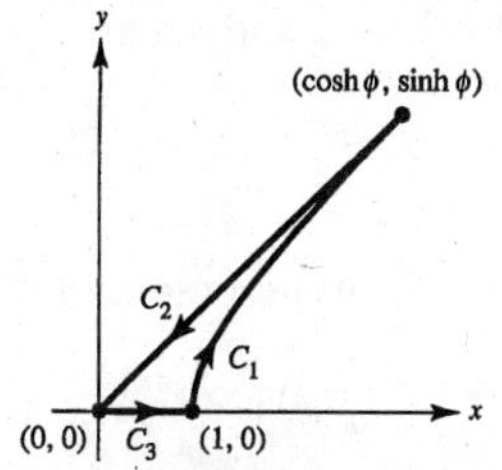


$$C_1: x = \cosh t,\ y = \sinh t,\ 0 \le t \le \phi.$$

Because $x\,dy - y\,dx = 0\,dt$ for any parameterization of straight lines through the origin,

$$A = \frac{1}{2}\int_C x\,dy - y\,dx = \frac{1}{2}\int_{C_1} x\,dy - y\,dx = \frac{1}{2}\int_{C_2} 0\,dt = \frac{1}{2}\int_{C_3} 0\,dt$$

$$= \frac{1}{2}\int_{C_3} \cosh^2 t\,dt - \sinh^2 t\,dt + 0 + 0 = \frac{1}{2}\int_0^{\phi} dt = \frac{1}{2}\phi.$$

(c) The area in the first quadrant to the right of the line

$$x = \frac{\cosh \phi}{\sinh \phi} y$$

and to the left of the curve $x = \sqrt{y^2 + 1}$ is

$$A = \int_0^{\sinh \phi} \left[\sqrt{y^2 + 1} - (\coth \phi)y\right] dy.$$

Using a tolerance of $1 \times 10^{-5}$, the numeric approximations $A = 0.5, 1, 2,$ and $5$ were obtained for $\phi = 1, 2, 4,$ and $10$, respectively.

(d) The hyperbolic functions $f(\phi) = \cosh \phi$ and $g(\phi) = \sinh \phi$ could be defined to be the coordinates $(\cosh \phi, \sinh \phi)$ of the point of intersection of the right branch of the hyperbola $x^2 - y^2 = 1$ with the straight line through the origin that (along with the $x$-axis) bounds a region of area $\frac{1}{2}|\phi|$.

## Chapter 14, Section 6, page 1072

(a) The ratio $b/a$ determines the shape of the surface. For a fixed value of $a$, larger values of $b$ suggest a (bottomless) stem vase and smaller values of $b$ suggest a pulley for a rope.

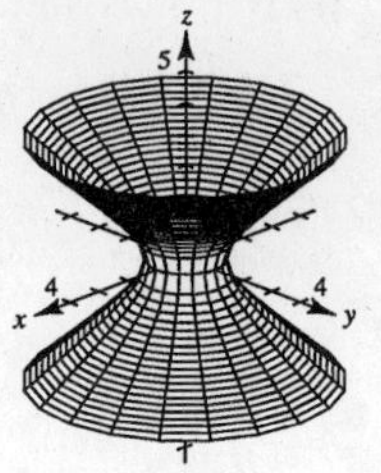


(b) $x^2 + y^2 = a^2 \cosh^2 u \cos^2 v + a^2 \cosh^2 u \sin^2 v = a^2 \cosh^2 u$

$$\frac{x^2}{a^2} + \frac{y^2}{a^2} - \frac{z^2}{b^2} = \cosh^2 u - \sinh^2 u = 1$$

(c) For each value of $u_0$, the curve is a circle of radius $a \cosh u_0$ centered about the $z$-axis in the plane $z = b \sinh u_0$.

(d) For each value of $v_0$, the curve is a hyperbola. In terms of cylindrical coordinates, it lies in the plane $\theta = v_0$ and satisfies

$$\frac{r^2}{a^2} - \frac{z^2}{b^2} = 1.$$

(e) The hyperboloid of one sheet is a level surface of

$$f(x, y, z) = \frac{x^2}{a^2} + \frac{y^2}{a^2} - \frac{z^2}{b^2}.$$

The gradient of $f$ is

$$\nabla f(x, y, z) = \frac{2x}{a^2}\mathbf{i} + \frac{2y}{a^2}\mathbf{j} - \frac{2z}{b^2}\mathbf{k} \text{ and } (x, y, z) = (a, 0, 0) \text{ if } (u, v) = (0, 0).$$

Therefore, a normal vector is $\nabla f(x, y, z) = \dfrac{2}{a}\mathbf{i}$.

## Chapter 14, Review Section, page 1089

(a)
$$\begin{aligned}\overrightarrow{OB} = \overrightarrow{OA} + \overrightarrow{AB} &= [x(t)\mathbf{i} + y(t)\mathbf{j}] + [L\cos\theta(t)\mathbf{i} + L\sin\theta(t)\mathbf{j}] \\ &= [x(t) + L\cos\theta(t)]\mathbf{i} + [y(t) + L\sin\theta(t)]\mathbf{j}.\end{aligned}$$

(b) Because $\theta(b) = \theta(a)$, $I_1 = \displaystyle\int_a^b \frac{1}{2}L^2\frac{d\theta}{dt}\,dt = \frac{1}{2}L^2\int_{\theta(a)}^{\theta(b)} d\theta = \frac{1}{2}L^2[\theta(b) - \theta(a)] = 0.$

The point $A$ traces the boundary $G$, say, of a region with no area. Therefore,

$$I_2 = \int_a^b \frac{1}{2}\left(x\frac{dy}{dt} - y\frac{dx}{dt}\right)dt = \frac{1}{2}\int_G x\,dy - y\,dx = 0.$$

(c) Because $\theta(b) = \theta(a)$, $x(b) = x(a)$, and $y(b) = y(a)$, we have

$$\begin{aligned}I_3 - I_4 &= \frac{1}{2}L\int_a^b\left[y(\sin\theta)\frac{d\theta}{dt} - (\cos\theta)\frac{dy}{dt} + x(\cos\theta)\frac{d\theta}{dt} + (\sin\theta)\frac{dx}{dt}\right]dt \\ &= \frac{1}{2}L\int_a^b \frac{d}{dt}(x\sin\theta - y\cos\theta)\,dt = \frac{1}{2}L\,(x\sin\theta - y\cos\theta)\Big|_{t=a}^{b} \\ &= \frac{1}{2}L\big[x(b)\sin\theta(b) - y(b)\cos\theta(b) - x(a)\sin\theta(a) + y(a)\cos\theta(a)\big] \\ &= \frac{1}{2}L\{[x(b)\sin\theta(b) - x(a)\sin\theta(a)] + [y(a)\cos\theta(a) - y(b)\cos\theta(b)]\} \\ &= \frac{1}{2}L(0 + 0) = 0.\end{aligned}$$

—CONTINUED—

## Chapter 14, Review Section, page 1089 —CONTINUED—

(d) If $\phi$ is the angle between the unit vectors **N** and **T**, then $\mathbf{N} \cdot \mathbf{T} = \cos \phi$. As the point $B$ moves through an element $ds$ of arc length in the direction of **T**, the wheel rolls a distance of $(\cos \phi)\, ds$ in the direction of **N** and skids a distance of $(\sin \phi)\, ds$ in the direction of $\overrightarrow{AB}$. The total distance rolled by the wheel is

$$D = \int_C (\cos \phi)\, ds = \int_C \mathbf{N} \cdot \mathbf{T}\, ds.$$

(e) If $X(t) = x(t) + L \cos \theta(t)$ and $Y(t) = y(t) + L \sin \theta(t)$, then the area of the region $R$ is

$$A = \frac{1}{2} \int_C X\, dY - Y\, dX$$

$$= \frac{1}{2} \int_a^b \left\{ (x + L \cos \theta)\left[\frac{dy}{dt} + L(\cos \theta)\frac{d\theta}{dt}\right] dt - (y + L \sin \theta)\left[\frac{dx}{dt} - L(\sin \theta)\frac{d\theta}{dt}\right]\right\} dt$$

$$= \frac{1}{2} \int_a^b \left\{ \left[L^2(\cos^2\theta)\frac{d\theta}{dt} + L^2(\sin^2\theta)\frac{d\theta}{dt}\right] + \left(x\frac{dy}{dt} - y\frac{dx}{dt}\right) + \left[yL(\sin \theta)\frac{d\theta}{dt} + xL(\cos \theta)\frac{d\theta}{dt}\right] \right.$$

$$\left. + \left[-L(\sin \theta)\frac{dx}{dt} + L(\cos \theta)\frac{dy}{dt}\right]\right\} dt$$

$$= I_1 + I_2 + I_3 + I_4 = 0 + 0 + 2I_4$$

Furthermore,

$$\mathbf{N} \cdot \mathbf{r}' = (-\sin \theta\, \mathbf{i} + \cos \theta\, \mathbf{j}) \cdot \left\{\left[\frac{dx}{dt} - L(\sin \theta)\frac{d\theta}{dt}\right]\mathbf{i} + \left[\frac{dy}{dt} + L(\cos \theta)\frac{d\theta}{dt}\right]\mathbf{j}\right\}$$

$$= -(\sin \theta)\frac{dx}{dt} + (\cos \theta)\frac{dy}{dt} + L\,\frac{d\theta}{dt}.$$

Therefore,

$$DL = L \int_C \mathbf{N} \cdot \mathbf{T}\, ds = L \int_a^b \mathbf{N} \cdot \mathbf{r}'\, dt = 2I_4 + 2I_1 = 2I_4 = A.$$